AF566926

Series Editor: **Rhodes W. Fairbridge**
Columbia University

Volume

1 ENVIRONMENTAL GEOMORPHOLOGY AND LANDSCAPE CONSERVATION, Volume I: Prior to 1900 / *Donald R. Coates*
2 RIVER MORPHOLOGY / *Stanley A. Schumm*
3 SPITS AND BARS / *Maurice L. Schwartz*
4 TEKTITES / *Virgil E. Barnes and Mildred A. Barnes*
5 GEOCHRONOLOGY: Radiometric Dating of Rocks and Minerals / *C. T. Harper*
6 SLOPE MORPHOLOGY / *Stanley A. Schumm and M. Paul Mosley*
7 MARINE EVAPORITES: Origin, Diagenesis, and Geochemistry / *Douglas W. Kirkland and Robert Evans*
8 ENVIRONMENTAL GEOMORPHOLOGY AND LANDSCAPE CONSERVATION, Volume III: Non-Urban / *Donald R. Coates*
9 BARRIER ISLANDS / *Maurice L. Schwartz*
10 GLACIAL ISOSTASY / *John T. Andrews*
11 GEOCHEMISTRY OF GERMANIUM / *Jon N. Weber*
12 ENVIRONMENTAL GEOMORPHOLOGY AND LANDSCAPE CONSERVATION, Volume II: Urban Areas / *Donald R. Coates*
13 PHILOSOPHY OF GEOHISTORY: 1785–1970 / *Claude C. Albritton, Jr.*
14 GEOCHEMISTRY AND THE ORIGIN OF LIFE / *Keith A. Kvenvolden*
15 SEDIMENTARY ROCKS: Concepts and History / *Albert V. Carozzi*
16 GEOCHEMISTRY OF WATER / *Yasushi Kitano*
17 METAMORPHISM AND PLATE TECTONIC REGIMES / *W. G. Ernst*
18 GEOCHEMISTRY OF IRON / *Henry Lepp*
19 SUBDUCTION ZONE METAMORPHISM / *W. G. Ernst*
20 PLAYAS AND DRIED LAKES: Occurrence and Development / *James T. Neal*
21 GLACIAL DEPOSITS / *Richard P. Goldthwait*
22 PLANATION SURFACES: Peneplains, Pediplains, and Etchplains / *George F. Adams*
23 GEOCHEMISTRY OF BORON / *C. T. Walker*
24 SUBMARINE CANYONS AND DEEP-SEA FANS: Modern and Ancient / *J. H. McD. Whitaker*
25 ENVIRONMENTAL GEOLOGY / *Frederick Betz, Jr.*
26 LOESS: Lithology and Genesis / *Ian J. Smalley*

27 PERIGLACIAL PROCESSES / *Cuchlaine A. M. King*
28 LANDFORMS AND GEOMORPHOLOGY: Concepts and History / *Cuchlaine A. M. King*
29 METALLOGENY AND GLOBAL TECTONICS / *Wilfred Walker*
30 HOLOCENE TIDAL SEDIMENTATION / *George deVries Klein*
31 PALEOBIOGEOGRAPHY / *Charles A. Ross*
32 MECHANICS OF THRUST FAULTS AND DÉCOLLEMENT / *Barry Voight*
33 WEST INDIES ISLAND ARCS / *Peter H. Mattson*
34 CRYSTAL FORM AND STRUCTURE / *Cecil J. Schneer*
35 OCEANOGRAPHY: Concepts and History / *Margaret B. Deacon*
36 METEORITE CRATERS / *G. J. H. McCall*
37 STATISTICAL ANALYSIS IN GEOLOGY / *John M. Cubitt and Stephen Henley*
38 AIR PHOTOGRAPHY AND COASTAL PROBLEMS / *Mohamed T. El-Ashry*
39 BEACH PROCESSES AND COASTAL HYDRODYNAMICS / *John S. Fisher and Robert Dolan*
40 DIAGENESIS OF DEEP-SEA BIOGENIC SEDIMENTS / *Gerrit J. van der Lingen*
41 DRAINAGE BASIN MORPHOLOGY / *Stanley A. Schumm*
42 COASTAL SEDIMENTATION / *Donald J. P. Swift and Harold D. Palmer*
43 ANCIENT CONTINENTAL DEPOSITS / *Franklyn B. Van Houten*
44 MINERAL DEPOSITS, CONTINENTAL DRIFT AND PLATE TECTONICS / *J. B. Wright*
45 SEA WATER: Cycles of the Major Elements / *James I. Drever*
46 PALYNOLOGY, PART I: Spores and Pollen / *Marjorie D. Muir and William A. S. Sarjeant*
47 PALYNOLOGY, PART II: Dinoflagellates, Acritarchs, and Other Microfossils / *Marjorie D. Muir and William A. S. Sarjeant*
48 GEOLOGY OF THE PLANT MARS / *Vivien Gornitz*
49 GEOCHEMISTRY OF BISMUTH / *Ernest E. Angino and David T. Long*

Benchmark Papers in Geology / 49

A BENCHMARK® Books Series

GEOCHEMISTRY OF BISMUTH

Edited by

ERNEST E. ANGINO

The University of Kansas

and

DAVID T. LONG

Michigan State University

Dowden, Hutchinson & Ross, Inc.

STROUDSBURG, PENNSYLVANIA

Benchmark Papers in Geology, Volume 49
Library of Congress Catalog Card Number: 78-24291
ISBN: 0-87933-234-4

81 80 79 1 2 3 4 5
Manufactured in the United States of America.

LIBRARY OF CONGRESS CATALOGING IN PUBLICATION DATA
Main entry under title:
Geochemistry of bismuth.
(Benchmark papers in geology; 49)
Includes indexes.
1. Bismuth—Addresses, essays, lectures.
I. Angino, Ernest E. II. Long, David T., 1947-
QE516.B57G46 553'.47 78-24291
ISBN: 0-87933-234-4

Distributed world wide by Academic Press,
a subsidiary of Harcourt Brace Jovanovich,
Publishers.

SERIES EDITOR'S FOREWORD

The philosophy behind the "Benchmark Papers in Geology" is one of collection, sifting, and rediffusion. Scientific literature today is so vast, so dispersed, and, in the case of old papers, so inaccessible for readers not in the immediate neighborhood of major libraries that much valuable information has been ignored by default. It has become just so difficult, or so time consuming, to search out the key papers in any basic area of research that one can hardly blame a busy person for skimping on some of his or her "homework."

This series of volumes has been devised, therefore, to make a practical contribution to this critical problem. The geologist, perhaps even more than any other scientist, often suffers from twin difficulties—isolation from central library resources and immensely diffused sources of material. New colleges and industrial libraries simply cannot afford to purchase complete runs of all the world's earth science literature. Specialists simply cannot locate reprints or copies of all their principal reference materials. So it is that we are now making a concerted effort to gather into single volumes the critical materials needed to reconstruct the background of any and every major topic of our discipline.

We are interpreting "geology" in its broadest sense: the fundamental science of the planet Earth, its materials, its history, and its dynamics. Because of training in "earthy" materials, we also take in astrogeology, the corresponding aspect of the planetary sciences. Besides the classical core disciplines such as mineralogy, petrology, strucuture, geomorphology, paleontology, and stratigraphy, we embrace the newer fields of geophysics and geochemistry, applied also to oceanography, geochronology, and paleoecology. We recognize the work of the mining geologists, the petroleum geologists, the hydrologists, and the engineering and environmental geologists. Each specialist needs a working library. We are endeavoring to make the task of compiling such a library a little easier.

Each volume in the series contains an introduction prepared by a specialist (the volume editor)—a "state of the art" opening or a summary of the object and content of the volume. The articles, usually some twenty to fifty reproduced either in their entirety or in significant extracts, are selected in an attempt to cover the field, from the key papers of the last century to fairly recent work. Where the original works are in foreign languages, we have endeavored to locate or commission translations. Geologists, because of their global subject, are often acutely aware of the oneness of our world. The selections cannot, therefore, be restricted to any one country, and whenever possible an attempt is made to scan the world literature.

To each article, or group of kindred articles, some sort of "highlight commentary" is usually supplied by the volume editor. This commentary should serve to bring that article into historical perspective and to emphasize its particular role in the growth of the field. References, or citations, wherever possible, will be reproduced in their entirety—for by this means the observant reader can assess the background material available to that particular author, or, if desired, he or she, too, can double check the earlier sources.

A "benchmark," in surveyor's terminology, is an established point on the ground, recorded on our maps. It is usually anything that is a vantage point, from a modest hill to a mountain peak. From the historical viewpoint, these benchmarks are the bricks of our scientific edifice.

RHODES W. FAIRBRIDGE

PREFACE

Bismuth has received little study by the geochemical community. Data, therefore, are scanty. Goldschmidt reviewed the early literature on this subject and presented a masterful summary of that information. Bismuth has both chalcophilic and lithophilic properties in the upper lithosphere. However, those bismuth minerals of economic importance are all chalcophilic in nature.

Our interest in compiling a review of bismuth geochemistry was triggered primarily by the striking absence of data on the element in Turekian and Wedepohl's review of elemental abundances in different units of the earth's crust.

Of all the elements, bismuth appeared to be the one for which little reliable data was available. Subsequently, we reviewed much of the literature on bismuth reported in chemical abstracts since 1960. From this review came the decision that a compilation of selected articles on the geochemistry of bismuth would be helpful in our work and perhaps to others. The present volume is a natural outgrowth of that assessment.

We expect there will be criticism and second guessing on the articles included in this collection. That is only natural. We do feel, however, that each paper included herein has contributed significantly to advancing our knowledge of the geochemistry of bismuth.

To the best of our knowledge, the Russian articles presented here appear in English translation for the first time. We encountered several problems with the Russian papers. Among them were the difficulty in obtaining clean copy for reproduction of figures and photographs. All of the figures have been redrafted. No easy solution to the photographic problem presented itself. In addition, even after securing copies of the original Russian articles, we found no affiliation for some of the authors was indicated.

The articles included in this collection have been grouped into five topics: Part I, General Geochemistry; Part II, Bismuth in Meteorities; Part III, Rock-Forming Processes; Part IV, Phase Equilibrium; and Part V, Economic Geology—Ore Deposits.

The element deserves far more attention and study than it has received to date. It is our hope that this situation will be changed by the publication of this volume. We feel that the articles selected herein will serve as a practical and useful introduction to the geochemistry of bismuth and possibly induce further studies of the natural distribution and occurrence of this element.

ERNEST E. ANGINO
DAVID T. LONG

CONTENTS

PART III: ROCK-FORMING PROCESSES

PART IV: PHASE EQUILIBRIUM

PART V: ECONOMIC GEOLOGY—ORE DEPOSITS

CONTENTS BY AUTHOR

GEOCHEMISTRY OF BISMUTH

INTRODUCTION

Bismuth is an element that has been little studied over the years by the geochemical profession. Several reasons exist for this situation, but two in particular seem clearly to be responsible for this state of affairs. Until about 1965, the analysis of bismuth in rocks and minerals was a rather difficult, time-consuming, and involved task. With the introduction of neutron activation and atomic absorption spectrometry, a major restriction on studies of bismuth distribution in geologic materials was removed. The second reason can be summarized by a statement sometimes heard among geochemists: "It is not an interesting element."

An awareness of these two facts helps to explain the dearth of material in the literature treating most facets of the geochemistry of bismuth. This literature gap has made the task of selecting articles for inclusion in this volume both easy and difficult at the same time—that is, the task was difficult because clearly the largest number of papers dealing exclusively with bismuth appears in the Russian literature with all the attendant problems of translation and obtaining copy.

If one considers the topics and papers covered in this compendium, it should become clear to anyone that considerable expansion of our knowledge of the geochemistry of bismuth is still to come. In brief, information on bismuth geochemistry is not extensive.

Bismuth has the properties (e.g., low melting point) that make it likely that it will become better known and more widely used in the future. It is not an element of great abundance, nor are any of its minerals very common. Bismuth as an element is frequently found in the native state, but it and its minerals are also quite dispersed in the matrix of many rocks and minerals.

The metal is commonly obtained commercially as a by-product of lead and copper smelting. Uses of bismuth and its componds are widespread but no one use requires great volume. Even in alloys, the metal is consumed in small quantities, but its unusual properties create important uses for it. It possesses curative medicinal properties, has a low melting point (271°C), forms alloys quite easily, and expands on solidifying. For example, Wood's metal (Bi-Sn-Cd) melts at 60°C, and other alloys melt at even lower temperatures, thus making bismuth a key metal in the use of automatic water sprinklers for fire control.

Its principal uses, however, are for medicinal and cosmetic preparations and in electrical and electronic apparatus. Other uses are in enameling, printing of fabrics, forming of special optical glass, making of fuses, locker safety plugs, and innumerable other minor applications. The potential of bismuth as a guide element in the geochemical exploration for other metals is beginning to be realized.

In spite of these uses and unusual properties, knowledge of the geochemistry of bismuth is still poorly disseminated or known. Thus, we felt the need to compile informative and key articles from the widely scattered literature on bismuth. These data will serve to make the geochemistry of bismuth better known and make information more available to new investigators.

As was noted in the Preface, extensive data on the geochemistry of bismuth is scarce. A review of the several parts of the Data of Geochemistry series (U.S. Geological Survey Prof. Paper 440) and of the now standard references of Taylor (1964) and Turekian and Wedepohl (1961) confirms this observation. Of the ten rock categories in Turekian and Wedepohl for which concentration values are normally given, entries for bismuth are listed under only two.

Accounting for this situation is difficult, for knowledge of the geochemistry of bismuth must be one of the keys to an understanding of the origin of our planet. Given the low melting point (271°C) of bismuth and its volatility, that any bismuth is present on the earth at all is surprising—if we accept the "hot" earth theory of our planet's origin. Why didn't all the bismuth boil off? As we know, "bismuth is lost progressively with temperature increase during the heating of geologic material, including primitive meteorites and basalt (BCR-1) for one (1) week at temperatures ranging from 500°–1000°C or more in a low-pressure (initially $\sim 10^{-5}$ atm H_2) system" (M. Lipschutz, pers. comm.).

The general geochemistry of bismuth is determined primarily from general principles and from mineralogical analyses. For upper lithospheric rocks and minerals, bismuth clearly has chalcophilic characteristics. In most deposits and occurrences known to contain bismuth, the element accompanies sulphide minerals or occurs in secondary alteration products of sulphides and their related compounds. Under certain conditions, the element also possesses lithophilic characteristics in the upper lithosphere.

Due to the lack of data, obtaining a well-authenticated value for the concentration of bismuth in the earth's crust is difficult. Taylor (1964) gives a crustal average for bismuth of 0.17 ppm. The correct value probably lies between 0.1 and 0.2 ppm. Bismuth is, however, likely to be concentrated in granitic magmas, and residual liquids formed therefrom.

Because of the recent studies of meteoritic data and recognition of the fact that bismuth may be a key element in tracing some of the cosmic reactions to which meteorites have been subjected, we now have a fair understanding of the geochemistry of bismuth in meteorites.

Knowledge of the behavior of bismuth in weathering processes is scarce. It is limited primarily to data on reactions in the oxidation zone of ore deposits. Except for its concentration in ferrugenous bauxites, little is known of the pathway of bismuth in soil formation or in the soils themselves. Bismuth is commonly present in most ranks of coal. Little reliable data exists on the concentration of bismuth in sedimentary rocks (Turekian and Wedepohl 1961).

Sulphide minerals of bismuth and sulphide minerals containing bismuth are commonly found associated with hydrothermal mineral deposits. As noted by Goldschmidt (1954, p. 482), bismuth minerals are also known as products of pneumatolysis associated with recent volcanic action. Bismuth is commonly associated with gold, selenium, and tellurium in high-temperature hydrothermal deposits. A common occurrence of bismuth is with (in) the mineral galena (PbS). From the literature, there appears to be a direct correlation between bismuth and temperature. The bismuth concentration of galenas increases with increasing temperature of mineral formation. Bismuth will very likely play a major role in the search for new mineral deposits in the United States and elsewhere. Knowledge of its role as a "pathfinder element" in geochemical exploration is still imperfectly understood, but clearly its use is in the development stage.

One area of the study of bismuth where we have a reasonable understanding of the relations involved is that of phase equilibria. Several published articles, including those in this compilation, summarize the phase relationships in bismuth systems. Levin, Robins, and McMurdie (1964) present data for Bi_2O_3–MoO_2–PbO; Bi_2O_3–PbO; and Bi_2O_3–PbO–WO_3. Van Hook (1960), Salanci (1965), and Craig (1967) have done extensive work on the Bi_2S_3–PbS system. Van Hook (1960) and Craig (1967) also carried out considerable work on the $AgBiS_2$–PbS system.

In sum, we hope that this volume on the geochemistry of bismuth will serve two purposes: first, to make information on the geochemistry of a rare and widely dispersed, but needed element more available and second, by collecting key articles on an interesting element into one volume, to stimulate further research on bismuth as another step in the endeavor to expand our knowledge of the geochemistry of the elements.

REFERENCES

Craig, J. R. 1967. Phase relations and mineral assemblages in the Ag–Bi–Pb–S system. *Miner. Deposita* **1**: 278–306.

Goldschmidt, V. M. 1954. *Geochemistry.* Oxford University Press, Oxford, pp. 480–84.

Levin, E. M., C. R. Robins, and H. F. McMurdie. 1964. Phase diagrams for ceramists. The American Ceramic Society, Columbus, Ohio. Annual supplements since 1964 have been published.

Salanci, B. 1964. Untersuchungen am system Bi_2S_3-PbS. *Neues Jahrb. Mineral. Monatsh.* **12**: 384–88.

Taylor, S. R. 1964. Abundance of chemical elements in the continental crust: A new table. *Geochem. Cosmochim. Acta* **38**: 1273–85.

Turekian, K. K., and K. H. Wedepohl. 1961. Distribution of elements in some major units of the earth's crust. *Bull. Geo. Soc. Am.* **73**: 175–92.

Van Hook, H. J. 1960. The ternary system Ag_2S–Bi_2S_3–PbS. *Econ. Geol.* **55**: 759–88.

Part I

GENERAL GEOCHEMISTRY

Editors' Comments on Papers 1, 2, and 3

Since little is known of the detailed geochemistry of bismuth and what is known is widely scattered in the literature, we decided to include three papers to serve as an introduction to the general geochemistry of bismuth.

Several brief summaries of the geochemistry of bismuth are available, including the two classic treatises on geochemistry by Rankama and Sahama (1950) and Goldschmidt (1954). The three papers in this section were selected because they provide a broad overview of our present knowledge of the geochemistry of the element bismuth. They include much of the pertinent material contained in the earlier reviews, in particular that from Goldschmidt and Rankama and Sahama.

In Paper 1, Angino gives a brief summary of the chemical physical properties of bismuth. This paper is provided as background material to introduce the articles that follow on the chemistry of bismuth in natural materials. Brief comments on some of the more common bismuth minerals are provided, followed by a more extensive discussion of the chemistry of bismuth in aqueous solution. From this information, it is obvious why the +3 valent state of bismuth is that seen in natural minerals. Although bismuth has many properties similar to those of antimony and arsenic, the common occurrence in nature of a +5 valent state is not one of them.

A review of the geochemistry and mineralogy of the element suggest that the crustal abundance of bismuth is around 0.2 ppm, although a considerable degree of uncertainty accompanies this figure. As is demonstrated, bismuth is a widely distributed element, but tends to become enriched in late magmatic differentiations. Mineralogically, bismuth is most commonly present in galenas. Paper 1 also presents some data on the marine geochemistry of bismuth.

Ernest E. Angino, the author of Paper 1, received his doctorate from the University of Kansas in 1961. After three years at the Department of Oceanography of Texas at A & M University, he joined the faculty of the University of Kansas in 1965 where he currently is professor of geology and civil engineering.

Papers 2 and 3 are from the section on bismuth in the *Handbook of Geochemistry* edited by K. H. Wedepohl. They provide valuable information on abundances, distribution, occurrences of bismuth in minerals, isotopes, and other physical data.

Knowledge of the crystal chemistry of bismuth is of considerable help in understanding the distribution of bismuth in minerals. In Paper 2, V. Kupčík presents an excellent summary of the geochemical occurrence of bismuth as controlled by crystallo-chemical relations. The main bismuth mineral groups are discussed. These groups include: (1) metallic bismuth and alloys; (2) sulfides, selenides, tellurides, and sulfosalts; and (3) oxides, oxosalts, and halogen-oxosalts. For many of the rarer bismuth minerals, formulae, Bi-O interatomic distances, structural data, interatomic distances, coordination shape, and references are tabulated.

Extensive data on bismuth are collected in the several tables presented in Paper 3 by Ahrens and Erlank. The long list of bismuth minerals presented in this paper will be of considerable help to anyone unfamiliar with the element, as many minerals containing bismuth are relatively rare. One should be cautioned, however, that the list is not complete. Some data on the phase equilibrium of bismuth is presented, and a brief summary and discussion of bismuth occurrences is given. An interesting observation in Paper 3 is the lack of qualitative data for bismuth in tektites.

From their review of bismuth geochemistry in igneous rocks, Ahrens and Erlank conclude that "the present data allow no firm conclusions to be drawn as to the variation of Bi in the various igneous rock types and with magmatic differentiation." They further conclude that "detailed knowledge regarding the mechanism of substitution of bismuth in sulfide minerals is lacking." These reviews show that many more data are required before we can state with confidence what the crustal abundance of bismuth is, how it moves through the magmatic processes, and what are its pathways to specific minerals.

L. H. Ahrens has contributed greatly over the years to expanding our knowledge of the geochemical cycles of the elements. In particular he has considerably extended our understanding of the part that ionic potentials play in trace element distributions in nature. A part of his early career was spent at the Massachusetts Institute of Technology. He has been associated for many years with the Department of Geochemistry of the Universtiy of Cape Town, South Africa. A. J. Erlank, a protégé of Ahrens, has also spent the largest part of his career at Cape Town.

In sum, although one may question whether summary articles can be considered benchmark papers, we have included these three because they provide a succinct introduction to the geochemistry of bismuth, a subject for which little data is available. Furthermore, they are stepping stones to an understanding of what follows in the other sections. In this sense, these papers, though introductory, are benchmarks.

REFERENCES

Goldschmidt, V. M. 1954. *Geochemistry.* Oxford University Press, London, 730 pp.

Rankama, K., and Th. G. Sahama. 1950. *Geochemistry.* University of Chicago Press, Chicago, 911 pp.

1

This article was prepared expressly for this Benchmark volume

BISMUTH: ELEMENT AND GEOCHEMISTRY

E. E. Angino

PHYSICAL PROPERTIES

Bismuth (Ger., Weisse Masse, white mass, later Wismuth), symbol Bi, atomic weight 208.980; atomic number 83; electronic configuration (Xe core) $4f^{14}5d^{10}6s^{2}6p^{3}$; melting point 271.3°C; boiling point 1560°C; specific gravity 9.75 (20°C); valence 3 or 5; ionic radius (+3) 1.16Å. Crystal modification: hexagonal metallic form (ditrigional scalenohedral class). Isotopic abundances: Bi^{209}, 100%; several other isotopes occur as daughter products in the decay chains of uranium and thorium. Bismuth is a brittle, very dia-magnetic, reddish white metal (often with brassy tarnish colors) and is a poor conductor of heat and electricity.

MINERALS AND COMPOUNDS

Bismuth is commonly found in the native state and often contains traces of arsenic (As), sulfur (S), and tellurium (Te). The most important bismuth mineral is bismuthinite (Bi_2S_3), which resembles stibnite (Sb_2S_3) very closely. Bismuth is not abundant and is usually found in veins associated with silver (Ag), cobalt (Co), lead (Pb), tin (Sn), or zinc (Zn) ores. Rarer minerals include bismuthotantalite ($BiTaO_4$), tetradymite (Bi_2Te_2S), guanajuatite [$Bi_2(S,Se)_3$], matildite ($AgBiS_2$), and aramayoite [$Ag(Sb,Bi)S_2$]. Important localities are Frieberg, Saxony; Joachimstal, Bohemia; Bolivia; Cornwall, England; and Great Bear Lake, Canada. The most important compounds are the trioxides and the subnitrates of medicinal use. Its soluble salts are characterized by the formation of insoluble basic salts on addition of water.

CHEMISTRY

The chemical properties of bismuth resemble those of antimony and arsenic. Chemical Bi is stable in air at room temperature, but burns, when heated, to yield Bi_2O_3. At elevated temperatures, it combines with the halogens, sulfur, and a variety of metals. Bi is unreactive toward mon-oxidizing acids. The best solvent for Bi is nitric acid.

$$Bi + H_2O \rightarrow BiO^+ + 2H^+ + 3e-$$
$$E° = -0.32V$$

Bi has a strong tendency to assume the +3 condition. The +5 state is very rare; Bi^2O^5 has not been prepared in the pure condition. Bi^{5+} is a very strong oxidizing agent in acid solution.

$$Bi_2O_5 + 6H^+ + 4e^- \rightarrow 2BiO^+ + 3H_2O$$
$$Eo \simeq 1.60V$$

Treatment of alkali metal bismuthates ($MBiO_3$) with HNO_3 gives a red-brown substance approximating Bi_2O in composition, which compound loses considerable oxygen, even at temperatures below 100°C. Orthobismuthates (M_3BiO_4) are obtained by heating the sesquioxide with appropriate amounts of alkali metal oxide (or perioxide) in air.

A complete series of Bi trihalides is known; all trihalides can be formed by direct combination of the elements. Bi trihalides are partially and reversibly decomposed with the formation of insoluble compounds containing BiO^+ (bismuthyl) cations.

$$BiX_3 + H_2O \rightarrow BiO^+ + X^- + 2H^+ + 2X^-$$

Complexes are known. Bi trihalides yield salts of the types $M[BiX_4]$, $M_2[BiX_5]$, and $M_3[BiX_6]$. Additional compounds, mainly of the 1:1 variety, are formed with organic donor molecules. Bi(III) chloride yields such compounds with tertiary amines. At least two crystalline forms of Bi(III) oxides are known. The compounds exhibit no acidic tendencies and are insoluble in alkalis but dissolve easily in acids to form Bi(III) salts.

The addition of OH^- to solutions of Bi salts gives $Bi(OH)_3^+$, which is completely basic. The hydrated cationic species Bi $(H_2O)^{3+}$ is present in solution when the sesquioxides are dissolved in concentrated solutions of strong oxyacids. Normal salts such as $Bi_2(SO_4)_3$ or $Bi(NO_3)_3 \cdot 5H_2O$ can be crystallized from such solutions, but they hydrolyze easily.

GEOCHEMISTRY

There is a dearth of geochemical information concerning Bi, and a general outline of the geochemistry of Bi is available only from general principles and mineralogical experience. Bismuth is a distinctly chalcophile element, but it also possesses certain lithophile characteristics. Concentrations in meteorites range from .02–2 ppm. Various estimates of the abundance of bismuth in the upper lithosphere have been made; however, the few data available cluster around a figure of 0.2 ppm for a crustal abundance. Only small amounts of bismuth are found in rocks and minerals belonging to the early and main stages of magmatic crystallization; it becomes strongly enriched in late magmatic differentiations. Little data is available on the concentration of Bi in the various members of the sequences of magmatic rocks. It is known to be captured in apatites and probably in other early calcium minerals. Bi tends to be concentrated in granitic magmas and is common in granitic pegmatities. In dolerite, concentrations of 0.01–0.80 ppm have been reported. Other rock types ands ranges are: silicic rocks (.02–.9), intermediate rocks (.02–.25), basaltic rocks (.01–.8), and ultramafic rocks (.01–1.2).

Bismuth is probably widely distributed in minerals of yttrium and yttrium lanthanides and in hydrothermal mineral deposits associated with granitic magmas. Bismuth minerals are also known in pneymatolytic deposits associated with volcanic action. Bismuthinite is the most important hydrothermal mineral of Bi. In high-temperature hydrothermal deposits, Bi is often assocaited with gold (Au), Te, and selenium (Se). It is most commonly present in galena.

Little is known of the behavior of bismuth during weathering. Observations are limited to the oxidation zones of ore deposits. Products of oxidation are bismite, hydrocarbonates, vanadates, and, rarely, silicates of Bi. No oxidation state in Bi minerals higher than 3 is known. Bismuth is enriched to some degree in sedimentary iron ores and in the hydrolyzates and can be detected in some reduzates. Residual sediments are essentially devoid of bismuth. In shales concentrations range from .1 to 2.1 ppm.

Bismuth occurs in seawater at a reported concentration of 2×10^{-5} mg/l (ranging from 1.5 to 4.2×10^{-5} mg/l) and has a residence time of 4.5×10^5 years. It has also been found in Mn nodules ranging in concentration from 2.5–24 ppm.

The radioactive isotope concentrations of Bi in seawater and in surface sediments are given in the following table (from Horne 1968):

Isotope	*Decay Time*	*Decay Mode*	*Est. Aver. Conc Seawater g/l*	*Est. Aver. Conc. Surface Sediments g/g*
Bi^{214}	19.7 min	β	2.1×10^{-21}	8.8×10^{-20}
Bi^{212}	60.5 min	β,α	2.2×10^{-22}	3.7×10^{-20}
Bi^{211}	2.16 min	α,β	$< 5.6 \times 10^{-34}$	1.0×10^{-21}
Bi^{210}	5.01 days	β	7.8×10^{-19}	3.1×10^{-17}

Little is known about the biogeochemistry of Bi, except that it appears to be one of the least toxic of the heavy metals.

ECONOMIC GEOLOGY

In all known deposits of economic import, Bi is found in sulfide minerals or in secondary alteration products of sulfides and related compounds. Commercial ore minerals of bismuth are native Bi, bismuthinite, and bismuth ochre. Few deposits are mined exclusively for Bi. Bi is commonly produced as a by-product of lead refining and is also derived from foreign and domestic ores of Sn, Copper (Cu), and Ag.

Bismuth minerals occur in fissure veins and in replacement deposits formed by hydrothermal solutions. Major producing countries are Bolivia, Canada, Japan, Mexico, Peru (about one-quarter of world's supply), South Korea, and the United States.

Extraction of Bi requires a complex process to eliminate other metals. Ores are reduced in furnaces with carbon (C) or iron (Fe) and soda ash. The Bi in Pb is dissolved in HCl as an oxychloride and smelted. Refinery slimes are fused with caustic soda and soda ash. Bismuth metal is consumed in small quantities. Its greatest uses are in metallurgical, chemical, and medical applications.

REFERENCES

Bateman, A. M. 1950. *Economic Mineral Deposits.* John Wiley & Sons, New York, 916 pp.

Brooks, R. R., and L. H. Ahrens. 1961. Some observations on the distribution of thallium, cadmium, and bismuth in silicate rocks and the significance of covalency on their degree of association with other elements. *Geochim. Cosmochim. Acta* **25**: 100–15.

Fleischer, M. Personal Communication, 1976.

Goldschmidt, V. M. 1954. *Geochemistry.* Oxford University Press, London, 730 pp.

Horne, R. A. 1969. *Marine Chemistry.* Wiley-Interscience, New York, 568 pp.

Howe, H. E. 1969. Bismuth. In the *Encyclopedia of the Chemical Elements,* Clifford A. Hamper, ed. Reinhold, New York, pp. 56–65.

Kleinburg, J., W. J. Argersinger, and E. Griswold. 1960. *Inorganic Chemistry.* D. C. Heath and Co., Lexington, Mass., 680 pp.

Rankama, K., and Th. G. Sahama. 1950. *Geochemistry.* University of Chicago Press, Chicago, 911 pp.

2

Reprinted from pages 83-A-1–83-A-7 and references of *Handbook of Geochemistry*, Vol. II/1, K. H. Wedepohl, ed., Berlin, Heidelberg, New York: Springer-Verlag, 1972

BISMUTH: CRYSTAL CHEMISTRY

V. Kupčík

Bismuth is crystallo-chemically very closely related to arsenic and antimony, as well as to lead. This follows from its position in the periodic table (group Vb), from the common electronic structure of Bi^{+3} and Pb^{+2}, and from the common geochemical occurrence. Bismuth is found mainly in sulfidic ore deposits and their oxidation zones. It occurs in rocks only as a trace element. The main groups of bismuth minerals include:

a) metallic bismuth and alloys (rare),
b) sulfides, selenides, tellurides, and sulfosalts,
c) oxides, oxosalts and halogen-oxosalts.

Bismuth, in contrast to arsenic and antimony, is trivalent in all its minerals. The ionic radius of Bi^{3+} is 1.20 Å according to WYCKOFF, and 0.96 Å according to AHRENS. The covalent radius (nowhere given precisely) is about 1.40 Å; the electron structure is 6 s^2p^3. The electro-positive character of Bi is the strongest within group Vb, but is only slightly stronger than that of antimony.

I. Metallic Bi and Alloys

Metallic bismuth occurs frequently in sulfidic ores. It is, like antimony and arsenic, trigonal of space group R3m. In the crystal structure (CUCKA and BARRET, 1962), each bismuth atom is coordinated by six neighbors arranged in a trigonal antiprism ($Bi^{[3+3]}$, Bi—Bi distances: 3×3.07 Å and 3×3.53 Å).

In alloy-like compounds, bismuth can, e.g., replace upto ~10% of the Ag atoms in chilenite (DANA, 1868), or it can form ordered structures, e.g., maldonite, $Au_2^{[6\,Au+6\,Bi]}\ Bi^{[12\,Au+4\,Bi]}_k$ (JURRIAANSE, SB 1933—1935, 315).

II. Sulfides, Selenides, Tellurides, and Sulfosalts

The structures of the S, Se, and Te compounds of Bi may be divided into two groups:

a) Bi_2S_3, with a small isomorphic addition of Se (upto a ratio of S:Se=2:1). This has a chain structure comparable with antimonite, $^{1}_{\infty}[Sb_2^{[3+2]}S_3]$ (cf. ŠĆAVNIČAR, SB 1960, 51).[1]

b) A trigonal layer structure, which occurs with higher Se content in the presence of Te. In this structure Bi is coordinated by S, Se, or Te in the form of a trigonal antiprism. The structures of other compounds of the type $Bi_n(S, Se, Te)_m$ are most probably also layer structures (see Fig. 83-A-1 for the structure of tetradymite, $^{2}_{\infty}[Bi^{[3+3]}]Te_2S$; HARKER, SB 1933—1935, 287).

---

[1] A very flat trigonal antiprism has been found in artificial $Bi^{[3+3]}(Bi_2S_3)_9I_3$, MIEHE and KUPČÍK (1971).

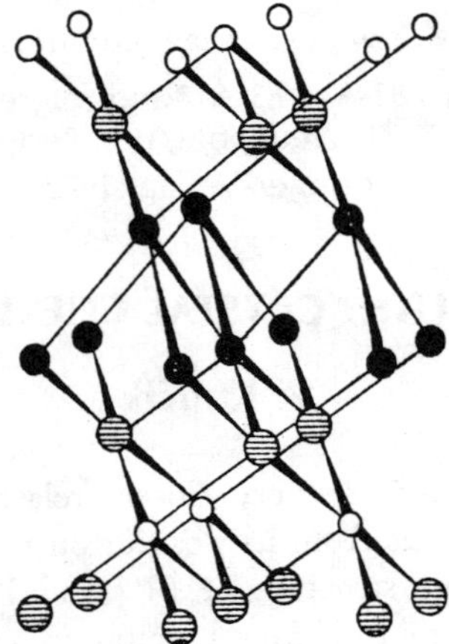

Fig. 83-A-1. Crystal structure of tetradymite, Bi_2Te_2S. Large hatched circles: Bi; medium black circles: Te; small white circles: S

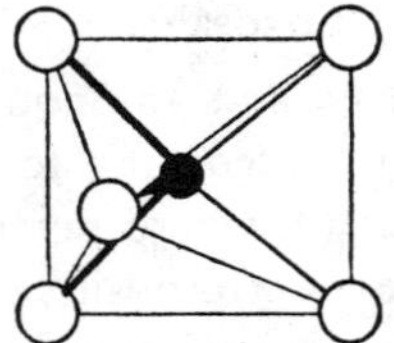

Fig. 83-A-2. Typical $[BiS_5]$ coordination pyramid

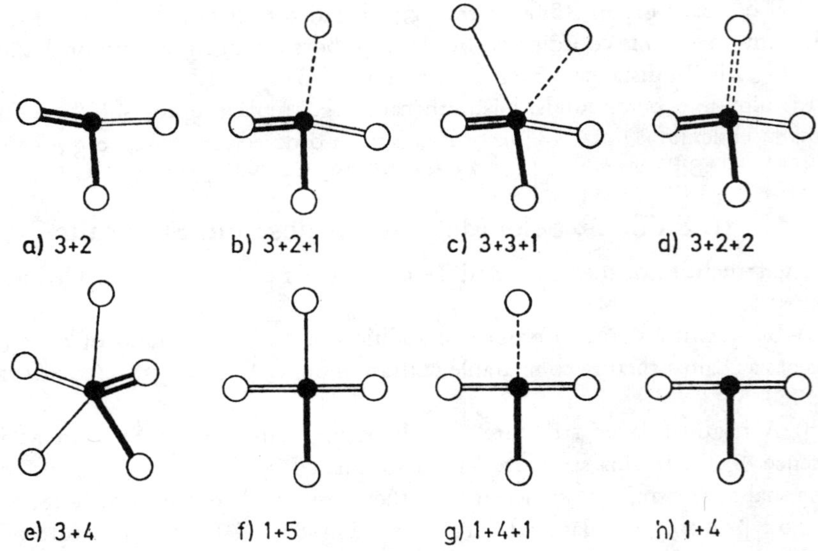

Fig. 83-A-3a—h. Different $[BiS_n]$ coordination polyhedra. See text for explanation. Small black circles: Bi; large white circles: S

Table 83-A-1

Chemical formula Total	Structural	Name	Shape of the coordination polyhedra and some interatomic distances (Å)	References
Bi_2S_3	$^{2}_{\infty}[Bi^{3+3}Bi^{[3+2+1]}S_3]$	bismuthinite	$c\begin{cases}1\times2.69\\2\times2.67\\1\times3.05\\2\times3.07\end{cases} + d\begin{cases}1\times2.54\\2\times2.76\\2\times2.97\end{cases}$	KUPČÍK- and VESELÁ-NOVÁKOVÁ (1970)
$CuBiS_2$	$Cu^{[4]}_{\frac{1}{2}}\,\infty[Bi^{[3+3]}_{\frac{1}{2}}S_4]$	emplectite	$a\begin{cases}1\times2.55\\2\times2.66\\2\times3.16\end{cases}$	HOFMANN SB 1933—35, 393: KUPČÍK (1965)
$Cu_4Bi_6S_{11}$	[a]	hodrushite	$4\times d + 1\times b$ + 1 octahedron	KUPČÍK and MAKOVICKÝ (1968)
$PbBi_2S_4$	$Pb^{[5]}\,^{3}_{\infty}[Bi^{[3+2+1]}_{2}S_4]$	galenobismuthite	$b\begin{cases}1\times2.63\\2\times2.73\\2\times2.99\\1\times3.12\end{cases} + e\begin{cases}2\times2.76\\1\times2.79\\1\times3.00\\2\times3.02\\1\times3.10\end{cases}$	ITAKA and NOWACKI (1962)
$Pb_2Bi_2S_5$	[a]	cosalite	$3\times b + 1\times g$	WEITZ and HELLNER, SR 1960, 63
$CuPbBiS_3$	$Cu^{[4]}\,Pb^{[5]}\,\infty[Bi^{[3+2]}S_3]$	aikinite	$b\begin{cases}1\times2.66\\2\times2.76\\2\times2.95\\1\times3.16\end{cases}$	OHMASA and NOWACKI (1970)
$Pb_3Bi_2S_6$	$Pb^{[6+2]}\,^{2}_{\infty}[Pb^{[6]}_{2}Bi^{[1+5]}_{2}S_6]$	phase III	$f\begin{cases}1\times2.80\\4\times2.90\\1\times2.92\end{cases}$	OTTO and STRUNZ (1968)
$(Cu, Fe)Pb_6(Bi, Sb)_7S_{17.5}$	[a]	kobellite	$2\times a;\ 2\times b;\ b;\ g;\ h$	MIEHE (1971)

[a] Structural formula too complicated, therefore omitted.

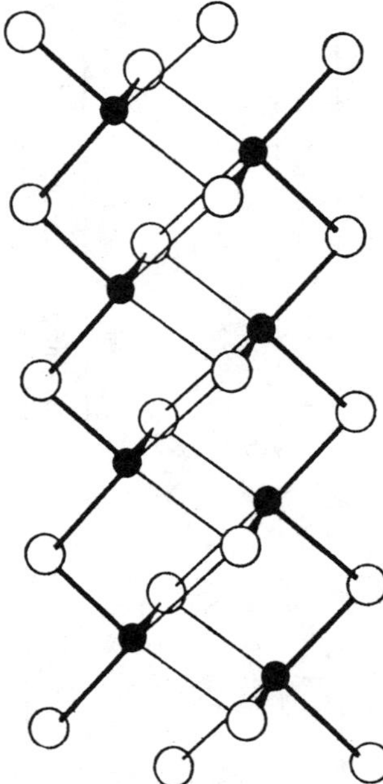

Fig. 83-A-4. Typical $[Bi_2S_4]$ double chain in emplectite, $CuBiS_2$

Bismuthinite, Bi_2S_3 (Kupčík, 1970), with its irregular coordination of sulfur atoms around bismuth, is representative of the whole group of Bi sulfosalts. The common coordination polyhedron is a tetragonal pyramid (Fig. 83-A-2) where the bismuth atom is located slightly above or below the base. One or two additional atoms are often located at the back side of the pyramid. There is only one known exception: galenobismuthite, $PbBi_2S_4$ (Iitaka and Nowacki, 1962), see Fig. 83-A-3e.

The Bi—S valences can be divided according to bond character and strength into the following groups:

a) Bi—S distances from 2.52 to 2.67 Å. These are p^3 bonds approximately rectangular to each other. The vector to the top of the coordination pyramid belongs always to this group of relatively short bonds (see Fig. 83-A-3; heavy lines).

b) Bi—S distances above 2.80 Å. These are hybrid *p-d* bonds (which mostly lie within the base of the pyramid; medium lines).

c) Bi—S distances from 2.94 to 3.17 Å. These are pure *d-γ* bonds lying within the base of the pyramid (light lines).

d) Bi—S distances above 3.27 Å. These are weak *d-γ* bonds (effect of the inner *s*-electron pair; broken lines).

In every structure there are several slightly differently coordinated Bi atoms. The coordination polyhedra polymerize by edge-sharing infinite single chains $\infty[BiS_2]$, or further to double or multiple chains, e.g., $\infty[Bi_2S_4]$, $\infty[Bi_4S_6]$ (see also Fig. 83-A-4).

Isomorphic replacement of As by Bi is unknown and of Sb by Bi possible but, for geochemical reasons, rare. A typical example for a wide-range substitution is the structure of kobellite (Cu, Fe)Pb_6(Bi, Sb)$_7S_{17.5}$ (Miehe, 1971).

The occurrence of a regular coordination (octahedron with Bi atom in the center) must be interpreted statistically. In such a case Bi may be isomorphically replaced by Pb as in hodrushite (Pb, Bi, Ag)$Cu_4Bi_5S_{11}$ (KUPČÍK and MAKOVICKÝ, 1968).

Replacement of Bi by Pb may also occur when each Bi atom replaced by a Pb results in an additional Cu atom occupying tetrahedral interstices of the bismuthinite structure. This explains the isomorphism of the bismuthinite-aikinite series, $(Bi_{2-x}, Pb_x)Cu_xS_3$ with $x=0$ to 1 (OHMASA and NOWACKI, 1970). It is assumed that the intermediate members have an ordered Bi—Pb, Cu distribution, which leads to multiple lattice constants and therefore to different mineral types. Data are given in Table 83-A-1.

III. Oxidic Compounds

In contrast to Bi sulfides and sulfosalts, which are very similar to the corresponding Sb compounds and in which Bi always has the same sulfur coordination, the oxidic compounds of Bi are much more complicated. E. g., α- and β-Bi_2O_3 are not related crystallo-chemically to the corresponding Sb compounds. The coordination around

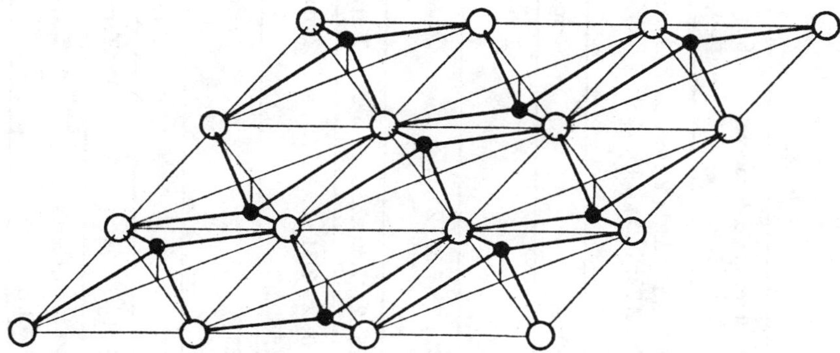

Fig. 83-A-5. $^2_\infty$ [BiO] nets formed from linked [BiO_4] pyramids in kettnerite, CaBi[OF|CO_3]. Small black circles: Bi; large white circles: O

Bi is very irregular (except in bismuthite and β-Bi_2O_3) and is difficult to describe in terms of common polyhedra types, though two frequently occurring basic arrangements may be distinguished:

a) $p^3\,d^{2-3}$ bonding (similar to sulfides) in a deformed and more or less incomplete octahedral coordination with 3 short bonds (tetragonal pyramid with Bi occupying the basis in α-Bi_2O_3 or trigonal antiprism in eulytite, $Bi_4Si_3O_{12}$);

b) extremely flat tetragonal pyramid with Bi occupying the apex.

Schematically, bond lengths can be related to the kind of hybridization:

a) ~2.09—2.30 Å	pure p^3 bonding
b) ~2.30—2.40 Å	$p^3\,d^n$ hybrid bonding
c) ~2.50—2.65 Å	pure d bonding
d) >2.70 Å	weak d bonding with repulsion effects from inert electron pair

Table 83-A-2

Chemical formula	Name	Bi—O interatomic distances [Å]	Shape of the coordination polyhedra	Reference
α-Bi_2O_3	bismite	Bi_1 2.08; 2.17; 2.21; 2.54; 2.63 B_2 2.14; 2.22; 2.29; 2.48; 2.54; 2.80	Tetragonal pyramid (3 + 2) Trigonal antiprism	Malmros (1970)
β-$Bi_2^{[6]} O_3$	—	6 × 2.39	Trigonal antiprism	Sillén SR 1937, 69
$Bi_2^{[2]} O_2[CO_3]$	bismutite	4 × 2.29	Tetragonal pyramid	Lagerkrantz and Sillén, SR 1947—1948, 308
$Bi^{[4+4]} [PO_4] \cdot 0.5 H_2O$	—	2 × 2.31; 2 × 2.33; 2 × 2.56;	Quadr.-planar + 2 × 2	Mooney and Slater (1962)
$Bi^{[2+2+1]} Mo^{[2+4]} O_6$	koechlinite	2 × 2.25; 2 × 2.39; 2.53	Tetragonal pyramid + 1	Zemann SR 1956, 449
$Bi^{[4+4]} V^{[4]} O_4$	pucherite	2 × 2.20; 2 × 2.31; 2 × 2.54; 2 × 2.73	Tetragonal prisma	Qurashi and Barnes SR 1953, 503
$CaBi_2^{[4]} O_2[CO_3]_2$	beyerite	4 × 2.29	Tetragonal pyramid	Lagerkrantz and Sillén, SR 1947—1948, 308
$Bi_4^{[3+3]} [SiO_4]_3$	eulytite	3 × 2.15; 3 × 2.62	Trigonal antiprism	Segal *et al.* (1966)
$Bi^{[4]} OCl$	bismoclite	4 × 2.32	Tetragonal pyramid	Sillén, SR 1947—1948, 312
$CaBi^{[4]} [OF/CO_3]$	kettnerite	4 × 2.29 + 4 from CO_3 undefined	Tetragonal pyramid	Syneček and Žák, SR 1960, 426

Data are given in Table 83-A-2.

There are several different ways of linking the coordination polyhedra, and the following types have been found:

a) isolated $[BiO_6]$ groups in eulytite, $Bi_4[SiO_4]_3$;

b) $^2_\infty[BiO]$ nets built of BiO_4 pyramids sharing edges (see Fig. 83-A-5) in bismoclite, BiOCl, bismutite, $Bi_2O_2[CO_3]$, and kettnerite, $CaBi[OF|CO_3]$;

c) probably $^1_\infty[BiO_2]$ chains with edge-sharing in koechlinite, $BiMoO_6$;

d) three-dimensional arrays $^3_\infty[Bi_2O_3]$, resulting from polyhedra sharing edges and corners in bismite, α-Bi_2O_3.

Isomorphous replacement of Bi by other elements is not known in oxidic minerals.

References: Section 83-A

Cucka, P., Barret, C. S.: The crystal structure of Bi and solid solutions of Pb, Sn, Sb and Te in Bi. Acta Cryst. **15**, 865 (1962).

Dana in: Strunz, H. Mineralogische Tabellen (4. ed.), p. 100. Leipzig: Akad. Verlagsgesellschaft 1966.

Iitaka, Y., Nowacki, W.: A redetermination of the crystal structure of galenobismutite. $PbBi_2S_4$. Acta Cryst. **15**, 691 (1962).

Kupčík, V.: Struktur des Emplektit $CuBiS_2$. Referate der 8. Diskussionstagung der Sektion für Kristallkunde der DMG, Marburg, 16 (1965).

— Makovický, E.: Die Kristallstruktur des Minerals (Pb, Ag, Bi)$Cu_4Bi_5S_{11}$. Neues Jahrb. Mineral., Monatsh. **236** (1968).

— Veselá-Nováková, L.: Zur Kristallstruktur des Bismuthinits, Bi_2S_3. Mineral. Petrog. Mitt. **14**, 55 (1970).

Malmros, G.: The crystal structure of α-Bi_2O_3. Acta Cryst. **24**, 384 (1970).

Miehe, G.: Crystal structure of kobellite. Nature Physical Science **231**, 133 (1971).

— Kupčík, V.: Die Kristallstruktur des $Bi(Bi_2S_3)_9J_3$. Naturwissenschaften **58**, 219 (1971).

Mooney-Slater, R. C. L.: Polymorphic forms of bismuth phosphate. Z. Krist. **117**, 317 (1962).

Ohmasa, M., Nowacki, W.: Note on the space group and on the structure of aikinite derivatives. Neues Jahrb. Mineral., Monatsh. **158** (1970).

— — A redetermination of the crystal structure of aikinite $[BiS_2|S|Cu^{IV}Pb^{VII}]$. Z. Krist. **132**, 7 (1970).

Otto, H. H., Strunz, H.: Zur Kristallchemie synthetischer Blei-Wismut-Spießglanze. Neues Jahrb. Mineral., Abhandl. **108**, 1 (1968).

Segal, D. J., Santoro, R. P., Newnham, R. E.: Neutron-diffraction study of $Bi_4Si_3O_{12}$. Z. Krist. **123**, 73 (1966).

3

Reprinted from pages 83-B-1–83-M-1, 83-0-1, and references of *Handbock of Geochemistry*, Vol. II/1, K. H. Wedepohl, ed., Berlin, Heidelberg, New York: Springer-Verlag, 1969

BISMUTH: ISOTOPES, ABUNDANCES, AND INTERELEMENT RELATIONSHIPS

L. Ahrens and A. J. Erlank

83-B. Isotopes in Nature

Only one bismuth isotope, Bi^{209}, is found in nature. The 1961 atomic weight of bismuth is 208.980.

83-C. Abundance in Cosmos, Meteorites and Tektites

I. Cosmos

In the absence of abundance data for the solar atmosphere, two methods of assigning a cosmic or solar system abundance value for Bi have been proposed. The first involves the use of abundances in the chondritic meteorites. The large difference in the various chondrite groups (Table 83-C-1) introduces uncertainty in this respect.

Table 83-C-1. *Bismuth contents of meteorites*

Name	Class symbol	References	ppb Bi[a]
Chondrites			
Beardsley	CH	1	3.3
		2	3.5
Forest City	CH	1	0.28
		2	1.0
Plainview	CH	1	3.6
Holbrook	CL	1	2.1
		2	7.5
		2	2.0
Modoc	CL	1	1.1
		2	0—6
Mighei	Cc_2	2	180
Orgueil	Cc_1	2	130
Abee	Ce	2	80
Achondrites			
Johnstown	Ab	1	2.8
Nuevo Laredo	Ap	2	<1
Meteoritic troilite from Iron Meteorites			
Cañon Diablo troilite		2	40
Toluca troilite		2	180

References: 1 = EHMANN and HUIZENGA (1959). 2 = REED, KIGOSHI, TURKEVICH (1960).
[a] All results by neutron activation analysis, 1 ppb $= 10^{-3}$ ppm.

The consensus view is that particular significance should be attached to the carbonaceous chondrites. In the most recent compilation of this type (CAMERON, 1968), the atomic abundance of Bi is given as 0.164 (relative to Si $= 10^6$ atoms).

The second approach attempts to predict cosmic abundances using modern theories of nucleosynthesis and other nuclear considerations. The latest calculations by SEEGER *et al.* (1965) give a Bi value of 0.75 (relative to Si $= 10^6$ atoms). The disparity between this value and that given by CAMERON (1967) requires further attention.

II. Meteorites

Table 83-C-1 lists all available neutron activation analyses for Bi in meteorites. It may be noted that the Bi concentration in the "ordinary" chondrites (class symbols CH and CL) varies rather widely, averaging about 3 ppb (0.003 ppm).

However, the Bi magnitude increases to 100—200 ppb (0.1—0.2 ppm) in the carbonaceous and enstatite chondrites (class symbols Cc and Ce), showing that Bi, in common with other heavy chalcophile elements, Hg and Tl in particular, is markedly fractionated in the principal chondrite groups (Reed *et al.*, 1960). The contents of Bi in the troilite phases of iron meteorites bears testimony to the chalcophile behaviour of Bi.

III. Tektites

No quantitative data for Bi in tektites are available. Taylor (1966) analysed five australites by means of a spark source mass spectrographic technique but could not detect Bi. However, he indicates Bi to be <10 ppb (0.01 ppm). If so, this probably implies a depletion of Bi in australites as compared with suggested parent materials, as two greywackes and an impact glass from Henbury, Australia, have 0.59, 0.65 and 0.37 ppm Bi respectively.

83-D. Abundance in Sulphides and in Rock Forming Minerals; Bismuth Minerals

Because of its chalcophile tendency, Bi occurs most commonly in sulphide minerals and hence most data are available on sulphides. Data for normal silicate minerals are virtually non-existent, as the abundance levels likely to be encountered can only be estimated by sensitive techniques such as neutron activation analysis or other techniques employing pre-enrichment procedures. Hence the geochemistry of Bi is dominated by its presence in sulphides, and it is necessary to consider this aspect in some detail. It should of course be borne in mind that sulphides are not normally rock forming minerals.

Ionic Radius of Bi^{3+}

V. M. Goldschmidt (1926) estimated the radius of Bi^{3+} as 1.20 A. Subsequently, Ahrens (1952) recommended a value of 0.96 A. According to Ahrens and Kable (in press) this value is too low and they recommend a radius of 1.16 A.

I. Chalcophile Tendency of Bismuth

A distinct indication based on meteorite evidence, that Bi is mainly chalcophile was provided by I. and W. Noddack (1930 and 1934), who reported the Bi contents in metal, sulphide (troilite) and silicate phases of meteorites as 0.5, 2 and 0.02 ppm, respectively.

The other two Group V-b elements, As and Sb, also tend to be chalcophile, but have distinctly greater siderophile tendencies than Bi. Although the techniques used by the Noddacks do not appear to be highly accurate, the relationships they observed between the three principal meteorite phases probably reflect the relative intensities of the respective tendencies. The chalcophile behaviour of bismuth is also borne out by the results reported by Reed *et al.* (1960) who report 0.04 and 0.18 ppm Bi in two samples of troilite from iron meteorites, as compared with a level of 0.003 ppm Bi in ordinary chondrites (see Table 83-C-1).

II. Chalcophile Behavior of Bismuth

Bi forms several sulphide minerals of its own (Table 83-A-1). The major proportion of "chalcophile" Bi in the earth's crust occurs, however, in trace quantities in various common sulphide minerals, notably galena which appears to be the principal host mineral for this element. In the discussions which follow on the chalcophile behaviour of Bi, galena will be considered first. For convenience, much of the available data for common sulphides is summarized in Table 83-D-1. It is not possible to calculate realistic averages for the individual minerals as many of the analysed samples are below the detection limits of the methods employed. Note also the variation in sensitivity given by the various authors, hence in Fleischer's compilation the category "not found" in some cases indicates a concentration > 10 ppm.

Table 83-D-1. *Summary of bismuth contents of common sulphide minerals*

Max. conc. % or ppm	No. of samples in each concentration range: 1% or more	5,000 to 10,000 ppm	1,000 to 5,000 ppm	500 to 1,000 ppm	200 to 500 ppm	100 to 200 ppm	50 to 100 ppm	10 to 50 ppm	< 10 ppm	not found	Total no. of samples	Sensitivity ppm	Reference	Method
Galena														
5%	23	1	28	18	26	19	8	43	36	125	327	1—100	1	S
3.9%	1	2		3		1		3	10	12	32	10	2	C
Sphalerite														
1,000						7	1	13	2	143	166	1—10	1	S
4,000			4	3	3	7	12	16	57	120	222	1	3	S
900				3	7	7	11	35	64	59	186	1—3	4	S
Chalcopyrite														
2,000			6	7	5	4	9	30	31	137	229	1	3	S
2,000			2	1	8	3	7	14	10	82	127	1—3	4	S
Pyrite														
100						1	—	4	1	11	17	10—100	1	S
780				4	9	21	27	63	49	422	605	10	5	S

References: 1 = Fleischer (1955), 2 = Nesterova (1958), 3 = Burnham (1959), 4 = Rose (1967), 5 = Cambel and Jarkovsky (1967).

In a detailed review on the occurrence of Bi in sulphide minerals FLEISCHER (1955) refers to several pre-1954 investigations on galena (PINA DE RUBIES and DOETSCH, LOPEZ DE AZCONA, WASSERSTEIN, OFTEDAL, WAHLSTROM, LEUTWEIN and HERRMANN: for reference details, see FLEISCHER's review), and on the basis of their observations it is quite clear that the Bi content varies enormously from <10 ppm up to 5%. Subsequent observations (NESTEROVA, 1958; HERTEL, 1966, for example) corroborate this conclusion. In 40 specimens of U.S.S.R. galena, Bi was found to vary from <10 ppm to almost 4%.

The mode of occurrence of Bi in galena is problematical. Some Bi, together with Ag, may occur structurally because of the possibility of the coupled substitution

$$Ag^{I}Bi^{III} \rightleftharpoons 2\,Pb^{II}.$$

If, however, the Bi and Ag concentrations are high, schapbachite (matildite) ($AgBiS_2$) may tend to exsolve (RAMDOHR, 1938 etc. — References in RAMDOHR, 1960; OFTEDAL, 1942). HOEHNE (1934) had earlier concluded that a substantial proportion of Bi occurred in galena as inclusion impurities. This view was also held by BETEKHTIN (1950) and later by NESTEROVA (1958) who maintained that the inclusions were mainly bismuthinite. The more complex Bi sulphides (benjaminite, tetradymite and matildite) are relatively rare.

The possibility that the Bi content of galena might be temparature dependent was investigated first by OFTEDAL (1941). American sphalerite data are given by SIMS and BARTON (1961).

CAMBEL and JARKOVSKY (1967) carried out a very detailed study of the abundances and distribution of several elements, including Bi, in pyrites from Czechoslovakia. Samples from various modes of origin and from a wide variety of localities were analyzed. The Bi concentrations in the great majority of specimens are below the detection limit (10 ppm) of the spectro-chemical method that was used. Several specimens, but nevertheless a minority, contain 10—40 ppm, and occasionally the level rises above 100 ppm reaching a maximum of $\sim$1,000 ppm. (one specimen only).

The concentrations of most elements investigated by CAMBEL and JARKOVSKY appear to follow a lognormal-type distribution in the pyrite specimens they studied. Data for Bi were, however, not sufficient to establish the statistical nature of the distribution of the concentration of this element in pyrite. The data of NESTEROVA (1958), BURNHAM (1959) and ROSE (1967), as indicated in Table 83-D-1, would, however, suggest that Bi does follow a lognormal distribution in galena, sphalerite and chalcopyrite.

In summary, two features are notable, (1) the preference of Bi for galenas, and (2) the large spread of concentration for all the sulphides discussed. In the latter respect, it is worth re-emphasizing that many of the higher concentrations encountered in the various sulphides may be explained by admixture of small amounts of foreign bismuth minerals. Despite considerable variations in the Bi concentration in galena from different modes of occurrence, there is a distinct indication that the Bi content tended to increase with increase in temperature. This possible temperature effect has been discussed by GOLDSCHMIDT (1954). SCHROLL (1955) is also of the opinion that relatively high Bi and Ag concentrations tended to occur in galenas which had been formed at high temperatures. Further discussion may be found in the section dealing with phase relationships in the Pb—Bi—S system (subsection D-VI).

The Bi concentration in other common sulphide minerals is usually much less than in galena. In his review of data up to 1954 FLEISCHER (1955) concluded that in the zinc sulphides (sphalerite and wurtzite), most specimens contain <1—10 ppm Bi; concentrations up to a few hundred ppm are occasionally found. According to the pre-1955 information, the Bi content of chalcopyrite is usually less (maximum, a few ppm) than in sphalerite and in pyrite, concentrations are still lower. The investigations by BURNHAM (1959) and ROSE (1967) on the composition of chalcopyrite and sphalerite, mainly from metallogenic provinces of the south western U.S.A. and northern New Mexico, show that where as the Bi levels are often low in these minerals, occasional high concentrations (up to $\sim$1,000 ppm) may be found in both sphalerite and chalcopyrite. It is quite clear from the detailed work of ROSE in particular that not only does the Bi content vary enormously — together with several other chalcophile elements — from specimen to specimen, whether from one area or not, but variation can be quite extreme in single specimens. Further data on the Bi contents of amounts of foreign bismuth minerals.

Detailed knowledge regarding the mechanism of substitution of Bi in sulphide minerals is lacking. Although bonding in sulphides is typically covalent, ionic size is apparently a consideration, as the ionic radius of Bi^{3+} (1.16 A) is closer to that of Pb^{2+} (1.20 A) than it is to Fe^{2+} (0.74 A), Cu^{2+} (0.72 A) or Zn^{2+} (0.74 A), hence providing some evidence for the higher concentration of Bi in galena. Covalent radii show the same relationships. Further discussion may be found in FLEISCHER (1955).

III. Lithophile Behavior of Bismuth

Although Bi is a distinctly chalcophile element (noted above), Bi^{III} exhibits some lithophile tendency. It will be recalled that Bi^{3+} is a relatively large cation ($r = 1.16$ A) and silicates the possibility of a Bi^{3+}—Ca^{2+} ($r = 1.01$ A) substitution exists, as mentioned by GOLDSCHMIDT (1954). If such substitution does take place to a significant extent, Bi would be expected to be enriched in plagioclase rich (basaltic) rocks, as compared with acidic rocks. BROOKS *et al.* (1960) and particularly BROOKS and AHRENS (1961) found no such relationship.

The possibility of a substitution of Bi^{3+} for Ca^{2+} in the mineral apatite has been stressed by GOLDSCHMIDT (1954) who stated that Bi is "... known to be captured in apatite." Neither the source of information nor any data are given. GURNEY and AHRENS (in press) have recently carried out a reconnaissance survey on 13 specimens of apatite from various localities. They estimate 5 of the samples to have Bi contents within the range 5—100 ppm with the highest Bi contents occurring in apatites from pegmatites. The other samples have Bi contents below the detection limit ($\sim$1 ppm) of the semi-quantitative spectrographic method employed.

Further data are presented by CLARK (1965) who reports the presence of Bi in several apatites from hydrothermal veins in Finland. Concentrations of up to 1,150 ppm Bi are recorded. Although the veins contain bismuth minerals, CLARK suggests that Bi occurs in the apatite structure, substituting for Ca.

Table 83-D-2 lists the meagre data available for silicate minerals. From this, and from the data available for silicate rocks (see next section), it appears that a bismuth content of <1 ppm can normally be expected in the common silicate minerals.

IV. Inter-Element Relationship Between Bismuth and the Rare Earth Elements

As the radius of Bi^{3+} is close to the radii of the trebly charged Rare Earth (RE) cations (0.88 A to 1.15 A) and as the crystal chemistry of several Bi^{III} compounds is similar to that of Rare Earth compounds (some are isostructural) some Bi—RE association might be expected.

Work on the Bi contents of RE minerals, notably gadolinite, has been carried out by I. and W. Noddack (1931) and by Gurney and Ahrens (in press); results are summarized in Table 83-D-3. Both studies indicate that of the minerals investigated gadolinite ($Y_2FeBe_2[O|SiO_4]_2$) contains the highest amounts of Bi. This suggests that the Bi^{3+}—Y^{3+} association may be closer than that existing between Bi and the other RE elements. In common with other authorities, we class Y as a Rare Earth element.

Table 83-D-2. *Bismuth contents of silicate minerals*

Mineral	No.	Method	ppm Bi average
Perthite, Reed *et al.* (1960)	1	N/R	0.011
Biotite, Brooks and Ahrens (1961)	1	S	0.21
Glauconite, Brooks and Ahrens (1961)	1	S	0.13
Crocidolite, Brooks and Ahrens (1961)	1	S	0.01

Table 83-D-3. *Bismuth contents of rare earth minerals*

Mineral	No.	Method	ppm Bi	
			range	average
Gadolinite, I. and W. Noddack (1931)	4	X	~1—20	
Gadolinite, Gurney and Ahrens (in press)	29	S	12—262	63
Fergusonite, Gurney and Ahrens (in press)	5	S[a]	~20—40	
Allanite, Gurney and Ahrens (in press)	6	S[a]	<1—10	
Monazite, Gurney and Ahrens (in press)	3	S[a]	≈ 5	
Euxenite, Gurney and Ahrens (in press)	2	S[a]	≈ 40	

[a] Semi-quantitative method.

V. Bismuth Minerals

Some bismuth minerals are listed in Table 83-D-4. Many of these are relatively rare and are mainly found in hydrothermal ore deposits where bismuth usually occurs in combination with either sulphur or other metals such as Au, Se and Te. The commonest bismuth minerals are bismuthinite (Bi_2S_3) and bismite (Bi_2O_3), the latter usually occuring as an alteration product. The only bismuth silicate minerals known are eulytite ($Bi_4(SiO_4)_3$) and bismutoferrite ($BiFe_2(SiO_4)_2(OH)$). However, bismuth oxides, carbonates, vanadates, arsenates, molybdates are also found. It should be noted that Table 83-D-4 does not contain a complete list of bismuth minerals, nor does it list those which are considered mixtures or which are discredited in the recent review by Fleischer (1966). Further details may be found in the references listed in Table 83-D-4.

Table 83-D-4. *Bismuth minerals*

Aikinite	$Cu_2S \cdot 2PbS \cdot Bi_2S_3$	Kettnerite	$CaBi(CO_3)OF$
Aramayoite	$Ag(Sb,Bi)S_2$	Kotulskite	$Pd(Te,Bi)_{1-2}$
Arsenobismutite	$Bi_2(AsO_4)(OH)_3$	Koechlinite	$(BiO_2)(MoO_4)$
Atelestite	$Bi_3(AsO_4)O_2(OH)_2$?	Laitakarite	Bi_4Se_2S
Benjaminite	$(Cu, Ag)_2Pb_2Bi_4S_9$	Lindstromite	$PbCuBi_3S_6$
Beyerite	$(Ca, Pb)Bi_2(CO_3)_2O_2$	Maldonite	Au_2Bi
Bismite	Bi_2O_3	Michenerite	$(Pd, Pt)BiTe$
Bismoclite	$BiOCl$	Mixite	$Cu_{11}Bi(AsO_4)_5(OH)_{10} \cdot 6H_2O$
(Daubreéite)	$BiO(OH,Cl)$	Moncheite	$(Pt, Pd)(Te, Bi)_2$
Bismuthinite	Bi_2S_3	Montanite	$(BiO)_2(TeO_4) \cdot 2H_2O$?
Bismutite	$Bi_2(CO_3)O_2$	Paraguanajuatite	$Bi_2(S,Se)_3$
Bismutoferrite	$BiFe_2(SiO_4)_2(OH)$	Parkerite	$Ni_3(Bi, Pb)_2S_2$
Bismutotantalite	$Bi(Ta,Nb)O_4$	Pavonite	$AgBi_3S_5$
Bonchevite	$PbBi_4S_7$	Perite	$PbBiO_2Cl$
Bursaite	$Pb_5Bi_4S_{11}$	Pucherite	$BiVO_4$
Cosalite	$2PbS \cdot Bi_2S_3$	Rooseveltite	$BiAsO_4$
Csiklovaite	$Bi_2Te(S,Se)_2$	Russellite	$(Bi_2, W)O_3$
Cuprobismutite	$CuBiS_2$	Schapbachite (Matildite)	$AgBiS_2$
Emplectite	$Cu_2S \cdot Bi_2S_3$		
Eulytite	$Bi_4(SiO_4)_3$	Sakharovaite	$(Pb, Fe) (Bi, Sb)_2S_4$
Froodite	$PdBi_2$	Sillenite	Bi_2O_3
Galenobismutite	$PbS \cdot Bi_2S_3$	Telluro-bismuthite	Bi_2Te_3
(Cannizzarite)	$Pb_3Bi_5S_{11}$?		
(Bismutoplagionite)	$5PbS \cdot 4Bi_2S_3$	Tetradymite	Bi_2Te_2S
Giessenite	$Pb_9CuBi_6Sb_{1.5}S_{30}$	Ustarasite	$Pb(Bi, Sb)_6S_{10}$
Gladite	$PbCuBi_5S_9$	Walpurgite	$Bi_{10}(UO_2)(AsO_4)_2O_4 \cdot 3H_2O$?
Grunlingite	Bi_4TeS_3	Waylandite	$(Bi, Ca)Al_3(PO_4,SiO_4)_2(OH)_6$
Guanajuatite	$BiSe_3$	Wehrlite	Bi_3Te_2?
Hammarite	$Pb_2Cu_2Bi_4S_9$?	Westgrenite	$(Bi, Ca)(Ta, Nb)_2O_6(OH)$
Healeyite	Bi_7Te_3	Wittichenite	Cu_3BiS_3
Ikunolite	$Bi_4(S, Se)_3$	Wittite	$Pb_5Bi_6(S, Se)_{14}$
Joseite	Bi_4TeS_2	Zavaritskite	$BiOF$

References: PALACHE, BERMAN and FRONDEL (1958), SHORT (1940), American Society for Testing and Materials, ASTM powder diffraction file (1966), FLEISCHER (1966), STRUNZ (1963).

Note: Minerals listed in brackets and not in alphabetical order are possibly varieties of the minerals listed immediately before them.

Sulphide minerals of bismuth are found widely distributed in hydrothermal deposits, in pyrometasomatic deposits alongside granite contacts with limestones, and in other pneumatolytic deposits. In the famous hydrothermal deposits of Tazna, Bolivia, bismuth is associated with tin deposits. Bismuth minerals are also found in granite pegmatites. GOLDSCHMIDT (1954) refers to crystals up to 1 inch wide and 6 inches long, in the Precambrian granite pegmatites from Southern Norway. Bismutotantalite is present in granite pegmatites of East Africa. GOLDSCHMIDT also mentions that he found major amounts of Bi in other tantalates and niobates in Norwegian pegmatites (no data given).

Metallic bismuth often occurs as a secondary mineral after bismuthinite, or in secondary zones of cementation. Further discussion on the occurrence of bismuth minerals and on the general geochemistry of Bi may be found in GOLDSCHMIDT (1954), RANKAMA and SAHAMA (1950), and RAMDOHR (1960).

VI. Phase Equilibria

A useful summary of the phase equilibriu mrelationships in bismuth systems: (Bi_2O_3—MoO_2—PbO; Bi_2O_3—PbO; Bi_2O_3—PbO—WO_3) may be found in LEVIN, ROBINS and MCMURDIE (1964).

VAN HOOK (1960), SALANCI (1965) and CRAIG (1967) have carried out extensive experimental work on the Bi_2S_3—PbS system. The latest data by CRAIG indicate the maximum solubility of Bi_2S_3 in galena to be 10 mol-% (19 weight-%) at 829 ± 6° C,

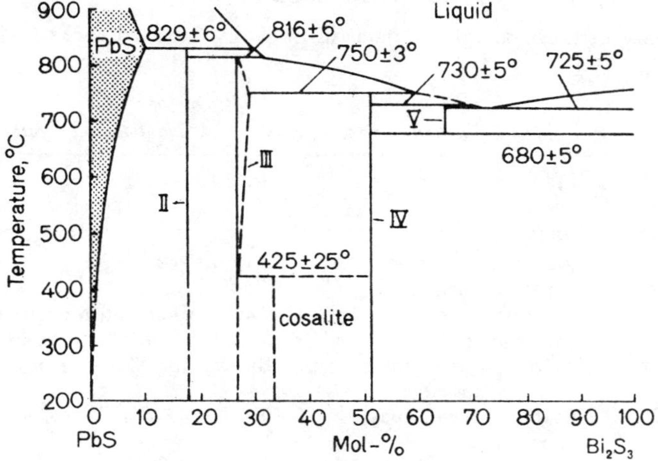

Fig. 83-D-1. System PbS-Bi_2S_3 after CRAIG (1967)

decreasing at lower temperatures (Fig 83-D-1). CRAIG suggests that the solid solution may vary sufficiently with temperature to provide a potential geologic thermometer.

VAN HOOK (1960) and CRAIG (1967) have also investigated the $AgBiS_2$—PbS system. The latter author's data show that solid solution is complete above 215 ± 15° C; below this temperature characteristic Widmanstätten structure-like textures are formed through exsolution. Even so, about 8% $AgBiS_2$ may remain dissolved in PbS at 200° C. Hence the solubility of bismuth in galena may be extensive, even at low temperatures, when silver is present.

83-E. Abundance in Common Igneous Rock Types

I. Bismuth in Standard Rocks, Granite G-1 and Diabase (tholeiitic basalt) W-1

A summary of the available Bi data (in ppm) in standard rocks G-1 and W-1 is given in Table 83-E-1.

Table 83-E-1. *Bismuth in standard rocks, granite G-1 and diabase (basalt) W-1*

G-1 ppm Bi	W-1 ppm Bi	Method	Source
0.09	—	ion exchange enrichment — spectrochemical	BROOKS *et al.* (1960)
0.25	—	polarographic	CARMICHAEL and McDONALD (1961)
0.1	—	spectrographic	SCHROLL and WENINGER (1965)
<0.9	<0.9	spark source — mass spectrometer	BROWN and WOLSTENHOLME (1964)
0.10	0.25	spark source — mass spectrometer	TAYLOR (1965)

Some of the data on G-1 agree reasonably well at a magnitude of 0.1 ppm Bi. Data on W-1 are much less satisfactory. These data emphasize the need to improve the analytical techniques for determining Bi in rocks.

II. Bismuth in Igneous Rocks

Table 83-E-2 lists all other available data for Bi in igneous rocks. In mafic rocks, mainly in Karroo doleritic basalt, the Bi content varies enormously (<0.01 ppm to 0.80 ppm) and averages at 0.15 ppm (BROOKS *et al.*, 1960; BROOKS and AHRENS, 1961). According to the data of the same workers, variations of the Bi content in acid rocks is also quite large. They report a range of 0.02 to 0.9 ppm Bi and an average of 0.18 ppm, close to the average for basaltic rocks. Alkaline rocks (syenite and foyaite) appear to contain less Bi (ave. $\sim$0.03 ppm) but as only five specimens were analysed, an accurate abundance estimate cannot be made.

Table 83-E-2. *Bismuth contents of igneous rocks*

Rock type	No.	Method	ppm Bi range	ppm Bi ave.
Ultramafic rocks				
Peridotite, Norway, BROOKS *et al.* (1960)	1	S[a]		1.20
Lherzolite, France, BROOKS *et al.* (1960)	1	S		0.03
Eclogites (see Table 83-M-1)				

Table 83-E-2 (Continued)

Rock type	No.	Method	ppm Bi range	ave.
Mafic rocks				
Norite, South Africa, BROOKS *et al.* (1960)	1	S		0.05
Dolerite, South Africa, BROOKS *et al.* (1960)	1	S		0.15
Karroo dolerite, South Africa, BROOKS and AHRENS (1961)	16	S	0.01—0.80	0.22
Basalt, S.W. Africa, BROOKS and AHRENS (1961)	1	S		0.02
Basalt, U.S.A., REED *et al.* (1960)	1	N/R		0.008
Intermediate rocks				
Trachyte, New Zealand, BROOKS *et al.* (1960)	1	S		0.09
Syenite, S.W. Africa, BROOKS and AHRENS (1961)	3	S	0.02—0.03	0.03
Syenite, Canada, INGAMELLS and SUHR (1963)	1	S		0.25
Foyaite, S.W. Africa, BROOKS and AHRENS (1961)	2	S	0.04, 0.04	0.04
Andesite, New Zealand, TAYLOR and WHITE (1966)	1	I		0.12
Carbonatites				
Soevite, Ondurakorume Complex, S.W. Africa, BROOKS and AHRENS (1961)	1	S		0.04
Mocaceous, Ondurakorume Complex, S.W. Africa, BROOKS and AHRENS (1961)	1	S		0.06
Dolomitic, Ondurakorume Complex, S.W. Africa, BROOKS and AHRENS (1961)	1	S		0.06
Acidic rocks				
Granites composite 14, Germany, PREUSS (1940)		S		2
Granite, S. Africa, BROOKS *et al.* (1960)	1	S		0.9
Granite, France, INGAMELLS and SUHR (1963)	1	S		0.15
Tonalite, Tanganyika, SCHROLL and WENINGER (1965)	1	S		0.5
Rhyolite, S.W. Africa, BROOKS and AHRENS (1961)	4	S	0.02—0.22	0.12
Acid porphyry, S.W. Africa, BROOKS and AHRENS (1961)	2	S	0.04, 0.04	0.04

[a] Data obtained by an ion-exchange enrichment spectrochemical procedure.

Hence, the present data allow no firm conclusions to be drawn as to the variation of Bi in the various igneous rock types and with magmatic diferentiation.

III. Crustal Abundance of Bismuth

Some estimated crustal abundances of Bi are given in Table 83-E-3.

Table 83-E-3. *Estimated crustal abundances bismuth (in ppm)*

Bi	Source
0.07	NODDACK and NODDACK (1934)
0.20	GOLDSCHMIDT (1937), RANKAMA and SAHAMA (1950)
0.009	VINOGRADOV (1962)
0.17	BROOKS and AHRENS (1961), TAYLOR (1964)

There is little doubt that many more data are required before the crustal abundance of Bi can be established accurately. The most recent estimate (TAYLOR, 1964) is based on the work of BROOKS, AHRENS and TAYLOR (1960) and BROOKS and AHRENS (1961). As TAYLOR points out, his estimate refers particularly to the *continental* crust, but as the Bi content of acid rocks and of basalts appears to be of the same general magnitude, the continental crust value might be similar to the value for the entire crust (continental and oceanic). Data on the Bi content of oceanic basalts are unfortunately not available.

83-F. Behavior in Magmatogenic Processes

In the previous sections it was noted that Bi is not easily accepted into common silicate minerals, the Bi abundance level usually being <1 ppm. Furthermore, as the average igneous rock has only 260 ppm S (TAYLOR, 1964), mainly in the form of pyrite and chalcopyrite where Bi is usually <10 ppm, a negligible proportion of Bi in igneous rocks is present in sulphides. These features suggest that Bi will tend to accumulate in residual melts during the magmatic differentiation process. Support for this suggestion is provided by the observation that Bi minerals are typically found in granite pegmatites and hydrothermal ore deposits. Quantitative data regarding the paragenesis of Bi minerals are apparently lacking. RANKAMA and SAHAMA (1950) mention that bismuth minerals separate and crystallize mainly from hot hydrothermal solutions, after arsenic minerals (pegmatitic stage), but before the crystallisation of antimony minerals.

83-G. Behavior during Weathering and Alteration of Rocks

Few data are available concerning the behaviour of Bi during weathering processes and hence it is not possible to discuss quantitatively the relationship existing between igneous and sedimentary rocks in terms of the geochemical cycle of Bi. To a certain extent, however, the geochemical cycle of Bi will be influenced by the stability of sulphide minerals such as bismuthinite and galena. Sulphides are readily attacked by acid solutions and hence Bi may be released for re-distribution in the secondary geochemical environment. Bismite (Bi_2O_3) is in fact often found as a secondary alteration product. There is a suggestion that some sedimentary rock types may be relatively enriched in Bi (see Table 83-K-1). GOLDSCHMIDT (1954) mentions that Bi concentrates in ferruginous bauxites and sedimentary oxidic iron ores; no data are given.

83-H. Behavior in Natural Waters

The characteristic oxidation state of Bi in the geological environment is Bi^{3+}. VINOGRADOV (1967) expects that Bi occurs in sea water in the form of the species BiO^+, $BiCl^-$ and BiOCl, while the residence time of Bi in the oceans is given by GOLDBERG (1965) as 4.5×10^4 years. The solubility of Bi_2S_3 in water is 0.18 ppm at 18° C; that of BiO(OH) is 1.44 ppm at 20° C (STEPHEN and STEPHEN, 1963).

RANKAMA and SAHAMA (1950) state that Bi is comparatively readily precipitated as basic carbonate which is formed during the hydrolytic decomposition of bismuth salts in aqueous solution. Hence, Bi should be enriched in hydrolyzate sediments. Bismutite [$Bi_2(CO_3)O_3$] is an important carbonate mineral of bismuth.

83-I. Abundance in Ocean and Lake Water

A few data on the concentration of Bi in ocean water are available. Most of this information, summarized in Table 83-I-1, refers to surface water.

The data indicate strongly that the magnitude of Bi in surface waters of the oceans is ~0.02—0.04 ppb (μg/l) and there is an indication also that the concentration is lower in deep water. Distinctly higher concentrations, several above 2 ppb (μg/l) (maximum 7.1 ppb (μg/l)), have been reported by BOSWELL, BROOKS and WILSON (1967) in Lakes of McMurdo oasis, Antarctica.

Table 83-I-1. *Bismuth concentration in ppb Bi in ocean water*

ppb Bi	Method	Location	Source
0.2	Spectrochemical after co-precipitation	Gulmarsfjord, Norway	NODDACK and NODDACK (1940)
0.02	Spectrochemical after ion exchange enrichment	Surface water, south Atlantic	BROOKS (1960)
0.024—0.026	Spectrophotometric, after dithizone extraction	Surface water, English Channel	PORTMANN and RILEY (1966)
0.035—0.042	Spectrophotometric, after dithizone extraction	Surface water, Irish Sea	PORTMANN and RILEY (1966)
0.033	Spectrophotometric, after dithizone extraction	Surface water, North Atlantic	PORTMANN and RILEY (1966)
0.015	Spectrophotometric, after dithizone extraction	Water at 2000 m, North Atlantic	PORTMANN and RILEY (1966)

83-K. Abundance in Common Sedimentary Rock Types

Bi has been estimated in very few sediments and little is known about its abundance in them. The available data is summarized in Table 83-K-1. The normal clastic types of sediment show a large spread in concentration (0.06—2 ppm), but even so, there is the suggestion that the abundance level in some sediments may be higher than in igneous rocks. More data are obviously required.

Table 83-K-1. *Bismuth contents of sedimentary rocks*

Rock Type	No.	Method	ppm Bi	
			range	average
A. *Clastic Sediments*				
Shales composite p. 36, Germany, PREUSS (1940)		S		1
Sandstones, composite of 23, Germany, PREUSS (1940)		S		0.3
Ocean Mud, 68° S, 32° E, BROOKS *et al.* (1960)	1	S		0.06
Shale, Table Mountain, S. Africa, BROOKS *et al.* (1960)	1	S		2.1
Shale, Malmesbury, S. Africa, BROOKS and AHRENS (1961)	1	S		0.10
Subgreywacke, Australia, TAYLOR (1966)	2	I	0.59, 0.65	0.62
Conglomerate, Eastern Alps, WENINGER (1965)	4	S	0.5—3.0	1.3
Kupferschiefer, Mansfeld, GOLDSCHMIDT (1954)		S		100
B. "*Organic*" *Sediments*				
Anthracite, Eastern Alps, WENINGER (1965)	2	S	1, 10	5.5
Coal and bitumen bearing sediments, Eastern Austria, BRANDENSTEIN *et al.* (1960)				
(1) Total rock samples	25	S	0.02—4.1	1.1
(11) Volatile, HF acid insoluble fraction	25	S	0.1—20	3.5
C. *Manganese Nodules*				
Atlantic, Indian, Pacific Oceans, AHRENS *et al.* (1967)	26	S	0.5—24	7
	6	S	<0.5	

With regard to the types classed as "organic", a distinct enrichment of Bi is noted, especially in the acid insoluble fractions of the coal and bitumen samples analysed by BRANDENSTEIN *et al.* (1960). However, lack of sample detail precludes further discussion. GOLDSCMHIDT (1954) also mentioned an enrichment of Bi in the ashes of coal seams; however, no data are given.

Both Goldschmidt (1954) and Rankama and Sahama (1950) mention the possibility of enrichment of Bi in hydrolyzate sediments, associated with ferric hydroxide fractions. Goldschmidt (1954) quotes 100 ppm Bi in the sulphide bearing Permian Kupferschiefer from Mansfeld and considers this to be due to precipitation of Bi_2S_3 by hydrogen sulphide from stagnant bottom waters. This figure is way in excess of others indicated in Table 83-K-1.

The available data are sufficiently interesting to merit further investigations of the distribution of Bi in sedimentary materials.

The Bi concentration in 26 *manganese nodules* has been estimated by Ahrens, Willis and Oosthuizen (1967). An ion exchange pre-enrichment — spectrochemical procedure was used. In common with several other elements, the Bi content was found to be much higher (Ave. = 7 ppm) in the nodules relative to common rocks and sediments. Variation of concentration is quite extreme (from <0.5 ppm to 24 ppm), a feature which is characteristic for several elements in manganese nodules.

83-L. Biogeochemistry

Gurney and Ahrens (in press), using a spectrochemical technique (detection limit ~1 ppm) were unable to detect Bi in 10 samples of calcium phosphate (shark's teeth, whale bones and phosphate nodules) dredged from the ocean floor.

I. and W. Noddack (1940), found from 0.025 to 0.6 ppm Bi in the dry substance of various marine animals, highest amounts being present in a sponge and a jelly-fish.

Bismuth in coals has been mentioned already under 83-K and in table 83-K-1.

83-M. Abundance in Common Metamorphic Rock Types

The few available data are summarized in Table 83-M-1. The Bi levels quoted by WENINGER (1965) for graphite bearing rocks are distinctly higher than found in common igneous·rocks and sediments, but are similar to the Bi contents of "organic" sediments (Table 83-K-1, B). However, their composition is unusual and hence too much significance should not be attached to the results. The few results cannot be used to study the distribution of Bi with increasing grades of metamorphism.

Table 83-M-1. *Bismuth contents of metamorphic rocks*

Rock type	No.	Method	ppm Bi	
			range	average
Graphite and talc bearing rocks, Eastern Alps, WENINGER (1965)				
Graphite schists	2	S	3, 10	6.5
Graphite phyllites	5	S	0.5—10	3.1
Quartz phyllites	2	S	0.5, 1.0	0.75
Talc phyllites	3	S	0.5—5	2.1
Grey Gneiss, Antarctica, BROOKS *et al.* (1960)	1	S		0.04
Eclogite, Almeklovdalen, Norway, BROOKS *et al.* (1960)	1	S		0.07
Eclogite, South Africa, BROOKS and AHRENS (1961)	3	S[a]	0.02—0.10	0.05

[a] Data obtained by ion exchange enrichment and spectrochemical procedure.

83-O Relations to Other Elements; Economic Importance etc.

I. Inter-Element Relationships

Possible associations existing between Bi and other elements are discussed in section D. However, for the sake of completeness, a short summary is provided here. The chalcophile behaviour of Bi dominates its geochemistry and, therefore, apart from bismuth minerals themselves, highest concentrations of Bi are found in common sulphides. This is particularly so for galena where a strong Bi^{3+}—Pb^{2+} association is observed.

In Rare Earth (RE) minerals, some Bi—RE coherence appears to exist, notably between Bi^{3+} and Y^{3+} in gadolinites. In common silicate minerals, the most likely association is between Bi^{3+} and Ca^{2+}; however, the available data do not support this suggestion and it appears that such an association might be non-existent. The abundance levels of Bi in normal silicate rocks are, moreover, extremely low.

II. Technical and Economic Importance

Rich deposits of bismuth are uncommon, and most of the world's supply is obtained from the treatment of lead and copper refinery slimes, and as a by-product from the mining of other ores. An exception is the Tasna mine, Bolivia, where quartz veins contain native bismuth, bismuthinite and bismutite. The ores carry from 20—30% bismuth.

Bismuth has been extensively used in pharmaceutical products where bismuth compounds are used in medicinal remedies, cosmetics, ointments and for internal X-ray examinations. Bismuth compounds are also used to give a glaze to porcelain, for enamelling, and making optical glass.

Bismuth forms various alloys which melt far below the melting points of the separate metals. For example, bismuth melts at 271° C, but Woods metal (Bi—Pb—Sn—Cd) melts at 72° C. These so-called fusible alloys play an important part in the manufacture of fire alarms, automatic sprinklers, boiler safety plugs, hand grenades and electrical apparatus. There are also innumerable other minor uses.

Bismuth is stated to play an important part in nuclear energy projects.

Further details may be found in Bateman (1942) and Johnstone (1954).

Acknowledgment

We are grateful to Miss C. C. Fullard for assistance with the compilation of data used in this study.

References: Sections 83-B to 83-M, 83-O

AHRENS, L. H.: The use of ionization potentials. Part I. Ionic radii of the elements. Geochim. Cosmochim. Acta 2, 155 (1952).

—, and E. J. D. KABLE: A revised set of cationic radii. (In press.)

— J. P. WILLIS, and C. O. OOSTHUIZEN: Further observations on the composition of manganese nodules, with particular reference to some of the rarer elements. Geochim. Cosmochim. Acta 31, 2169 (1967).

American Society for Testing and Materials. Index to the Powder Diffraction File, ASTM Publication PD1S—16i, Philadelphia (1966).

BATEMAN, A. M.: Economic mineral deposits. London: Chapman & Hall Ltd., 1942.

BETEKHIN, A. G.: Mineralogiya (Mineralogy). Moscow: Gosgeolizdat 1950.

BOSWELL, C. R., R. R. BROOKS, and A. T. WILSON: Some trace elements in lakes of McMurdo Oasis, Antarctica. Geochim. Cosmochim. Acta 31, 731 (1967).

BRANDENSTEIN, M., I. JANDA u. E. SCHROLL: Seltene Elemente in österreichischen Kohlen- und Bitumengesteinen. Tschermaks Mineral. Petrog. Mitt. 7, 260 (1960).

BROOKS, R. R.: The use of ion-exchange enrichment in the determination of trace elements in sea water. Analyst 85, 745 (1960).

—, and L. H. AHRENS: Some observations on the distribution of Tl, Cd, and Bi in silicate rocks and the significance of covalency on their degree of association with other elements. Geochim. Cosmochim. Acta 23, 100 (1961).

— —, and S. R. TAYLOR: The determination of trace elements in silicate rocks by a combined spectrochemical — anion exchange technique. Geochim. Cosmochim. Acta 18, 162 (1960).

BROWN, R., and W. A. WOLSTENHOLME: Analysis of geological samples by spark source mass spectrometry. Nature 201, 598 (1964).

BURNHAM, C. W.: Metallogenic Provinces of the south-western United States and Northern Mexico. Bull. 65, State Bureau of Mines and Mineral Resources, New Mexico Inst. Mining and Technology, Socorro, New Mexico (1959).

CAMBEL, B., u. J. JARKOVSKÝ: Geochemie der Pyrite einiger Lagerstätten der Tschechoslowakei. Vydavatefstvo Slovenskej akadémie vied Bratislava (1967).

CAMERON, A. G. W.: A new table of abundances of the elements in the solar system. Intern. Ser. of Monographs in Earth Sciences, Origin and distribution of the elements" (Paris, 1967). Oxford: Pergamon Press 1968.

CARMICHAEL, I., and A. MCDONALD: The colorimetric and polarographic determination of some trace elements in the standard rocks G-1 and W-1. Geochim. Cosmochim. Acta 22, 87 (1961).

CLARK, A. H.: The mineralogy and geochemistry of the Ylöjärvi Cu—W deposit, southwest Finland. Bull. Comm. Géol. Finlande 218, 195 (1965).

CRAIG, J. R.: Phase relations and mineral assemblages in the Ag—Bi—Pb—S system. Mineral. Deposita 1, 278 (1967.)

EHMANN, W. D., and J. R. HUIZENGA: Bismuth, thallium and mercury in stone meteorites by activation analysis. Geochim. Cosmochim. Acta 17, 125 (1959).

FLEISCHER, M.: Minor elements in some sulphide minerals. Econ. Geol. Fiftieth Anniversary Volume, 970 (1955).

— Index of new mineral names, discredited minerals, and changes of mineralogical nomenclature in volumes 1—50 of The American Mineralogist. Am. Mineralogist 51, 1247 (1966).

GOLDBERG, E. D.: Minor elements in sea water. In: Chemical oceanography (Edit. J. P. RILEY and G. SKIRROW). London and New York: Academic Press 1965.

GOLDSCHMIDT, V. M.: Geochemische Verteilungsgesetze der Elemente VII. Skrifter Norske Videnskaps-Akad. Oslo, Mat.-Naturv. Kl. No. 2 (1926).

Goldschmidt, V. M.: Geochemische Verteilungsgesetze der Elemente IX. Skrifter Norske Videnskaps-Akad. Oslo, Mat.-Naturv. Kl. No. **4** (1937).
Geochemistry (edited by Alex Muir). Oxford: Clarendon Press 1954.

Gurney, J. J., and L. H. Ahrens: The Bi contents of some rare earth minerals, notably gadolinite. (In press.)

Hertel, G.: Die Fremdelementführung der Bleiglanze als Hilfe zur Bestimmung der Bildungstemperatur. Erzmetall **19**, 632 (1966).

Hoehne, K.: Quantitative chemische und erzmikroskopische Bestimmung von As, Sb, Zn und Bi in vorwiegend schlesischen Bleiglanzen. Chem. Erde **9**, 219 (1934/35).

Hook, H. J. van: The ternary system Ag_2S—Bi_2S_3—PbS. Econ. Geol. **55**, 759 (1960).

Ingamells, C. O., and N. H. Suhr: Chemical and spectrochemical analysis standard silicate samples. Geochim. Cosmochim. Acta **27**, 897 1963).

Johnstone, S. J.: Minerals for the chemical and allied industries. London: Chapman & Hall Ltd. 1954.

Levin, E. M., C. R. Robins, and H. F. McMurdie: Phase diagrams for ceramists. Ohio: The American Ceramic Society 1964.

Nesterova, Yu. S.: The chemical composition of galena. Geochemistry No. 7, 835 (1958).

Noddack, I., u. W. Noddack: Die Häufigkeit der chemischen Elemente. Naturwissenschaften **18**, 757 (1930).

— — Die Geochemie des Rheniums. Z. Physik. Chem. A **154**, 207 (1931).

— — Die geochemischen Verteilungskoefficienten der Elemente. Svensk. Kem. Tidskr. **46**, 173 (1934).

— — Die Häufigkeiten der Schwermetalle in Meerestieren. Arkiv Zool. No.4, 32 A (1940).

Oftedal, I.: Untersuchungen über die Nebenbestandteile von Erzmineralien Norwegischer zinkblendeführender Vorkommen. Skrifter Norske Videnskaps-Akad. Oslo, Mat.-Naturv. Kl. No. 8 (1941).

— Om betingelsane for oktaedrisk delbarheit hos vismutrik blyglans. Norsk Geol. Tidsskr. **22**, 61 (1942).

Palache, C., H. Berman, and C. Frondel: The system of mineralogy of James Dwight Dana and Edward Salisbury Dana, 7th edit. New York: John Wiley & Sons 1958.

Portmann, J. E., and J. P. Riley: The determination of Bi in sea and natural waters. Anal. Chim. Acta **34**, 201 (1966).

Preuss, E.: Beiträge zur spektralanalytischen Methodik II. Bestimmung von Zn, Cd, Hg, In, Tl, Gn, Sn, Pb, Sb, und Bi durch fraktionierte Distillation. Z. Angew. Mineral. **3**, 8 (1940).

Ramdohr, P.: Die Erzmineralien und ihre Verwachsungen, 3. Aufl. Berlin: Akademie-Verlag **1960.**

Rankama, K., and Th. G. Sahama: Geochemistry. Chicago: Chicago University Press 1950.

Reed, G. W., K. Kigoshi, and A. Turkevich: Determination of concentrations of heavy elements in meteorites by activation analysis. Geochim. Cosmochim. Acta **20**, 122 (1960).

Rose, A. W.: Trace elements in sulfide minerals from the Central district New Mexico and the Bingham district, Utah. Geochim. Cosmochim. Acta **31**, 547 (1967).

Salanci, B.: Untersuchungen am system Bi_2S_3—PbS. Neues Jahrb. Mineral. Monatsh. **12**, 384 (1965).

Schroll, E.: Über das Vorkommen einiger Spurenmetalle Blei-Zink-Erzen der ostalpinen Metallprovinz. Tschermaks Mineral. Petrog. Mitt. **5**, 183 (1965).

—, u. M. Weninger: Eine empfindliche spektrochemische Analysenmethode zur Bestimmung von Germanium und Zinn unter Verwendung sulfidierender thermochemischer Reagenzien. Mikrochim. Acta **2**, 378 (1965).

Seeger, P. A., W. A. Fowler, and D. D. Clayton: Nucleosynthesis of heavy elements by neutron capture. Astrophys. J., Suppl. II, No. 97 121, (1965).

Short, M. N.: Microscopic determination of the ore minerals. U.S.G.S. Bull. 914, 2nd edit. (1940).

Sims, P. K., and P. B. Barton: Some aspects of the geochemistry of sphalerite, Central City District, Colorado. Econ. Geol. **56**, 1211 (1961).

STEPHEN, H., and T. STEPHEN: Solubilities of inorganic and organic compounds, vol. 1. London: Pergamon Press 1963.

STRUNZ, H.: Homöotypie Bi_2Se_2—Bi_2Se_3—Bi_2Se_4—Bi_4Se_5 usw. (Platynit, Ikunolith, Laitakarit.) Neues Jahrb. Mineral. Monatsh. **7**, 154 (1963).

TAYLOR, S. R.: Abundance of chemical elements in the continental crust: a new table. Geochim. Cosmochim. Acta **28**, 1273 (1964).

— Geochemical application of spark source mass spectrometry. Nature **205**, 34 (1965).

— Australites, Henbury impact glass and subgreywacke: a comparison of the abundance of 51 elements. Geochim. Cosmochim. Acta **30**, 1121 (1966).

—, and A. J. R. WHITE: Trace element abundances in Andesites. Bull. Volcanol. **29**, 177 (1966).

VINOGRADOV, A. P.: Average contents of chemical elements in the principal types of igneous rocks of the earth's crust. Geochemistry No. 7, 641 (1962).

— Introduction to geochemistry of the oceans [in Russian]. Moscow: Science Publishing House 1967.

WENINGER, M.: Über Gehalte an Germanium, Zinn und einigen anderen Spurenelementen in ostalpinen Graphit- und Talkgesteinen. Tschermaks Mineral. Petrog. Mitt. **10**, 475 (1965).

Part II

BISMUTH IN METEORITES

Editors' Comments on Papers 4 Through 8

4 **EHMANN and HUIZENGA**
Bismuth, Thallium and Mercury in Stone Meteorites by Activation Analysis

5 **REED, KIGOSHI, and TURKEVICH**
Determinations of Concentrations of Heavy Elements in Meteorites by Activation Analysis

6 **ANDERS**
Chemical Processes in the Early Solar System, as Inferred from Meteorites

7 **LAUL et al.**
Bismuth Content of Chondrites

8 **SANTOLIQUIDO and EHMANN**
Bismuth in Stony Meteorites and Standard Rocks

Fascinating is one word we can use to describe the studies of the bismuth content of meteorites. Since bismuth is unique in being the heaviest of the stable elements and also being highly volatile, understanding the chemical properties of Bi within meteorites is of considerable help in understanding cosmochemistry. To paraphrase a statement by Edward Anders (Paper 4), meteorites can be regarded as the poor man's space probe. We now have an important sensing device within the probe—bismuth.

The papers in this section were chosen to provide data on the distribution of bismuth in various types of meteorites, to give insights into the geochemical behavior of bismuth, and to show how the bismuth contents can be used to unravel the processes involved in the chemical and physical evolution of meteorites. The papers presented, though not in chronological order, were chosen to give the reader a feeling for the evolution of thought on the significance of the geochemistry of bismuth in meteorites and the many questions that still remain to be answered.

One of the earliest determinations of the bismuth content of meteorites is in the work of Noddack and Noddack (1934). Using X-ray and optical emission spectroscopy, they reported the atomic abundance of bismuth to be 0.144 (Si = 10^6). Suess and Urey (1956) used this value in their work on the cosmic abundances of the elements. From the distribution of the elements in the cosmos, Suess and Urey speculated on the theories of the origin of the elements. Burbidge et al. (1957) based much of their theory of the origin of the elements involving nucleosynthesis on the research of Suess and Urey.

Since the work of the Noddacks, neutron activation techniques have become the preferred method of analysis for heavy metals in meteorites. With the technique continually being refined, more reliable data are being generated. Papers 4 and 5 are benchmark papers in that they present some of the first data on the bismuth contents in meteorites using activation analyses.

The cosmic abundance of Bi determined by Ehmann and Huizenga (Paper 4) of 0.0016 (Si = 10^6) supported the value previously determined by Reed et al. (1958) of 0.005 (Si = 10^6). These values were much lower than the previous values assumed and predicted (Suess and Urey 1956; Cameron 1959). The conclusion that one might reach is that the theories of nucleosynthesis of the elements particularly as related to bismuth need to be modified. The problem however is that these abundance values were determined as averages from various meteoritic classes. Most researchers on meteorites now agree that the Type 1 carbonaceous chondrite is the most chemically primative meteorite and therefore should be used to estimate the atomic abundances of the elements. Using the other classes of meteorites to compute averages results in low values, as these papers show. However, both Papers 4 and 5 demonstrate that the Bi contents of ordinary chondrites are much lower than the Bi contents of the enstatite and carbonaceous chondrites, thus setting the stage for future studies to elucidate the meaning of the differences.

Reed et al. (Paper 5) determined that carbonaceous chondrites are probably of low temperature origin. Comparing bismuth abundances from carbonaceous chondrites to Bi abundances predicted by nucleosynthesis theory resulted in a fairly close match. This important observation suggested that since carbonaceous chondrites are of lower temperature origin, they may have retained their primordial elemental abundances. The atomic abundances computed in this paper are still lower than the predicted values, however, since a less primative chondrite (C2) was used in computing the average.

William Ehmann, who was working at the Argonne National Laboratories when he wrote Paper 4, has been at the Department of Chemistry of the University of Kentucky for most of his career. Subsequent to obtaining his Ph.D. from Carnegie Institute of Technology in 1957, Ehmann has held a variety of positions with groups such as the Atomic Energy Commission and the National Research Council. J. R. Huizenga received his Ph.D. in physical chemistry from the University of Illinois in 1949. He was a senior scientist at Argonne National Laboratories when the work on Bi reported herein was done. Since 1962, he has been a professor of chemistry and physics in the department of chemistry at the University of Rochester.

George Reed received his Ph.D. from the University of Chicago in 1952 in chemistry and has since been at Argonne National Laboratories where he is a research associate chemist. Anthony Turkevich has a Ph.D. in chemistry from Princeton (1940). He is presently a professor of chemistry at the Enrico Fermi Institute, the University of Chicago.

From previous studies, bismuth was recognized as one of the strongly depleted elements (Anders 1964). In order to account for the abundance patterns of the elements in chondrites including bismuth, several models have been proposed and are summarized in a paper by Keays et al. (1971). These models are: (1) the two-component condensation model, (2) the three-component condensation model, (3) the metamorphic model, and (4) the multi-component model. One of the more popular is the two-component model proposed by John Larimer and Edward Anders. They suggested that the various chondrites are produced as mixtures of a low temperature fraction and a high temperature fraction (Anders 1964; Larimer 1967; Larimer and Anders 1967). Paper 6 by Edward Anders, which has become a frequently cited paper, was chosen to summarize his concept. Although the original paper is not bismuth dedicated, we are including it because we think the concepts are pertinent to the understanding of the geochemistry of bismuth in meteorites.

Edward Anders has been at the forefront of our knowledge of the chemistry and origin of meteorites. His concepts account for much of our understanding of the geochemistry of bismuth in meteorites. He worked on his ideas concerning the chemical composition of meteorites during a senior postdoctoral fellowship sponsored by the National Science Foundation at the University of Berne, 1963–1964. Having received his Ph.D. (1954) in chemistry from Columbia, he has since taught and been a guest scientist at various institutions including the University of Chicago, University of Illinois, University of Berne, Brookhaven National Laboratories, and The Goddard Space Flight Center. Since 1964, he has been at the Enrico Fermi Institute, University of Chicago.

To complete our brief review of the present knowledge of the bismuth contents in stony meteorites, we have chosen Papers 7 and 8. These papers show that Bi concentration decreases with increasing petrologic grade among ordinary chondrites. Paper 8 demonstrates how the Bi content of meteorites can be used to compute accretion temperatures. The paper by Keays et al. (1971) noted earlier also treats this subject.

Anders (1964) pointed out the regularity of the depletion curves for the strongly depleted elements in carbonaceous chondrites. Although the bismuth data at the time were not complete, he predicted, using the two-component model, that Bi too would follow this regularity. The mean relative abundances predicted were 1.0, 0.55, and 0.32 for C1, C2, and C3 meteorites respectively. A test of this prediction came from the work of Laul et al. (Paper 7) during a more complete study of bismuth contents of chondrites. Their measured abundances closely matched the relative abundances predicted by the two-component model. Paper 7 demonstrates from the trends in the Bi concentrations that equilibrated ordinary chondrites (EOC) and unequilibrated ordinary chondrites (UOC) must have different processes controlling their formation.

Paper 7 is important because it shows that bismuth losses from a meteorite cannot be the result of late volatilization. The authors suggest that the loss must be the result of some primordial process. An insight into what this process might be was given by Keays et al. (1971). Using the two-component model and accretion temperatures of L-chondrites based on Bi and Tl contents, they demonstrated that abundances of Bi in meteorites can be accounted for by simple equilibrium condensation at falling temperatures. Subsequent work tends to support this observation (e.g., Binz et al. 1974 and Kurimoto et al. 1973). In their summary paper Kurimoto et al. (1973) compared predictions made by the two-component condensation model plus the other three models listed previously using inter-element relationships of carbonaceous chondrites and unequilibrated ordinary chondrites. The results are interesting in that three models with *ad hoc* assumptions can account for the general pattern of inter-element relationships in unequilibrated ordinary chondrites while the distribution pattern in carbonaceous chondrites is better explained in a modified two-component model.

J. C. Laul received his Ph.D. from Purdue University and Paper 7 is part of his dissertation. Born in Bandung, Indonesia, Friedrich

Schmidt-Bleek received his doctorate from the University of Maine in 1960 and was in the department of chemistry at the University of Tennessee.

M. E. Lipschutz has contributed much to the understanding of the contents of trace elements, including Bi, in meteorites. Spending a year at Berne just after Anders left as a National Science Foundation Fellow, he is presently a professor of chemistry at Purdue University where he has been since 1965.

REFERENCES

Anders, E. 1964. Origin, age and composition of meteorites. *Space Sci. Rev.* **3**:583–714.

Binz, C. M., R. K. Kurimoto, and M. E. Lipschutz. 1974. Trace elements in primitive meteorites. V: Abundance patterns of thirteen trace elements and interelement relationships in enstatite chondrites. *Geochim. Cosmochim. Acta* (preprint).

Burbidge, E. M., G. R. Burbidge, W. A. Fowler, and F. Hoyle. 1957. Synthesis of the elements in stars. *Rev. Mod. Phys.* **20**:586–650.

Cameron, A. G. W. 1959. A revised table of abundances of the elements. *Astophys. J.* **129**: 676–99.

Keays, R. R., R. Ganapathy, and E. Anders. 1971. Chemical fractionations in meteorites—IV. Abundances of fourteen trace elements in L-chondrites: Implication for cosmothermometry. *Geochim. Cosmochim. Acta* **35**:337–363.

Kurimoto, R. K., I. Z. Pelly, J. C. Laul, and M. E. Lipschutz. 1973. Interelement relationships between trace elements in primitive carbonaceous and unequilibrated ordinary chondrites. *Geochim.Cosmochim. Acta* **37**:209–24.

Larimer, J. W. 1967. Chemical fractionations in meteorites in condensation of the elements. *Geochim. Cosmochim. Acta* **31**:1215–38.

Larimer, J. W., and E. Anders. 1967. Chemical fractionations in meteorites. II: Abundance patterns and their interpretation. *Geochim. Cosmochim. Acta* **31**:1239–70.

Lipschutz, M. E. 1971. Bismuth (83). *In* B. Mason, ed., *Handbook of Elemental Abundances in Meteorites,* N.Y.: Gordon & Breach, pp. 511–15.

Noddack, I., and W. Noddack. 1934. Die geochemischen Verteilung Koeffizunten der Element. Svensk Kem. Tidskr. **46**:173.

Santoliquido, P. M. and E. D. Ehmann. 1972. Bismuth in stone meteorites and standard rocks. *Geochim. Cosmochim. Acta* **36**:897–902.

Suess, H. E., and H. C. Urey. 1956. Abundances of the elements. *Rev. Mod. Phys.* **28**:53.

4

Reprinted from *Geochim. et Cosmochim. Acta* **17**:125–135 (1959)

Bismuth, thallium and mercury in stone meteorites by activation analysis*

W. D. EHMANN† and J. R. HUIZENGA
Argonne National Laboratory, Lemont, Illinois

Abstract—The radiochemical procedures for the assay of bismuth, thallium, and mercury in stone meteorites following neutron activation are described in detail. A Bi^{209} abundance of $2\cdot2 \times 10^{-9}$ gramme/gramme meteorite leading to a cosmic abundance of 0·0016 (per 10^6 Si atoms) was determined from the analysis of six stone meteorites. A Tl^{203} abundance of $0\cdot40 \times 10^{-9}$ g/g meteorite and a Hg^{202} abundance of 30×10^{-9} g/g meteorite corresponding to cosmic abundances of 0·00030 and 0·023 (per 10^6 Si atoms), respectively, were determined from analysis of five stone meteorites. Implications of these data as pertaining to the life history of meteorites are discussed.

INTRODUCTION

THE heavy elements Hg, Tl, Pb and Bi are of particular interest to those working in geochemistry and cosmochemistry, since they are the highest atomic number nonradioactive elements in the periodic table. In this work the abundances of three of these, Hg, Tl and Bi, were determined by use of the very sensitive analytical tool of neutron activation.

The advantages and disadvantages of neutron activation analysis have been discussed at length in the literature (BOYD, 1949; MEINKE, 1955; PLUMB and LEWIS, 1955; JENKINS and SMALES, 1956), and will not be dealt with in detail here. It is sufficient to say that the two major advantages of the method are its high sensitivity (in this work abundances down to $\sim 10^{-10}$ g/g) and its freedom from contributions to an elemental abundance due to reagent contamination during chemical processing. The latter is of extreme importance for low abundance elements especially in the case of Hg owing to the generally high prevalence of Hg contamination in chemistry laboratories.

Previous data on heavy element abundances in stone meteorites at the time this investigation began were very sparse. SUESS and UREY use an interpolated value of 0·284 (Si = 10^6) for the atomic abundance of mercury (SUESS and UREY, 1956). They state, however, that all early analyses are suspect because of possible contamination in the laboratory. The NODDACK's report the atomic abundance of thallium as 0·108 (Si = 10^6) (NODDACK and NODDACK, 1934), while SHAW (1952) reports thallium as less than 0·007 (Si = 10^6). EL BADRY and KOHMAN (1957) using microchemical techniques found the thallium abundance of the Plainview chondrite to be 0·0005 (Si = 10^6) based on one determination. SUESS and UREY adopt 0·144 (Si = 10^6) for the atomic abundance of bismuth based on the work of the NODDACKS (NODDACK and NODDACK, 1934). Recently REED *et al.* (1958)

* Based on work performed under the auspices of the United States Atomic Energy Commission.
† Permanent address: Department of Chemistry, University of Kentucky, Lexington, Kentucky.

have reported atomic abundances (Si = 10^6) by neutron activation analysis using low level β-counting of 0·00013, 0·027 to 0·178, and $\leq$0·005 for Tl^{203}, Pb^{208} and Bi^{209}, respectively.

Data obtained to date using the neutron activation technique, details of the various procedures, and the general significance of the data are reported in this paper.

Sample and Flux Monitor Preparation

The meteorite samples selected for analysis were obtained wherever possible from the interior of a freshly fractured large individual specimen. In this manner the effects of weathering or other post-fall contamination are minimized. Descriptions of the meteorites used may be found in Prior and Hey (1953). Ideally, dated falls would provide the most useful specimens. However, specimens of dated falls are difficult to obtain and one of the meteorites used in this work, the Plainview chondrite, is a "find" of unknown terrestrial age.

The selected interior sample was pulverized in a steel vibration ball mill. Appropriate precautions were observed at all times to prevent contamination of the samples before irradiation. As a matter of practice, the preparation of the meteorites prior to irradiation was done in the office area of the laboratory, as distinct from the areas in which the later chemistry was done.

The powdered specimen was placed for irradiation in a quartz vial (4 mm internal diameter and 30 mm long), having a ground quartz stopper. Approximately 1 g of meteorite was used for each determination.

Flux monitor packages for each of the elements being determined were wrapped tightly around the vial containing the sample. Each flux monitor package consisted of a "sandwich" of four $\frac{3}{4}$ in. diameter aluminium foils, each 0·00075 in. thick. An aliquot of a standard solution of bismuth, thallium or mercury was evaporated, along with the pipette washings, on one of these foils. Evaporation was accomplished under a heat lamp at a distance of about 2 ft, so that the foil is heated only slightly above room temperature. An identical foil was placed on top of the foil containing the evaporated salt to act as a cover. Another pair of foils, one placed on each side of the flux monitor and its cover, served as monitor blanks.

The sandwich of foils was crimped together tightly and wrapped in ordinary commercial aluminium foil. Each package was also held in its position around the quartz sample vial by a firm wrapping of commercial aluminium foil.

Previous work in this laboratory (Bate *et al.*, 1958) has indicated the absence of self-shadowing effects in the use of physical flux monitors of this type in activation analysis for thorium in stone meteorites. This is reasonable, considering that the known composition of chondrites indicates low abundances of nuclides having very large neutron capture cross-sections and the small physical size of the sample. These investigators also found a 4 per cent per in. flux gradient along the axis of the irradiation can. Therefore the placement of the flux monitor foils directly around and covering the section of the quartz vial containing the sample is an important consideration.

While the validity of neutron activation analysis of the type described has not been vigorously proven for bismuth, thallium and mercury at the concentration levels in which these elements are found in meteorites, there are no reasons at present to discount the method. The use of flux monitors of the type described to determine the absolute amount of the element in question by means of direct comparison of the flux monitor activity to that found in the sample is the best control available for the accuracy of the activation method.

Irradiation and Counting Procedures

The slow neutron activation of the meteorite samples was carried out in the "Thimble" of the Argonne CP-5 heavy water moderated reactor. The flux density for the positions used was of the order of 3×10^{13} neutrons-cm^{-2}-sec^{-1}. The Cd ratio for In activation (which is related to the ratio of thermal neutrons to fast neutrons) was approximately 10 to 1. The irradiations were usually for a period of from 10 to 15 days.

In this work bismuth was determined by counting the 5·298 MeV α-particles of Po^{210} (138·40 days), the active daughter of Bi^{210}, a 5 day β-emitter produced by an (n, γ) process on 100% natural abundance Bi^{209}. The chemical yield for polonium was determined by use of a "spike" of a known amount of cyclotron produced Po^{208} introduced at the start of the chemistry. The spike also contained small amounts of Po^{209} and Po^{210}. The Po^{210} to Po^{208} ratio for the sample polonium plate was determined by use of an α-ionization chamber connected to a multi-channel pulse-height analyser. The sample Po^{210} counting rate was usually 1–5 α-counts/min. A correction was applied for the small amount of Po^{210} in the spike, and the total Po^{210} produced in the irradiation was then calculated. The appropriate decay corrections for the spike were, of course, applied during the course of this work.

The use of α-counting of Po^{210} has at least three major advantages over direct β-counting of Bi^{210}. First, the α-energy, which is easily determined with the α-pulse-height analysis system, provides a firm indentification of the nuclide being counted. Second, less radiochemical purification is necessary owing to the low background of α-emitters formed by the irradiations and the ease of discriminating against impurities by α-energy measurements in a pulse-height analysis. The natural background of the counter in this region was very low. Third, the longer half-life of Po^{210} as compared to Bi^{210} allows a longer "cooling" time for the sample and as a result, less remote handling. Also, decay corrections in processing the data are less severe.

The amount of Po^{210} activity produced in the sample as determined above is compared with the total Po^{210} α-activity of the flux monitor to determine the absolute amount of bismuth in the sample. The foil containing the evaporated bismuth salt, the cover foil and the two blank outside foils were all counted directly without chemistry in a 2π internal sample Bradley-type α-counter. This same counter was used to determine the total counting rate of Po spike added at the start of the chemistry.

A 25 λ aliquot of a standard solution of bismuth nitrate containing 10·31 μg of bismuth per 25 λ was used to prepare the flux monitors. With the usual irradiation time this resulted in a Po^{210} counting rate for the flux monitor of 10,000 to

15,000 α-counts/min in the 2π Bradley counter. Normally, 75–90 per cent of the activity was found on the flux monitor foil itself with 10–25 per cent found transferred to the cover foil. The blank outside foils always gave a negligible counting rate (1 — 5 α-counts/min) by comparison. The total flux monitor counting rate used was the sum of that of the flux monitor foil and the cover foil.

Thallium was determined by β-counting ($E_{max} = 0{\cdot}765$ MeV) 4·0 year Tl^{204} produced by an (n, γ) reaction on 29·50% natural abundance Tl^{203}. The chemical yield of the thallium was determined by use of 13·95 mg of inert thallium nitrate carrier added at the start of the chemistry. The thallium from both the flux monitors and the samples was counted as a precipitate of TlI. The sample TlI was counted at a geometry of approximately 35 per cent in a low background proportional counter. This counting system used bulk shielding and a ring of anticoincidence tubes to achieve a reproducible background of 1·5 counts/min. The flux monitor TlI was counted under nearly the same geometry conditions with a standard end-window proportional counter having a background of approximately 10 counts/min. A correction was made for the slight difference between the efficiences of the two counters by calibration with a Tl^{204} standard source.

Unlike bismuth, the thallium flux monitors could not be counted directly owing to the very high β-background from the aluminium foils themselves. The flux monitor for thallium was prepared by evaporation of 25 λ of a standard thallium nitrate solution, containing 2·12 μg of Tl per 25 λ, on an aluminium foil. After irradiation the flux monitor foil and its cover were rinsed in portions of warm 1 M H_2SO_4 containing thallium nitrate carrier until additional washings removed no more activity. After reduction with SO_2, TlI was precipitated from this solution in the same manner used for the sample. Washings of the two blank foils from the flux monitor package in a similar manner yielded a negligible counting rate as compared to that from the flux monitor foil and cover. Under the irradiation conditions used, the Tl^{204} flux monitor counting rate was generally from 20,000 to 30,000 β-counts/min on shelf 1 of the standard proportional counter system.

Aluminium absorption curves on the flux monitor Tl^{204} were identical to those for a standard Tl^{204} source. Absorption curves were also taken for the TlI from each meteorite sample and compared to the standard Tl^{204} source. Using the chemical and irradiation procedures of this paper the counting rates obtained for the sample TlI were usually from 3 to 30 counts/min. This allowed a reasonably accurate β-absorption curve to be run in 24–48 hr. In all cases the curve agreed closely with that for the standard source. The absence in all samples of a γ-"tail" on the absorption curve indicated a pure β-emitter, although in the case of the lower activity samples the counting statistics for the points near the range of the Tl^{204} β are necessarily poor. Identical slopes were obtained for all samples over the initial portions of the absorption curve where statistics were good.

Mercury was determined by counting the 279 keV γ-ray associated with the decay of 45·8 day Hg^{203}. Hg^{203} is formed in the neutron irradiation of mercury by an (n, γ) reaction on 29·8% natural abundance Hg^{202}. The chemical yield of mercury was determined by the addition of 15·2 mg of inert mercuric nitrate carrier at the start of the chemistry.

The counting equipment consisted of a 3 in. × 3 in. NaI crystal scintillation detector coupled to a 200-channel pulse-height analyser. The Hg^{203} γ-ray was identified by both energy calibration and half-life. The counting rate of the mercury samples was about 1000 counts/min. The flux monitor and its cover foil were counted as a unit without chemistry. The spectrum obtained was compared to that obtained by counting the two blank foils from the same package. The foil background in the area of the 279 keV photopeak, as represented by counting the two blank foils, was negligible compared to the Hg^{203} flux monitor foil and its cover. The mercury flux monitor was prepared by the evaporation on a foil of a 25 λ aliquot of a standard solution of mercuric nitrate containing 7·60 μg of mercury per 25 λ. This yielded an integrated counting rate for the flux monitor in the 279 keV photopeak of approximately 50,000 counts/min for a geometry of 3·7 per cent and an irradiation time of about 10 days.

Radiochemical Procedures

Solution and general ion exchange separation. The irradiated meteorite powder was transferred to a small platinum dish containing 5 ml of concentrated H_2SO_4, the Po^{208} spike, and the mercury and thallium carriers. The slurry was warmed under a heat lamp with occasional stirring for from 15 to 20 min. Concentrated HF was then added dropwise with stirring until effervescence ceased. An additional 10 ml of HF were then added, and the mixture was gently heated under a heat lamp at a distance of at least 20 in. until the residue was reduced to a paste. Care was taken to prevent the mixture from going to dryness at any time. Additional HF was added, and the evaporation repeated.

Concentrated HNO_3 (5 ml) was added to the resulting paste, and the slurry was transferred to a glass beaker. A small amount of concentrated HCl was added, and after any evidence for reaction ceased, the solution was made up to 50 ml with 9 M HCl. Usually, the solution at this point was clear. In some cases, however, a very small black deposit settled out on standing. Where this was observed, the residue was treated with aqua regia, diluted with 9 M HCl, and again added to the main portion.

This solution was then introduced into a 20 to 25 ml resin volume anion exchange column. The resin bed was about $\frac{1}{2}$ in. in diameter and 5 in. in length. Dowex 1, X-10, 100 to 200 mesh resin was used in all this work. After introduction of the sample the column was washed successively with 75 ml of 9 M HCl, 75 ml of 0·5 M HCl, and 25 ml of distilled H_2O. This treatment elutes the gross amount of activity from the column, which may now be treated in a "low-level' laboratory (column is $\ll$5 mr/hr at 2 in.).

Polonium chemistry. Polonium is eluted from the column with 75 ml of 1 M H_2SO_4. The eluted solution is brought to boiling and SO_2 bubbled through it to reduce any high valency ions present, which tend to reduce the polonium plating yield.

Polonium is plated directly in two batches from 1 M H_2SO_4 solution on a grease-free silver disk ($1\frac{1}{8}$ in. diameter). A plating time of 3 hr was found to give plating yields in excess of 90 per cent. No potential was necessary to plate polonium on silver. The plate was washed with benzene and acetone and submitted for α-pulse analysis.

Thallium chemistry. Thallium is eluted from the column by reduction of Tl^{3+} to Tl^{1+}. Specifically, the thallium is eluted with 75 ml of 1 M H_2SO_4 freshly saturated with SO_2. Carriers for aluminium, antimony, cadmium, copper, indium, iridium, tellurium and tin (generally in the form of nitrate or chloride solutions) were added to the eluant. The pH of the solution was adjusted to 6·5 with NaOH. The combined precipitates formed were removed by centrifugation and discarded. The supernatant was brought to pH 4 with H_2SO_4, iridium carrier again added, and TlI precipitated by the addition of excess NaI.

The TlI was separated by centrifugation, washed with H_2O, and dissolved in a few drops of aqua regia. Iridium carrier was again added, and the solution was made up to 30–40 ml with 4 M HCl. Several drops of $KMnO_4$ were added to assure oxidation of Tl^{1+} to Tl^{3+}.

The Tl^{3+} was extracted into 25 ml of equilibrated ethyl ether in a separatory funnel. The ether layer was washed three times with 10 ml portions of 4 M HCl, all aqueous layers being discarded. The ether layer was then evaporated over 10 to 15 ml of 1 M H_2SO_4 containing iridium carrier. Tl^{3+} was reduced to Tl^{1+} with SO_2, and the solution was again scavenged with Al^{3+} carrier at pH 7.

The supernatant from the $Al(OH)_3$ precipitation containing thallium was again treated with NaI to precipitate TlI. The TlI was filtered on a weighted paper, washed with 1 M H_2SO_4, H_2O and ethyl alcohol, dried at 110°C, weighed and mounted for counting. Chemical yields for thallium averaged about 70 per cent.

Mercury chemistry. Mercury was eluted from the column with 75 ml of 4 M HNO_3. The eluant was neutralized to pH 6·5 with solid NaOH and HgS precipitated by addition of H_2S gas. The HgS was separated by centrifugation, washed with H_2O and dissolved in a few drops of aqua regia. Iridium carrier was added, and the solution evaporated with low heat to near dryness to remove excess HNO_3 after addition of 5 ml of concentrated HCl.

The small volume of solution was diluted to 10 ml with H_2O, and mercury was precipitated by the addition of several drops of freshly prepared $SnCl_2$ solution. The mercury precipitate was separated by centrifugation, washed with H_2O, dissolved in aqua regia, and again precipitated as above using iridium carrier and $SnCl_2$.

This final mercury precipitate was again separated and washed and dissolved in several drops of aqua regia. The resulting solution was diluted to 10 ml with H_2O, brought to pH 7 with NaOH and HgS precipitated with H_2S gas.

The HgS was filtered on weighed paper, washed with H_2O and ethyl alcohol, dried at 110°C, weighed and mounted for counting. Chemical yields generally ranged from 50 to 80%, although some early runs resulted in considerably lower yields.

RESULTS AND DISCUSSION

The results obtained in this study to date are presented in Table 1 and summarized in Table 2.

It should be noted that sample no. 1 in each case represents an individual sample extracted from the larger meteorite individual and irradiated alone. Samples 2 and 3 represent portions of a separate homogeneous sample from the larger meteorite individual irradiated simultaneously, but with separate flux monitors in the same irradiation can. Therefore, comparison of the results for no. 2 and no. 3 may be looked upon as a measure of the reproducibility of the method, and a comparison of no. 1 to the average of no. 2 and no. 3 as a measure of the homogeneity of the sample. In the case of the Holbrook chondrite two entirely different specimens were used for run no. 1 and no. 2. Run no. 1 was a portion of a small Holbrook pebble. This specimen was found to yield erratic results for several elemental analyses performed in this laboratory. Therefore, the results for Holbrook no. 1, while listed in the tables, are not included in the averages computed. Holbrook no. 2 in contrast was a portion of a large Holbrook fragment and this specimen has given reproducible results in other studies in this laboratory. Therefore, the results of the Holbrook no. 2 specimen are included in the computation of the averages.

The statistical errors due to counting are of the order of 5 per cent for bismuth, 3 per cent for thallium, and 1 per cent for mercury. In general, it was attempted to record a minimum of 10^3 counts of Po^{210}, Tl^{204} and Hg^{203} for each determination. In practice, over 10^4 counts were usually recorded for Hg^{203}, while in some cases less than 10^3 counts for Po^{210} were recorded, owing to demands upon the α-pulse analysis system.

Bismuth. The results obtained for bismuth are reasonably consistent for each individual meteorite with the exception of the Johnstown and Modoc specimens. Run no. 1 for these specimens exhibits an abundance greater than that for the second irradiation group. There is no experimental reason evident for this discrepancy, and it may be due to a true non-homogeneous distribution of bismuth in these specimens. Johnstown is described by some observers as an achondrite,

Table 1. Bismuth, thallium and mercury content of stone meteorites (10^{-9} g of nuclide/g meteorite)

Sample	No.	Weight irradiated (g)	Bi^{209}	Tl^{203}	Hg^{202}
Beardsley	1	1·2513	2·7	0·57	22
	2	1·2000	3·8	1·4	—
	3	1·2446	3·3	1·3	—
Forest City	1	1·1489	0·18	0·15	24
	2	1·0114	0·50	0·12	—
	3	1·2840	0·15	0·089	—
Holbrook	1	0·6808	(0·49)*	(0·27)*	—
	2	1·1032	2·1	0·35	52
Johnstown	1	1·2614	5·6	0·29	36
	2	1·0245	1·2	0·12	—
	3	1·0343	1·5	0·24	—
Modoc	1	1·4496	3·0	(<1·23)*	—
	2	1·1184	0·07	0·071	—
	3	0·8427	0·33	0·12	—
Plainview	1	1·2994	4·1	—	16
	2	1·0975	3·0	—	—
	3	1·0281	3·8	—	—
Average			2·2	0·40	30

* Omitted from average.

while all the other specimens used are chondrites. In this work the abundances in Johnstown were not significantly different from the major group of chondrites and hence. were included in the average values computed.

It will be noted that the bismuth abundance obtained in this work is significantly less than that adopted by SUESS and UREY. This is an interesting result, since it might be expected that element formation by successive neutron capture on a slow time scale would result in a pile-up at bismuth, and hence a relatively high abundance. It is evident that the bismuth abundance is quite relevant to theories of nucleogenesis.

Thallium. the thallium result for Modoc no. 1 is listed as a limit, since a low chemical yield resulted in a high uncertainty. This value is not included in the averages. The results show some "scatter," which is not unexpected considering the extremely low levels involved. The results for Forest City show quite good internal consistency. Only Beardsley appears to be greatly out of line (almost an order of magnitude greater than Forest City.) BATE *et al.* (1958) found the thorium content of this same Beardsley specimen to be about 25 per cent higher than the average of four other chondrites. It is possible this specimen may have been subjected to post-fall alteration. However, there is no reason to ascribe this result to experimental error, and the Beardsley values are included in the determination of the average. If they were not included, the Tl^{203} abundance average listed in

Table 2. Cosmic abundances of the heavy nuclides Bi^{209}, Tl^{203} and Hg^{202} on the basis of neutron activation analysis on stone meteorites

Nuclide	Abundance (atoms/10^6 Si)	
	SUESS and UREY	This work
Bi^{209}	0·144	0·0016
Tl^{203}	0·0319	0·00030
Hg^{202}	0·0846	(0·023)*

* Preliminary result based on five determinations.

Table 1 would be 0·17 rather than 0·40 (10^{-9} g Tl^{203}/g meteorite). This would result in an atomic abundance (Si $= 10^6$) for Tl^{203} of 0·00013 instead of 0·00030 as listed in Table 2. Both of these values are significantly lower than that listed by SUESS and UREY.

Mercury. The results reported here for mercury must be regarded as preliminary, since they are based on only five determinations. Each of the values listed in Table 1 are for a single meteorite sample irradiated alone in an irradiation can. Additional determinations in which two meteorite specimens were irradiated in the same can, as was the case for specimens no. 2 and no. 3, gave variations in the counting rates of the mercury flux monitors from the same can and mercury contents, which were spurious and larger than those reported in Table 1.

A possible explanation may be that the irradiations, in which two meteorite specimens were in the same irradiation can, were exposed to sufficiently different cooling conditions to cause loss of mercuric nitrate by volatilization from the flux monitor foils. There is also an indication that $Hg(NO_3)_2$ may have been physically lost from several of the run no. 2 and no. 3 flux monitors, owing to formation of a flaky non-adhering deposit. This type of deposit did not appear in the preparation of run 1 flux monitors and is not as yet explained. To what extent volatilization may have occurred in the single sample irradiations is unknown, but the consistency of the flux monitor counts and calculated mercury abundances in this group suggests a certain confidence in the values listed in Table 1. Additional work is planned in order to better establish these mercury abundances.

Table 2 summarizes the results of this study and compares them to the literature values accepted by SUESS and UREY (1956). In this table the results are in terms of atomic abundance of the nuclide per 10^6 atoms of Si, using an average Si content of 0·185 by weight in primitive solar non-volatile material. The results of this work are in reasonably good agreement with those of REED *et al.* (1958) for Bi^{209} ($\leq$0·005) and Tl^{203} (0·00013).

These values together with the cosmic abundances of uranium ($1{\cdot}14 \times 10^{-8}$ g U/g meteorite) reported by HAMAGUCHI *et al.* (1957) and thorium ($3{\cdot}96 \times 10^{-8}$ g Th/g meteorite) reported by BATE *et al.* (1958) are plotted in Fig. 1.

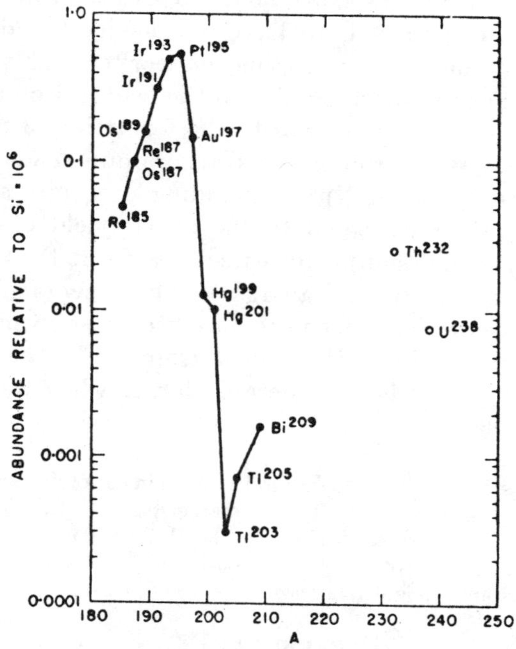

Fig. 1. Cosmic abundances of heavy nuclides based on analyses of chondrites; (●) odd A nuclides; (○) even A nuclides. Re, Os, Ir, Pt and Au, SUESS and UREY (1956); Hg, Tl and Bi, This work; Th, BATE *et al.* (1958); U, HAMAGUCHI *et al.* (1957).

The known abundances of thorium and uranium in chondrites are large compared to the abundances of thallium and bismuth. Processes, such as neutron capture on a fast time scale (BURBIDGE *et al.*, 1957; CAMERON, 1957), which have been invoked to explain the abundances of thorium and uranium, are also expected to make copious amounts of the progenitors of Bi^{209}. The fact that chondrites contain at present only a very small amount of bismuth relative to thorium and uranium strongly suggests that bismuth has been fractionated from these heavier elements sometime in the past.

The longest lived member of the $4n + 1$ radioactive series is Np^{237} which has a half-life of only 2·2 million years. With existing estimates of the time interval between element production and the formation of the solar system (WASSERBURG and HAYDEN, 1955) it is possible that the Bi^{209} progenitors decayed to very low concentrations prior to the formation of the solar system. Fractionation of elements with extremely low concentrations in chondrites such as mercury, thallium, lead, bismuth and indium (SCHINDEWOLF and WAHLGREN, 1958) may have occurred therefore prior to or during the formation of the solar system (including meteoritic materials). It is interesting to note that present estimates of the lead content in the sun reported by ALLER (1957) is a factor of at least 10^4 greater than the thallium and bismuth concentrations of chondrites reported in Table 2. A comparison of the isotopic content of the lead in iron meteorites and stone meteorites like Nuevo Laredo (PATTERSON, 1955) indicates that the fractionation of lead from thorium and uranium in the meteorites has not been important during their $4 \cdot 5 \times 10^9$ year lifetime. On this basis one would expect the lead in the sun, if its concentration is as high as reported, to have a considerably different isotopic composition from lead in the earth and stone meteorites. This solar lead may possibly have an isotopic composition similar to the lead in iron meteorites, if the fractionation did take place during or prior to the formation of the solar system.

If the process or processes forming the transuranium elements (BURBIDGE *et al.*, 1957) give an initial yield of Np^{237} (including progenitors decaying by α-emission) which is approximately equal to the initial yield of Th^{232} (including progenitors decaying by α-emission), one calculates from the present bismuth and thorium contents in chondrites and an age of the elements of $6 \cdot 6 \times 10^9$ years that more than 95 per cent of the bismuth has been lost from the chondrites. This figure includes only the bismuth formed from Np^{237} and its progenitors decaying by α-emission and will be considerably larger when the other modes of producing Bi^{209} are included.

Acknowledgements—The assistance of a number of individuals in supplying samples for this work is gratefully acknowledged. The sources of all these meteorites with the exception of the Holbrook no. 2 sample are listed by BATE *et al.* (1958). The Holbrook no. 2 sample was obtained through the courtesy of Dr. U. SCHINDEWOLF. We also wish to thank Mr D. J. HENDERSON for his assistance with the α-pulse-height analyses.

REFERENCES

ALLER L. H. (1957) *The Abundance of the Elements Determined from the Sun and Main-sequence Stars* Chap. 2. Interscience, New York. (Preprint).

BATE G. L., HUIZENGA J. R. and POTRATZ H. A. (1958) Thorium in stone meteorites by neutron activation analysis. In press.

BOYD G. E. (1949) Method of activation analysis. *Analyt. Chem.* **21,** 335.

BURBIDGE E. M., BURBIDGE G. R., FOWLER W. A. and HOYLE F. (1957) Synthesis of the elements in stars. *Rev. Mod. Phys.* **29,** 547–650.

CAMERON A. G. W. (1957) *Stellar Evolution, Nuclear Astrophysics, and Nucleogenesis.* Chalk River Laboratory Report No. 41.

EL BADRY H. and KOHMAN T. P. (1957) Private communication.

HAMAGUCHI H., REED G. W. and TURKEVICH A. (1957) Uranium and barium in stone meteorites *Geochim. et Cosmochim. Acta* **12,** 337.

JENKINS E. N. and SMALES A. A. (1956) Radioactivation analysis. *Quart. Rev. Chem. Soc.* **10,** 83.

MEINKE W. W. (1955) Trace-element sensitivity: Comparison of activation analysis with other methods. *Science* **121,** 177.

NODDACK I. and NODDACK W. (1934) Die geochemischen Verteilungkoeffizunten der Element. *Svensk Kem. Tidskr.* **46,** 173.

PATTERSON C. C. (1955) The Pb^{207}/Pb^{206} ages of some stone meteorites. *Geochim. et Cosmochim. Acta* **7,** 151.

PATTERSON C. C., TILTON G. and INGHRAM M. (1955) Age of the Earth. *Science* **121,** 69.

PLUMB R. C. and LEWIS J. E. (1955) How to minimize errors in neutron activation analysis. *Nucleonics* **13,** 42.

PRIOR G. T. and HEY M. H. (1953) *Catalogue of Meteorites* (2nd Ed.). British Museum, London.

REED G. W., KIGOSHI K. and TURKEVICH A. (1958) Activation analysis for heavy elements in stone meteorites. *Second United Nations International Conference on the Peaceful Uses of Atomic Energy* Paper 953.

SCHINDEWOLF U. and WAHLGREN M. A. (1958) Private communication.

SHAW D. M. (1952) The geochemistry of Tl. *Geochim. et Cosmochim. Acta* **2,** 118.

SUESS H. E. and UREY H. C. (1956) Abundances of the elements. *Rev. Mod. Phys.* **28,** 53.

WASSERBURG G. J. and HAYDEN R. J. (1955) Time interval between nucleogenesis and the formation of the meteorites. *Nature, Lond.* **176,** 130.

5

Reprinted from *Geochim. et Cosmochim. Acta* **20**:122–140 (1960)

Determinations of concentrations of heavy elements in meteorites by activation analysis*

G. W. REED, K. KIGOSHI† and A. TURKEVICH

Argonne National Laboratory, Lemont, Illinois and Enrico Fermi Institute for Nuclear Studies, University of Chicago, Chicago, Illinois

Abstract—Barium, mercury, thallium, lead, bismuth and uranium contents and isotopic composition of stone and the troilite phase of iron meteorites have been determined by neutron activation analysis. Two groups of chondrites were observed: the one containing enstatite and carbonaceous chondrites has orders of magnitude more of the chalcophilic elements studied than other chondrites. Barium and uranium do not show this variation. The lead in this group is primordial in its 208/204 isotopic composition as might be expected from the Pb/U ratio. Some of the nuclear and geochemical implications of the results are suggested.

INTRODUCTION

NEUTRON activation analysis has recently been used to measure the abundance of trace heavy elements in meteorites (REED and TURKEVICH, 1955; HAMAGUCHI *et al.*, 1957; REED, *et al.*, 1958a; BATE *et al.*, 1957, 1958; EBERT *et al.*, 1957; HERNEGGER and WÄNKE, 1957; REED *et al.*, 1958b). This paper reports results obtained by this method for the barium, mercury, thallium, lead, bismuth and uranium concentrations in samples of meteoritic matter. Some of the results of these analyses have already been reported (REED *et al.*, 1958b). These early data indicated that ordinary chondrites, such as Forest City, Modoc, Holbrook, etc., which have often been used for estimating the cosmic abundances of heavy elements, are unexpectedly low in thallium, lead and bismuth. This result was confirmed by EHMANN and HUIZENGA (1959) who also showed that mercury was low in such meteorites.

The present paper reports more data on a larger variety of non-metallic meteorites. It has been found that samples of enstatite and carbonaceous chondrites have much larger amounts of thallium, lead and bismuth than the other types of chondrites studied. The abundances of these elements in these particular stony meteorites are closer to what might be expected on the basis of nuclear abundance regularities and current theories of nucleogenesis.

EXPERIMENTAL PROCEDURES

The nuclear reactions used, the effective cross-sections, and the characteristics of the radioactive species produced, have already been presented (REED and TURKEVICH, 1955; HAMAGUCHI *et al.*, 1957; REED *et al.*, 1958) except for the case of mercury. In this case the radiations from the 65 hr Hg^{197} and 48 day Hg^{203} formed by (n, γ) reactions on stable mercury isotopes were measured. The large neutron capture cross-section (2300 barns) for the formation of Hg^{197} more than

* This work was supported in part by the U.S. Atomic Energy Commission.

† Present address: Gakushuin University, Mejiro, Tokyo, Japan.

compensates for the low isotopic abundance of Hg^{196} (0·15 per cent) and makes the shorter-lived activity very prominent in neutron irradiated mercury.

The meteorite samples studied, as well as their physical form and source are

Table 1. Description of meteoritic samples

Samples	Irradiation	Physical form	Source	Reference†
Forest City	6, 7 12	Powder fragments	C. C. Patterson	a, b, c, d
Modoc	6, 8, 10 13, 14	Powder fragments	H. C. UREY	a, b, c
Holbrook	8, 10 13 11, 23	Powder fragments Fragments	R. R. Marshall H. C. Urey	a, c, d
Beardsley	11 19, 23	Fragments Fragments	R. R. Marshall G. Bate	a, c
Nuevo Laredo	7, 19, 20	Powder	C. Patterson	a, b, c
Abee*	12, 14, 24	Fragments	F. Begemann	e
Indarch Mighei Orgueil	14, 18 18 19, 20	Fragments	H. C. Urey	f
Troilite from:				
Canyon Diablo Toluca	15, 16, 21 16	Fragments fragments	C. Patterson H. Nininger	g
Basalt Perthite	9 9	Powder	G. R. Tilton	h i

* K. R. DAWSON, Geological Survey of Canada, describes Abee as a "black, polymict chondrite breccia". H. RAMBERG, Department of Geology, University of Chicago, in a cursory examination was only able to identify an orthorhombic pyroxene which appeared to be enstatite.

† References to Table 1 are as follows:

(a) HAMAGUCHI *et al.* (1957).
(b) PATTERSON (1955).
(c) BATE *et al.* (1957).
(d) MARSHALL and HESS (1958).
(e) BEGEMANN *et al.* (1959).
(f) WIIK (1956).
(g) PATTERSON *et al.* (1953).
(h) PATTERSON *et al.* (1955).
(i) TILTON *et al.* (1955).

listed in Table 1. In addition we have indicated some of the known previous analytical work on the same samples. As insurance against the introduction of terrestrial contamination, many samples were irradiated, without crushing, as freshly broken fragments.

The samples include four chondrites (Forest City, Modoc, Holbrook and Beardsley) that have frequently been taken as representing what we shall refer to as "ordinary chondrites". Nuevo Laredo is an achondrite. These five meteorites have been studied previously in our laboratories (HAMAGUCHI *et al.*, 1957; REED *et al.*, 1958b). Abee and Indarch are examples of enstatite chondrites: Orgueil and Mighei are carbonaceous chondrites. (See footnote to Table 1 for classification of Abee.) Two samples of troilite were studied: one from the Canyon Diablo iron meteorite and a sample from the Toluca iron meteorite. Some uncertainty exists in the identification of this latter sample as troilite. The material did not dissolve in acid. In polished sections it was revealed to be an intimate mixture of graphite and troilite. This was confirmed by chemical analysis by R. BANE of the Argonne National Laboratory.

The irradiation technique, chemical processing and counting have already been described in some detail (REED *et al.*, 1958b). Briefly, 1–2 g samples of meteorite were sealed in quartz vials and irradiated along with monitors of meteorite powder spiked with known amounts of the elements to be measured. Most of the irradiations were carried out inside a special fuel element at the Argonne CP–5 reactor so that the fast neutrons available could be used to produce Pb^{203} from Pb^{204} by $(n, 2n)$ reactions*. This, along with Pb^{209} from neutron capture by Pb^{208}, permitted a determination of the Pb^{208}/Pb^{204} isotopic ratio as well as the lead concentration.

With few exceptions, barium, thallium, lead, bismuth and uranium were all measured in the same sample. Mercury was determined in only a few experiments; because of possible losses due to volatilization, a different procedure (see below) was used in the initial solution of the meteorite when a determination of mercury was to be made.

After the exposure to neutrons, the samples of stone meteorites were usually fused with sodium peroxide in the presence of carriers in a nickel crucible. A clear melt was considered as evidence of complete equilibration between added carriers and radioactive isotopes. When mercury was to be determined, the samples and added carriers were digested with $HClO_4$, HNO_3 and HF until only a small residue remained. In one case this residue was shown not to contain any mercury; this will be discussed later. The troilite samples were dissolved in HCl; only the residues, if any, from this acid treatment were fused.

The same radiochemical purification procedure was used for lead, bismuth, thallium, uranium and barium as described in the previous reports (HAMAGUCHI *et al.*, 1957; REED *et al.*, 1958b). The mercury, when determined, was precipitated with the bismuth as the sulphide from an acid solution after the separation of the lead and barium sulphates. It was separated from the bismuth when the latter was precipitated as the oxychloride and was then brought down as sulphide. The purification cycles consisted of reduction with $SnCl_2$, precipitation of Hg_2Cl_2 with HPO_3 from dilute HCl solution, AgCl scavenging, $Al(OH)_3$ scavenging with the mercury complexed as iodide, reduction in alkaline solution with hydroxylamine sulphate, and ether extraction of the iodide complex.

As in all experiments of this type, the monitors were separated and purified in

* The effective thermal cross-section for the $Pb^{204}(n, 2n)Pb^{203}$ reaction for a sample irradiated inside an enriched uranium fuel rod is 0·002 barns and not the value given in our earlier report (REED *et al.*, 1958b).

parallel procedures. Usually the fusion step was not performed, since complete recovery of the monitor radioactivity was possible by digestion in acid; nor was the extreme radioactive decontamination necessary.

Counting was done routinely in methane-flow proportional counters at efficiencies usually approximating 50 per cent. The very low activity samples frequently encountered were measured on a proportional counter of the same type but

Table 2. Experimental results on lead contents of meteorites

Meteorite	Expt. no.	Pb^{204} (g/10^8 g**)	Pb^{208} (g/10^6 g)	"Total Lead Content" (g/10^6 g†) via Pb^{204}	"Total Lead Content" (g/10^6 g†) via Pb^{208}	Isotopic ratio § Pb^{208}/Pb^{204}
Fotest City	6*	<0·7	<0·2			
	7*	<0·64	0·09 ± 0·02		0·17 ± 0·04	
	12	0·20 ± 0·02	0·05 ± 0·01‡	0·15 ± 0·02	0·10 ± 0·02‡	
Modoc	6*	<0·2	0·05 ± 0·01		0·10 ± 0·02	
	8*	<0·14	0·02 ± 0·005		0·04 ± 0·01	
	10*	0·08 ± 0·03	<0·15	0·06 ± 0·02		
	13	0·08 ± 0·03		0·06 ± 0·02		
	14	<2·3	<0·20			
Holbrook	8*	0·55 ± 0·05	0·23 ± 0·01	0·41 ± 0·04	0·44 ± 0·02	
	10*	0·80 ± 0·20	0·19 ± 0·07	0·60 ± 0·16	0·36 ± 0·14	
	11*	0·45 ± 0·03	0·16 ± 0·01	0·34 ± 0·04	0·31 ± 0·02	
	13	0·13 ± 0·02		0·10 ± 0·02		
Beardsley	11*	0·23 ± 0·02	0·07 ± 0·01	0·17 ± 0·02	0·13 ± 0·02	
	19	<0·36	<0·15			
Nuevo Laredo	7*	0·56 ± 0·06	0·33 ± 0·03	0·42 ± 0·04	0·63 ± 0·06	59 ± 8
	20		>0·17			
Abee	12	5·6 ± 0·1	1·5 ± 0·2‡	4·20 ± 0·08	2·9 ± 0·4‡	
	14	>1·8	>0·2			
Indarch	18	3·43 ± 0·05	0·93 ± 0·03	2·56 ± 0·04	1·78 ± 0·06	26·8 ± 1·2
Mighei	18	2·34 ± 0·06	0·71 ± 0·03	1·75 ± 0·05	1·36 ± 0·06	31·1 ± 1·2
Orgueil	20	4·53 ± 0·32	1·31 ± 0·03	3·38 ± 0·25	2·50 ± 0·06	30·3 ± 1·2
Canyon Diablo troilite	15	9·2 ± 0·1	2·67 ± 0·02	6·86 ± 0·08	5·10 ± 0·04	28·8 ± 1·0
	16	10·5 ± 0·2	2·93 ± 0·03	7·84 ± 0·15	5·60 ± 0·06	27·6 ± 1·0
Toluca troilite	16	10·8 ± 0·2	3·11 ± 0·06	8·06 ± 0·15	5·94 ± 0·12	29 ± 3
Basalt	9*	9·0 ± 0·2		6·7 ± 0·2		
Perthite	9*	9·1 ± 0·2		6·8 ± 0·2		

* Experiments reported previously (REED *et al.*, 1958b), but data re-evaluated.

† The "lead content" for this column was calculated on the assumption of normal lead isotopic composition in the meteorite, even though in many of the cases the data indicate different Pb^{208}/Pb^{204} ratios.

‡ Experiment no. 12 was performed in a different part of the pile where the activation of Pb^{208} in the monitor appeared to be slightly (~20 per cent) anomalous.

§ On basis of 208/204 = 38·9 in monitor. This had been determined mass spectrometrically (REED *et al.*, 1958b).

** The units in this column (g/10^8 g) differ from the units in columns (4), (5) and (6) (g/10^6 g).

shielded with anti-coincidence counters so that its background was only 1 c/min. Thallium X-rays arising from both Pb^{203} electron capture decay and Hg^{197} β-decay were measured on a single channel scintillation counter. These data gave results in agreement with those obtained from the proportional counter measurements.

EXPERIMENTAL RESULTS

Lead

Table 2 gives a summary of our experimental data on the lead contents of meteorites. Most of the results on ordinary chondrites (Forest City, Modoc, Holbrook and Beardsley) have been reported earlier (REED *et al.*, 1958b). (These results are accompanied by symbol* in the table, and are reproduced for completeness.) In a few cases the numbers have been changed slightly as a result of re-evaluation of the experimental data. The first two columns identify the meteorite and the experiment number. The third and fourth columns indicate the results for the Pb^{204} and Pb^{208} contents of the meteorite as deduced from the Pb^{203} and Pb^{209} activity, respectively. In some cases, only limits (usually upper limits) are indicated because of the possibility that the isolated radioactivity was contaminated. The fifth and sixth columns give the gross lead contents of the meteorites calculated on the assumption that the isotopic composition of the lead in the meteorites was the same as that of terrestrial lead, even though, as will be seen, this assumption is not generally valid.

Finally, in the last column, we give the Pb^{208}/Pb^{204} ratio in the meteorites from our data for those cases in which we have sufficient confidence.

The new experiments on the lead contents of ordinary chondrites confirm the earlier results, but unfortunately the activity levels are still too low to make an appreciable improvement in the data. Forest City would appear to have about 0·15 p.p.m. of lead; Modoc about 0·06 p.p.m., Holbrook about 0·40 p.p.m. and Beardsley about 0·15 p.p.m. Thus, a lead content of 0·15 p.p.m. in these chondrites appears to be a median value with a range of a factor of 3 each way.

The levels of activity in the case of these chondrites were too low in most cases to give more than an indication of the isotopic composition. In the cases where both isotopes were determined in the same irradiation, there is some indication that the lead in these meteorites may be less radiogenic than terrestrial lead.

The lead concentrations for the ordinary chondrites reported here and in our previous paper are confirmed by the mass spectrometry measurements of MARSHALL and HESS (1958, 1959) who observed the same range of concentrations. Their values for Forest City and Holbrook are about 0·09 and 0·3 p.p.m. (MARSHALL and HESS, 1958). Richardton, a chondrite, not studied here, has a lead content of about 0·1 p.p.m.

As reported earlier (REED *et al.*, 1958), the results on Nuevo Laredo are in satisfactory agreement with PATTERSON (1955) in both the absolute amounts of Pb^{204} and Pb^{208} and in the very radiogenic nature of this lead ($Pb^{208}/Pb^{204} = 59 \pm 8$). PATTERSON (1955) reports 68 for this ratio.

The meteorites Abee, Indarch, Mighei and Orgueil stand out in having an order of magnitude more lead than the ordinary chondrites, varying between 1 and 5 p.p.m. Furthermore, the Pb^{208}/Pb^{204} ratios (27–31) are low relative to those in terrestrial lead ($\sim$39). That the lead in these meteorites should be relatively non-radiogenic is reasonable in view of their low ($\sim 10^{-8}$ g/g) uranium contents (see below). If the thorium to uranium ratios are near normal (3–4) radioactive decay in 4·5Æ could not seriously affect the isotopic composition of lead present in p.p.m. quantities. Also, these high lead results would appear to be free of suspicion from contamination since it would be hard to contaminate with primordial lead.

Finally, the lead contents of the troilite from two iron meteorites (Canyon Diablo and Toluca) are in the 5–8 p.p.m. range. This is appreciably lower than the 18 p.p.m. reported by PATTERSON *et al.* (1953). Our best value for the Pb^{208}/

Pb^{204} ratio in such material comes from the two irradiations of Canyon Diablo troilite. Fig. 1 shows the decay curve of lead from Canyon Diablo and that of a lead sample with Pb^{208}/Pb^{204} ratio of 38·9. The average of two such runs indicates a Pb^{208}/Pb^{204} ratio in Canyon Diablo of 28·2 $\pm$ 1·0. The quoted error includes an estimate of possible systematic effects; the statistical errors and agreement between the two determinations are appreciably smaller. The result is not far from the value of 29·2 reported by PATTERSON *et al.* (1953), for the lead from the troilite of this meteorite, although perhaps significantly lower. The isotopic

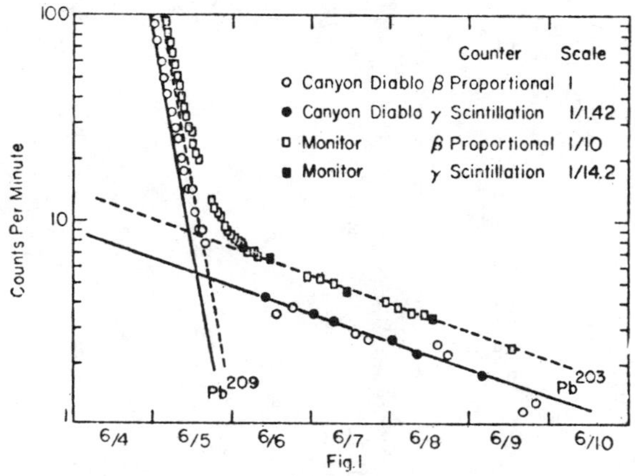

Fig. 1. Decay curve of lead from Canyon Diablo troilite in comparison with that of monitor lead. The straight lines indicate the decomposition of the two components, Pb^{203} and Pb^{209}. At a time when the two species contribute equally in the case of the monitor (6/5 1400), it is seen that the sample has more Pb^{203} than Pb^{209}.

ratio (29) deduced for the troilite from the Toluca meteorite is similar to that found in Canyon Diablo. Unfortunately, in this case a slight amount of radioactive contamination prevents us from assigning an error of less than 10 per cent to the ratio.

The absolute amounts of lead found by this method in the terrestrial basalt and perthite, about 7 p.p.m. have been shown (REED *et al.*, 1958b) to be in satisfactory agreement with results obtained by PATTERSON *et al.* (1955) and TILTON *et al.* (1955), respectively, by isotope dilution methods.

Barium

Table 3 gives a summary of our experimental data on the barium contents of meteorites. These data are based primarily on the 11·6-day Ba^{131} activity induced in the samples. The correction from the β-activity of Ba^{140} from fission, was made as described by HAMAGUCHI *et al.* (1957), by La^{140} milking, and amounted to about 20 per cent in most cases.

The first column of Table 3 indicates the meteorite studied; the second gives the irradiation number; the third gives the individual results for barium contents in grammes per 10^6 g of meteorite. The probable error in these numbers is considered to be about 15 per cent except where indicated. The results on meteorites analysed previously are in excellent agreement with the values reported by

Table 3. Experimental results on the barium and "uranium" contents of meteorites

Meteorite	Irradiation no.	Barium content (g/10 g)	Average*	"Uranium" content (g/10^8 g)	Average†
Forest City	6	4·5	3·3	1·6	1·5
	7				
	12	2·5		1·4	
Modoc	6	4·8	3·8	1·8	1·4
	8	2·7		1·2	
	10	3·2		1·1	
	14	2·6‡		1·5	
Richardton§			3·2		
Holbrook	8	3·6	3·6	1·6	1·6
	10	3·6		2·1	
	11	3·2		1·6	
	23			1·2**	
Beardsley	11	3·8	3·0	1·7	1·2
	19	2·2		0·9	
	23			1·1**	
Nuevo Laredo	19	41·9	44	14	15
	20	44·3		17	
Abee	12	2·0	1·8	1·1	1·1
	14	1·5		1·0	
	23			1·1**	
Indarch	18	1·9		1·6	
Mighei	18	2·5		1·6	
Orgueil	19			0·8	0·8
	20	2·4		0·8	
Canyon	15	0·4		0·4	0·4
Diablo	16	<0·006		0·4	
Troilite	21	<0·3			
Toluca					
Troilite	16	<0·1		<1·3	
Basalt	9	905		84	
Perthite	9	1010		34	

* The average barium content has been calculated including the results of HAMAGUCHI *et al.* (1957).
† The averages listed here are from the present work only. We suspect that they are probably systematically too high by about 30 per cent (see text).
‡ An error of ~50 per cent must be assigned to the results of this measurement.
§ Richardton was not examined in this study. The results are from HAMAGUCHI *et al.* (1957).
** Irradiation in thermal column of Oak Ridge National Laboratory pile.

HAMAGUCHI *et al.* (1957). The fourth column gives the average barium content of the meteorite in question based on both present and earlier results.

The scatter of the results from individual determinations is of the order of 20 per cent in the cases of all the stone meteorites. This would appear to be real and may be a reflection of the difficulty of getting a representative sample in a 1 g sample. The average of the first five stone meteorites (ordinary chondrites) is 3·4 p.p.m. The meteorites Abee, Indarch, Mighei and Orgueil would appear to be in a slightly lower range—about 2·2 p.p.m. The present results confirm the earlier ones (HAMAGUCHI *et al.*, 1957) in indicating 44 p.p.m. for the barium content of Nuevo Laredo.

Information on the relative abundances of the light isotopes of barium (Ba^{130}, Ba^{132}, Ba^{134}) can be obtained from the amount of short-lived radioactivity (29–35 hr) present in the samples as compared to the 11·6-day Ba^{131} radioactivity. All of the stone meteorites examined to date show decay curves consistent with terrestrial barium isotopic ratios to within an accuracy of 5 per cent.

The barium content in the troilite phase of the two iron meteorites studied is substantially lower than that of stone meteorites. One sample of Canyon Diablo troilite gave 0·4 p.p.m.; the other two samples of troilite studied gave only limits of 0·3 and 0·06 p.p.m.

Our results for the barium concentrations in basalt (905 p.p.m.) and in perthite (1010 p.p.m.) appear to be reasonable in comparison with data in the literature on the concentration of the element in basalts and granites (FAIRBAIRN, 1951; ENGEL and ENGEL, 1958) which give values between 500 and 2500 p.p.m. although a greater difference between basalt and perthite appears statistically to be probable.

Uranium

Table 3 also shows, in column (5), the experimental data on the uranium contents of the meteorites studied. This was calculated from the Ba^{140} contribution to the radioactivity of the isolated barium. As in our earlier work (HAMAGUCHI *et al.*, 1957), this contribution was determined by lanthanum "milking" of the barium after the La^{140} had grown in. The experiments thus analyse for fissionable isotopes. On the assumption that the main contribution is from the fission of U^{235} and assuming a normal isotopic ratio for the uranium—as established in our earlier work on chondrites (HAMAGUCHI *et al.*, 1957)—we can calculate the uranium contents from the La^{140} radioactivity.

Examination of the results listed in the table and comparison with those reported by HAMAGUCHI *et al.* (1957) show that the present results confirm the level of the uranium content of stone meteorites found by these authors. The scatter of the present results is a little greater, however, and the absolute amounts are somewhat higher than those reported previously. Although at present we do not completely understand the reason for this, we feel that the conditions of irradiation in most of the experiments reported here were much less suitable for a good uranium analysis than in our previous work. For example, in the present work there is some possibility of contribution from the fast fission of thorium. This is supported by the fact that irradiations of two chondrites (Holbrook and Beardsley) in the thermal flux at the Oak Ridge Pile (irradiation no.23) gave uranium contents consistent with the lower numbers reported by Hamaguchi *et al.* (1957). Thus we feel that the uranium contents of all ordinary chondrites, including Beardsley, which had not been studied previously, are all the same—close to 1·1 $\times 10^{-8}$ g/g.

The present study, moreover, establishes that the uranium contents of Abee, Indarch, Mighei and Orgueil are just about the same as those of other stony meteorites. The approximately ten times higher value for Nuevo Laredo has been noted before (Hamaguchi *et al.*, 1957). On the other hand, the troilite from Canyon Diablo has only about one-third the uranium that typical stony meteorites have. Patterson *et al.* (1953) have reported 0·9 $\times 10^{-8}$ g/g for the uranium content of Canyon Diablo troilite—thus a little more than twice the amount we find.

The results on the terrestrial materials agree moderately well with values obtained by Tilton *et al.* (1955). He found 80 $\times 10^{-8}$ g/g and 22 $\times 10^{-8}$ g/g in the basalt and perthite as compared to our results of 84 $\times 10^{-8}$ and 34 $\times 10^{-8}$ g/g. In view of the possible contribution of thorium fission to our studies and the possible variation due to sampling at the 1 g level, this agreement must be considered satisfactory.

Bismuth

Table 4 summarizes our experimental results on bismuth. Column (3) gives the results of individual experiments on the meteorites in units of grammes of bismuth in 10^9 g of meteorite. The experiments whose results have been reported previously (in a few cases slightly re-evaluated) are indicated with a symbol.* In the case of the ordinary chondrites (the first four entries in the table) the results indicate very low and, in several cases, varying amounts of bismuth in these meteorites. In addition to the usual criteria of satisfactory decay and absorption characteristics of the β-rays, in three of the experiments the isolated radioactivity was established to be pure by further chemical purification. We therefore feel that the bismuth content of these meteorites is variable, but in all cases, very low, having a median value of 3 $\times 10^{-9}$ g/g. The meteorite Nuevo Laredo seems to be in the same class, although only limits could be established to its bismuth content.

These low values for the bismuth content of ordinary chondrites have recently been confirmed by Ehmann and Huizenga (1959), who studied several of the same meteorites using a slightly different neutron activation method. Their average bismuth content is 2·2 $\times 10^{-9}$ g/g, which is close to our median value. In addition, their results on Beardsley check our value very well. They, too, find varying results for Modoc and lower values than the average for Forest City—in this case significantly lower than the results reported here.

In marked contrast to these low figures are the bismuth contents of Abee, Mighei and Orgueil, which lie in the range 80 to 180 $\times 10^{-9}$ g/g. Indarch also has a high value in this range, although an exact number could be deduced due to radioactive contamination. Likewise, the two samples of troilite gave bismuth contents of this same magnitude (40 to 180 $\times 10^{-9}$ g/g).

Thallium

Thallium samples separated from irradiated ordinary chondrites usually

Table 4. Experimental results on the bismuth, thallium and mercury contents of meteorites

Meteorite	Irrad. no.	Bismuth content (g/10⁹ g)	Ave. or range	Thallium content (g/10⁹ g)	Mercury content (g/10⁹ g)
Forest City	6*	<2·7		0·51	
	7*	1·1	1·0	0·39†	
	12	0·7		0·14†	
Modoc	6*	1·2		0·83†	
	8*	<3·4		0·14†	
	10*	0·4	0–6	0·23†	
	13	1·0		0·03†	
	14	6·2		—	
Holbrook	8*	<8	7·5	0·45†	
	10*	7·6		0·56	
	11*	1·7	2·0	3·8†	
	13	2·6		0·42†	
Beardsley	11*	3·0†	3·5	0·36†	
	19	4·1		—	
Nuevo Laredo	7*	<1		0·75	
	19	<1·6		—	0·3‡
					78·0§
					0·04**
	20	⩽1·4		0·57	
Abee	12	93		71	
	14	68	80	96†	
	24	—		—	4·0§
Indarch	14	high		—	
	18	high		125	
Mighei	18	180		97	
Orgueil	19	—		—	11,000‡
					6260§
	20	130		141	
Canyon Diablo troilite	15	40		13†	7·5§
	16	⩽65	40	7·8	5·9§
Toluca troilite	16	180		198	<34
Basalt	9*	7·6		72·5	
Perthite	9*	11		2300	

* The bismuth and thallium results of these experiments have been reported previously (Reed *et al.*, 1958b). Some of the data here have been re-evaluated.

† Sample sent through extra chemistry and found to be radiochemically pure. Almost all of the thallium samples were recycled and found to be radiochemically pure.

‡ Cold trap sample. § Acid treated sample. ** Fused residue.

contained no more than 2 c/min of activity. The long half-life of Tl^{204} made the observation of decay, which is one of the criteria for radioactive purity, impractical. As a result, chemical repurifications were carried out on most of the low activity samples. Constant specific activities (c/min per g) as well as correct absorption behaviour of the radiations were required to make the data acceptable. The spread in the data from successive purifications was in general less than 20 per cent and within the range expected from the estimated uncertainty in counting, etc.

The thallium results are compiled in column (5) of Table 4, and include our earlier reported results, in some cases slightly revised. As in the case of bismuth, the thallium contents show a tendency to scatter from sample to sample of the same meteorite. The mean thallium content of ordinary chondrites is about 4×10^{-10} g/g. This is the lowest concentration yet reported for non-volatile elements in chondrites. Nuevo Laredo, likewise, contains only about $6{\cdot}6 \times 10^{-10}$ g Tl/g. Ehmann and Huizenga (1959) find an average thallium content of ordinary meteorites (including four of those studied here) of 13×10^{-10} g/g. They also used neutron activation. Even though our levels of activity were lower, due to the shorter neutron exposure times, we feel confident of our results because of the tests of radiochemical purity which were made, and because, due to the possibility of radioactive contamination in the case of such a long-lived species as Tl^{204}, the lower results are more likely to be correct.

The carbonaceous and enstatite chondrites again have on the average 300 times more thallium than the ordinary chondrites; the concentrations range from (700–1400) $\times 10^{-10}$ g/g. Troilite has thallium concentrations in the range (100–2000) $\times 10^{-10}$ g/g.

Mercury

Analyses were conducted for mercury in stone meteorites in only the few irradiations where special control over the dissolving conditions was possible. This was because it was felt that the standard fusion dissolving procedure, geared for the short-lived Pb^{209}, might not be ideal for mercury in terms of equilibrating the radioactive mercury with the added carrier before possible vaporization losses. Moreover, in the case of two of the stones that were studied in single irradiations (Nuevo Laredo and Orgueil) an attempt was made to see if any of the radioactive mercury came off when the sample was heated to approximately 100°C for 2 hr while pumping with a mechanical pump. The pumping was done past a glass finger cooled to −78°C. Less than 1 per cent of the mercury activity was found on the cold finger in the case of Nuevo Laredo; more than 60 per cent of the activity was found in the cold trap in the case of Orgueil. Moreover, the Nuevo Laredo sample was dissolved in two stages after heating. It was first treated several times over a period of 2 days with an HF, $HClO_4$ and HNO_3 mixture, in the presence of mercury carrier. Most of the sample went into solution. The residue was fused in the presence of additional mercury carrier. This residue had less than $\frac{1}{10}$ per cent of the mercury activity of the sample. The other stone meteorites were treated only by the acid decomposition procedure.

The standard dissolving procedure in the case of the troilite samples was by acid treatment, and here the mercury analyses were conducted in parallel with the others reported in this paper.

The mercury analyses were based on the Hg^{197} and Hg^{203} radioactivity induced in the samples. The former was often measured by both β and X-ray counting. Usually the results, calculated on the basis of the terrestrial isotopic ratios, agreed to within a few per cent.

The results of these experiments on the mercury contents of meteorites are given in column (6) of Table 4. The result of the Toluca troilite is indicated as an upper limit because in this case the X-ray counting of the induced Hg^{197} activity gave a 40 per cent lower mercury content. Likewise the results for Abee in one irradiation were so inconsistent internally as to be completely discarded. It is seen that in the only case where check experiments were performed (Canyon Diablo troilite) the agreement is satisfactory. This value for Canyon Diablo troilite, $\sim 7 \times 10^{-9}$ g/g is much lower than that of NODDACK and NODDACK (1931) for this meteorite (200×10^{-9} g/g).

DISCUSSION

The analytical results reported in this paper can be examined from three relatively different standpoints. First, we can consider the results in the light of geochemical regularities observed in terrestrial rocks. This is particularly suitable in the cases of thallium and barium. Second, we can see if the results provide any new information about the nature or origin of the meteoritic samples. Finally, the nuclear implications of the data can be considered.

The geochemistry of thallium appears to be governed by the similarity of the ionic radii of K^+, Rb^+ and Tl^+ (1·33, 1·49 and 1·49 Å). SHAW (1952) has recently emphasized this point of view in a survey of the thallium contents of terrestrial rocks. These comparisons can now be extended to include meteorites. The average concentrations of thallium, rubidium and potassium in meteorites and composite terrestrial basalts and granites are given in Table 5. The meteoritic thallium data is from this paper. All the rubidium values and the terrestrial potassium values are from GAST (1959) and the meteoritic potassium data is from EDWARDS and UREY (1955), and EDWARDS (1955).

It can be seen from Table 5 that the absolute thallium concentration spreads over a much larger range (10^4) than does rubidium (350) or potassium (67) and this is reflected in the large variation in the Rb/Tl and K/Tl ratios. The meteorites show a greater variation in these ratios than do basalts and granites. It thus appears that factors other than ionic radii and charge have played a role in establishing the distribution of these three elements in meteoritic matter.

In spite of the type variations pointed out above, a very systematic trend exists in both the K/Tl ratios and Rb/Tl ratios. For each the values decrease in the following order: (1) ordinary chondrites, (2) Nuevo Laredo (basaltic achondrite), (3) basalts, (4) granites, (5) enstatite and carbonaceous chondrites. No data exists for the Rb concentration in the latter group so no check can be made of the Rb/Tl ratio.

Barium is another element for which some geochemical regularities have been noted. The range of barium contents of all stone meteorites except Nuevo Lared is remarkably small. All the results reported here and by HAMAGUCHI *et al.* (1957) lie between 1·8 and 4·0 p.p.m. The ordinary chondrites are even more homogeneous, having an average value of 3·5 p.p.m. The slightly lower barium content in the carbonaceous chondrites (2·4 p.p.m.) is close to this average value if calculated on a water- and carbon-free basis. A similar relationship between some

Table 5. Potassium, rubidium, and thallium contents of meteoritic and terrestrial matter

Material	Tl	Rb*	K	K/Tl ($\times 10^3$)	Rb/Tl
	(concentrations in p.p.m.)				
Ordinary chondrites	~0·0003	2·8	850†	2800	9300
Nuevo Laredo	0·0006	0·37	470†	800	600
Orgueil	0·17		730†	4·3	
Mighei	0·11		530†	4·8	
Indarch	0·125		880†	7·0	
Abee	0·084				
Basalts (composite)	0·13	30	9600*	74	231
This work	0·073				
Granites (composite)	3·1	130	31,500*	10	42
Perthite This work	2·3				
Igneous average	1·3				

* GAST (1959).
† EDWARDS and UREY (1955), EDWARDS (1955).

carbonaceous and ordinary chondrites has been noted before in the case of their sodium and potassium contents (EDWARDS, 1955).

The barium contents of the two enstatite chondrites examined are lower than those of ordinary chondrites by a factor of 2. In the enstatite chondrites there is not enough volatile material to explain the deviation. The deviation of Nuevo Laredo in the opposite direction is much larger, since it has more than 10 times the barium content of the ordinary chondrites.

In terrestrial rocks, the variation in barium content has been correlated with potassium content (ENGEL and ENGEL, 1958). Our limited data suggest that in meteorites the barium follows the calcium rather than the potassium. This is illustrated by the data of Table 6. Unfortunately, the calcium content of Nuevo Laredo has not been determined; the value in the table is an average for basaltic achondrites, with which Nuevo Laredo has been grouped (PATTERSON, 1955).

It is, however, the variation in the abundances of thallium, lead and bismuth in the different classes of stony meteorites studied, that appears to be especially significant as regards the nature and origin of these objects. The ordinary chondrites have been shown (REED *et al.*, 1958b) to have very low contents of bismuth and thallium, and even a relatively small amount of lead. The new results on ordinary chondrites presented here confirm these conclusions as do the recent results of EHMANN and HUIZENGA (1959). The latter authors also show that

Table 6. Barium, potassium, and calcium contents of meteorites

Type of meteorite	All concentrations in p.p.m.		
	Ba	K*	Ca
Enstatite chondrites	2·4	850	9000†
Ordinary and carbonaceous chondrites	3·5	880	16,000 24,000†‡
Basaltic achondrites‡			70,000
Nuevo Laredo	44	470	

* EDWARDS (1955).
† WIIK (1956).
‡ UREY and CRAIG (1953).

mercury is relatively rare in the chondrites they investigated. Our mercury content of 80×10^{-9} g/g for Nuevo Laredo is close to their average for chondrites. The other stone meteorites studied in this investigation, however (Indarch, Abee, Mighei and Orgueil) are much different in their contents of thallium, bismuth, lead and, in one case, mercury. Thallium is higher by a factor of about 200, lead by a factor of 5, bismuth by a factor of 50. The result for mercury, at least in Orgueil, is 80 times higher than the values of EHMANN and HUIZENGA (1959) for chondrites and our result on Nuevo Laredo. The mercury content of Abee, however, is a factor of 20 lower. In contrast with these data, the uranium and barium contents of these meteorites are very nearly the same as in all the stone meteorites (except Nuevo Laredo) examined to date.

Such a striking variation in the content of trace elements in different chondrites is very unusual. One may first try to understand it on the basis of different relative amounts of the three basic components of agglomerated meteorites, i.e. metal, sulphide and silicate. For example the amounts of thallium, lead and bismuth in the ordinary chondrites are consistent with these elements being associated with the few per cent troilite these meteorites contain, if it is assumed that this troilite is similar to that examined in this investigation. The simplest version of this picture is untenable, however, for the carbonaceous and enstatite chondrites. The sulphide content of the stone meteorites having the small thallium, lead and bismuth contents is only a factor of from 2 to 3 less than that of Indarch, Orgueil and Mighei. Moreover, these latter meteorites have amounts of Tl^{203}, Pb^{204} and Bi^{209} that are comparable to those in the separated sulphide phase of Canyon Diablo and Toluca. Thus one cannot attribute the variation in the amounts of these elements in the stone meteorites in a simple way to the sulphide contents. On the other hand, the limited data (NICHIPORUK and CHODOS, 1959) indicates that perhaps the sulphide component of meteoritic matter is more variable in composition than either the silicate or metal components.

Similarly, the variation in metal content cannot be responsible. For example, Mighei and Orgueil contain no metal whereas Indarch has about 24 per cent metal; all three are similar in their thallium, lead and bismuth contents. Forest City, on the other hand, has nearly as much metal as Indarch, yet contains orders of magnitude less thallium and bismuth.

The contents of these elements is not strongly correlated with water content. It is true that meteorites containing the larger quantities of these elements contain significant amounts of water (WIIK, 1956). However, within this group the water content varies by a factor of 10 (Orgueil and Mighei cf. Indarch) without affecting the thallium, lead and bismuth concentrations.

From these remarks it seems reasonable to consider that these elements are associated with the silicate component of the meteorites. It has recently been reported that indium (SCHINDEWOLF and WAHLGREN, 1960) is more than a factor of 10 lower in a typical chondrite than was expected. The chemical similarity of indium to thallium and bismuth seems significant and may be a clue to the chemical processes that appear to have removed these elements from ordinary chondrites.

The mercury content of meteorites is particularly interesting in providing information on the thermal history of these objects. Mercury and its compounds are readily lost at temperatures no greater than 350°C. Thus the variation in mercury content can be taken as an indication of thermal histories in the temperature range above that covered by the presence or absence of water.

The analytical results on mercury presented here are only fragmentary, but there appear to be three levels of mercury concentration. Nuevo Laredo, a basaltic achondrite, and the ordinary chondrites (Forest City, etc.) reported by EHMANN and HUIZENGA (1959) all appear to have mercury contents of about 80×10^{-9} g/g. However, the chondrite Abee, which may be typical of enstatite chondrites, falls in a class, along with the troilite phase of iron meteorites, having mercury contents in the range of $\sim 5 \times 10^{-9}$ g/g. Orgueil, a carbonaceous chondrite, contains about 6000×10^{-9} g/g. Thus, it is in the mercury concentration that a striking difference between the enstatite and carbonaceous chondrites is seen. This variation appears to be consistent with the apparent thermal history of the objects since the carbonaceous chondrites must have been at a low temperature throughout their history, having retained quite considerable amounts of water and easily decomposed organic compounds. On the other hand, polished section examination of Abee (also Indarch) reveals an almost continuous network of metal and sulphide in a silicate matrix suggesting high temperature exposure after accumulation.

We turn now to the nuclear evidence represented by these results. It has been remarked that the low abundances in ordinary chondrites of thallium, lead, bismuth and the single result on indium, if interpreted as cosmic abundances of these nuclides, are unsatisfactory in that they indicate abrupt discontinuities in elemental abundances. The results for the heavy elements in the carbonaceous and enstatitic chondrites are much more consistent with expectations. The situation is represented in Fig. 2. The ordinate represents, on a logarithmic scale, the abundance of the odd mass nuclides, as abscissae, in the region investigated. The reference abundance is the usual one of 10^6 atoms of silicon. The dashed curve gives the abundances as deduced by SUESS and UREY (1956). The solid curve is the prediction of CAMERON (1958) from calculations based on current theories of nucleogenesis. The open and closed circles represent data obtained

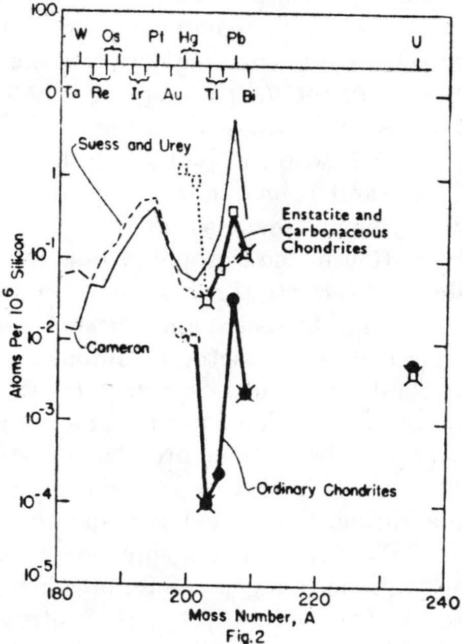

Fig. 2. Abundances of the heavy nuclei as deduced from activation analyses on meteorites in relation to abundance curves of SUESS and UREY and CAMERON.

from recent activation analysis experiments. Terrestrial isotopic composition has been assumed for mercury and thallium. The Pb^{207} point has been calculated on the basis of the Pb^{204} data and the primordial lead isotopic composition of PATTERSON *et al.* (1953). The point at mass number 235 is the abundance of U^{235} 4·5 Æ ago based on the U^{235} content today.

The experimental data on ordinary chondrites are presented as closed circles; the data for carbonaceous chondrites are shown as open squares. Except in the case of mercury, these open squares also represent the abundance curve for enstatite chondrites. The data for mercury at present is so sparse that its significance should not be overemphasized.

It is seen that the data for the ordinary chondrites in the mass region above 200 all lie far below the curves of both SUESS and UREY (1956) and CAMERON (1958). The data on carbonaceous and enstatite chondrites lie rather close to the Suess–Urey curve. They show the peak at Pb^{207} predicted by CAMERON, although the absolute amount of this nuclide is an order of magnitude less than predicted.

The agreement between the abundances of the heavy elements (except mercury) in these special meteorites with the figures presented by SUESS and UREY (1956) is most intriguing. In the case of the carbonaceous chondrites it suggests that the low-temperature history indicated by the physical character and presence of volatile compounds may have preserved a better record of elemental abundances at early times than some of the other types of meteorites previously used for this purpose.

In addition to the data on the abundance of the elements, our results also give some information on isotopic compositions. There has, of course, been considerable data reported indicating that the isotopic composition of even the heavy elements in meteorites (where not affected by radioactive decay) is the same as in terrestrial samples. For example, HAMAGUCHI *et al.* (1957) showed that to within 10 per cent the uranium in stone meteorites is the same as on the earth. The present data on the similarity of the Hg^{196}/Hg^{202} ratio in Nuevo Laredo, Orgueil, Abee and the troilite phase of Canyon Diablo and Toluca to that in terrestrial samples represents a check on the next heaviest element made so far.

In the case of lead, a variation in the isotopic ratio is to be expected. The age determination of meteorites and terrestrial rocks based on this variation has been widely accepted as one of the most trustworthy of such methods. A datum of great importance in this method is the isotopic composition of primordial lead. It has been assumed that the lead found in the troilite phase of iron meteorites represents such primordial lead since it has the least amount of radiogenic isotopes (Pb^{206}, Pb^{207}, Pb^{208}) relative to Pb^{204}. The present investigation confirms the findings of PATTERSON *et al.* (1953) in this respect, although only the Pb^{208}/Pb^{204} ratio was determined. In addition, the data shows that the enstatite and carbonaceous chondrites have Pb^{208}/Pb^{204} ratios in the same range (27–31) as the assumed primordial lead. Since the uranium (and, therefore, presumably thorium) contents of these meteorites are too low to have affected the isotopic composition of the lead, our results support the idea of a primordial lead common to many meteorites with isotopic composition close to that reported by PATTERSON *et al.* (1953).

Finally, the implications of these analytical results for time scales for meteoritic processes can be considered. In spite of intensive recent work, there is still only one borderline case of a meteorite in which the absolute amounts of lead, thorium and uranium are consistent, within experimental error, with an observed isotopic composition of the lead. This one case is Nuevo Laredo. Here there has been observed only 60–70 per cent of the uranium and thorium needed to produce in 4·5 Æ the amounts of radiogenic lead deduced from the isotopic composition of PATTERSON and our results on the absolute amount of lead present. Probably the experimental errors are sufficient to bridge this gap and so the age deduced can be trusted. In other meteorites for which Pb–Pb ages have been quoted, there is a question whether there is enough uranium and thorium to make the radiogenic lead, and the ages in our opinion must be considered suspect.

Nuevo Laredo is interesting also in giving approximate data on another time scale. Our experiments could set only an upper limit of about 2×10^{-9} g/g for the Bi^{209} content. In the same sample the U^{235} content is such that 4·5 Æ ago it was about 10^{-7} g/g. Np^{237} decays to Bi^{209} with a $2{\cdot}2 \times 10^{6}$ yr half-life. If the assumption is made that at the time the elements of this meteorite were formed, the Np^{237} was present in amounts equal to those of U^{235} at least $1{\cdot}6 \times 10^7$ years must have elapsed between the formation of the elements and their segregation into the matter that makes up the meteorite. This lower limit is shorter than the $\geqslant 4 \times 10^8$ year deduced by similar arguments from the Xe^{129} content of meteorite (WASSERBURG and HAYDEN, 1955; REYNOLDS and LIPSON, 1957) but has perhaps a firmer basis since it does not depend on the retention of a noble gas.

Acknowledgements—We wish to thank Professor H. C. UREY and Drs. F. BEGEMANN and R. MARSHALL for providing us with some of the meteorite samples not previously studied; Mr. R. BANE of the Argonne National Laboratory for several quantitative chemical analyses; and Dr. D. HESS of the Argonne for the mass spectrometric analysis of our lead monitor. We are indebted to Professors H. C. UREY, E. ANDERS, N. SUGARMAN, and Dr. R. MARSHALL for many illuminating discussions.

REFERENCES

BATE, G. L., HUIZENGA, J. R., and POTRATZ, H. A. (1957) "Thorium Content of Stone Meteorites". *Science* **126,** 612.

BATE, G. L., POTRATZ, H. A., and HUIZENGA, J. R. (1958) "Thorium in Iron Meteorites". *Geochim. et Cosmochim. Acta* **14,** 118–125.

BEGEMANN, F., EBERHARDT, P., and HESS, D. C. (1959) "He^3-H^3-Strahlungalter eine Steinmeteoriten". *Z. Naturf.* **14,** 500–503.

CAMERON, A. G. W. (1959) "A Revised Table of Abundances of the Elements". *Astrophysical Journal* **129,** 676–699.

EBERT, K. H., KÖNIG, H., and WÄNKE, H. (1957) "Eine neue Methode zur Bestimmung kleinster Uranmengen und ihre Anwandung auf die Urananalyse von Steinmeteoriten". *Z. Naturf.* **12a,** 763–765.

EDWARDS, G. (1955) "Sodium and Potassium in Meteorites". *Geochim. et Cosmochim. Acta* **8,** 285–294.

EDWARDS, G., and UREY, H. C. (1955) "Determination of Alkali Metals in Meteorites by A Distillation Process". *Geochim. et Cosmochim. Acta* **7,** 154–168.

EHMANN, W. D., and HUIZENGA, J. R. (1959) "Bismuth, Thallium and Mercury in Stone Meteorites by Activation Analysis". *Geochim. et Cosmochim. Acta* **17,** 125–135.

ENGEL, A. E. G., and ENGEL, C. G. (1958) "Progressive Metamorphism and Granitization of the Major Paragneiss, Northwest Adirondack Mountains, New York". *Bull. Geol. Soc. Amer.* **69,** 1369–1414.

FAIRBAIRN, H. W. (Editor) (1951) "A Cooperative Investigation of Precision and Accuracy in Chemical, Spectrochemical and Modal Analysis of Silicate Rocks". *Bull. U.S. Geol. Survey* H980.

GAST, P. W. (1960) "Alkali Metals in Stone Meteorites". *Geochim. et Cosmochim. Acta* **19,** 1–4.

HAMAGUCHI, H., REED, G. W., and TURKEVICH, A. (1957) "Uranium and Barium in Stone Meteorites". *Geochim. et Cosmochim. Acta* **12,** 337–347.

HERNEGGER, F., and WÄNKE, H. (1957) "Uber der Urangehalt der Steinmeteorite und deren 'Alter' ". *Z. Naturf.* **12a,** 759–762.

MARSHALL, R. R., and HESS, D. C. (1958) "Lead in Some Stone Meteorites". *J. Chem. Phys.* **28,** 1258–1259.

MARSHALL, R. R., and HESS, D. C. (1959) private communication.

NICHIPORUK, W., and CHODES, A. A. (1959) "The Concentration of Vanadium, Chromium, Iron, Cobalt, Nickel, Copper, Zinc and Arsenic in the Meteoritic Iron Sulfide Nodules". *J. Geophys. Res.* **64,** 2451–2463.

NODDACK, I., and NODDACK, W. (1931) "Die Geochemie des Rheniums". *Z. Phys. Chem.* **154,** 223–244.

PATTERSON, C. C. (1955) "The Pb^{207}/Pb^{206} Ages of Some Stone Meteorites". *Geochim. et Cosmochim. Acta* **7,** 151–153.

PATTERSON, C. C., BROWN, H., TILTON, G. R., and INGHRAM, M. G. (1953) "Concentration of Uranium and Thorium and the Isotopic Composition of Lead in Meteoritic Material". *Phys. Rev.* **92,** 1234–1235.

PATTERSON, C. C., TILTON, G. R., and INGHRAM, M. G. (1955) "Age of the Earth". *Science* **121,** 69–75.

REED, G. W., HAMAGUCHI, H., and TURKEVICH, A. (1958a) "The Uranium Contents of Iron Meteorites". *Geochim. et Cosmochim. Acta* **13,** 248–255.

REED, G. W., and TURKEVICH, A. (1955) "The Uranium Contents of Two Iron Meteorites". *Nature, Lond.* **176,** 794–795.

REED, G. W., KIGOSHI, K., and TURKEVICH, A. (1958b) "Activation Analysis for Heavy Elements in Stone Meteorites". *Proceedings of the Second International Conference on the Peaceful Uses of Atomic Energy, Geneva*, 1958, *Vol.* 28, *Paper* 953. United Nations, New York.

REYNOLDS, J. H., and LIPSON, J. I. (1957) "Rare Gases in the Nuevo Laredo Stone Meteorite". *Geochim. et Cosmochim. Acta* **12,** 330–336.

SCHINDEWOLF, U., and WAHLGREN, M. (1960) "The Rhodium, Silver and Indium Content of Some Chondritic Meteorites". *Geochim. et Cosmochim. Acta* **18,** 36–41.

SHAW, D. M. (1952) "The Geochemistry of Thallium". *Geochim. et Cosmochim. Acta* **2,** 118–154.

SUESS, H. E., and UREY, H. C. (1956) "Abundances of the Elements". *Rev. Mod. Phys.* **28,** 53–74.

TILTON, G. R., PATTERSON, C. C., BROWN, H., INGHRAM, M. G., HAYDEN, R. J., HESS, D. C., and LARSEN, E. S. (1955) "Isotopic Composition and Distribution of Lead, Uranium and Thorium in Precambrian Granites". *Bull. Geol. Soc. Amer.* **66,** 1131–1148.

UREY, H. C., and CRAIG, H. (1953) "The Composition of Stony Meteorites and the Origin of the Meteorites". *Geochim. et Cosmochim. Acta* **4,** 36–82.

WASSERBURG, G. J., and HAYDEN, R. J. (1955) "Time Interval Between Nucleogenesis and the Formation of Meteorites". *Nature, Lond.* **176,** 130–131.

WIIK, H. B. (1956) "The Chemical Composition of Some Stony Meteorites". *Geochim. et Cosmochim. Acta* **9,** 279–289.

6

Reprinted from *Accounts Chem. Research* **1**:289–298 (1968)

Chemical Processes in the Early Solar System, as Inferred from Meteorites

EDWARD ANDERS

Enrico Fermi Institute and Department of Chemistry, University of Chicago, Chicago, Illinois 60637

Received June 10, 1968

Meteorites are the only extraterrestrial samples of matter accessible to man. They are also the oldest. Radioactive dating methods have consistently shown them to be 4.5–4.7 AE (= 10^9 years) old, as old as Earth itself and at least 1 AE older than the oldest rocks found on Earth thus far. Meteorites thus contain an ancient archeological record whose counterpart has been destroyed on Earth by geologic processes. The challenge to the cosmochemist is to decipher this record. The clues available to him are chemical and isotopic composition, structure, and mineralogy of meteorites.[1,2] From these he must deduce environmental variables, such as temperature, pressure, oxidation potential, pH, etc., and their variation with time. Meteorites may thus be regarded as the "poor man's space probe."[3]

A great stimulus for this work has come from the recent realization that some meteorites are even more primitive than had been supposed. Some of their properties seem to date back to the time when meteoritic and planetary matter was still dispersed in the solar nebula, a disk of gas and dust from which the planets accreted. Thus there is hope of reconstructing the very earliest history of planetary matter, up to the time when its evolution diverged from that of meteoritic matter.

Classification

Meteorites are divided into three broad classes, differing in their ratio of metal to silicate: irons, stones, and stony irons (Figure 1). Stones comprise 92% of the 778 known falls and are divided into two subclasses: chondrites, characterized by millimeter-sized globules of silicate (= chondrules, after the Greek work χονδροσ, grain), and achondrites, chondrule-free stones differing from chondrites in both texture and composition. Each of these classes is further subdivided on the basis of various secondary criteria, such as structure and Ni, Ca, or Fe^{2+} content (Figure 1).

Chondrites occupy the highest place in the hierarchy of meteorites, being not only the most abundant (84% of all known falls), but also the most primitive in composition. If a sample of solar matter were freed of its volatile elements (H, He, C, and the excess O not required to form metal oxides) the residue would closely resemble the chondrites in composition. On paper, at least, the other classes of meteorites can be derived from chondrite-like material by chemical differentiation. Partial melting would give a metal layer corresponding to the irons and two silicate fractions, corresponding to the Ca-rich and Ca-poor achondrites: a low-melting fraction enriched in Ca and Al and a high-melting residue enriched in Mg. The composition of the products would also be affected by redox conditions. Under strongly reducing conditions, iron would go largely into metal, giving a nickel-poor metal phase (hexahedrites) and iron-free silicates (aubrites). Oxidizing conditions would yield more Ni-rich metal and Fe-rich silicates. Age determinations, though incomplete, support this hierarchy: chondrites seem to be 10^8 years older than achondrites or irons.[4] Hence we shall focus our attention on chondrites.

Chondrites are divided into five subclasses on the basis of their degree of oxidation, as reflected in the chemical state of their iron (Figure 1). All chondrites contain 20–30% Fe, partitioned between *metal* (Ni, Fe), *troilite* (FeS), and the silicates *olivine* [$(Fe,Mg)_2SiO_4$] and *pyroxene* [$(Fe,Mg)SiO_3$]. At one end of the spectrum, *enstatite chondrites* are highly reduced, containing iron mainly as metal or troilite. Their principal silicate, enstatite ($MgSiO_3$), has less than 0.1 mol % Fe^{2+} in

(1) The following books and the review in ref 2 provide further background information: (a) F. Heide, "Meteorites," University of Chicago Press, Chicago, Ill., 1963; (b) B. Mason, "Meteorites," John Wiley & Sons, Inc., New York, N. Y., 1962; (c) J. A. Wood, "Meteorites and the Origin of Planets," McGraw-Hill Book Co., Inc., New York, N. Y., 1968.

(2) E. Anders, *Space Sci. Rev.*, **3**, 583 (1964).

(3) E. Anders, paper presented at the 126th Annual Meeting of American Association for the Advancement of Science, Chicago, Ill., Dec 30, 1959.

(4) C. M. Hohenberg, M. N. Munk, and J. H. Reynolds, *J. Geophys. Res.*, **72**, 3139 (1967).

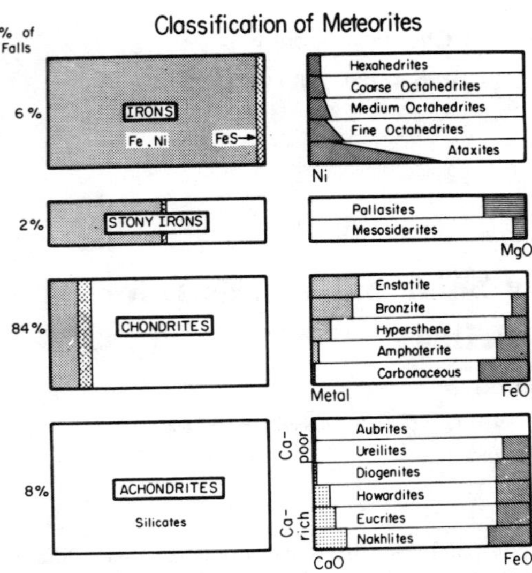

Figure 1. Meteorites are divided into four broad classes (left) on the basis of their ratio of metal (shaded) to silicate (white). Further subclassification (right) is based on various compositional or structural criteria.

solid solution. At the other end of the spectrum, *carbonaceous chondrites* are highly oxidized. They contain sulfur largely as S_8 or SO_4^{2-} and little or no metal. Most of their iron is present as Fe^{2+} (silicates) or even Fe^{3+} (Fe_3O_4, Fe_2O_3(aq)). Both enstatite and carbonaceous chondrites are quite rare, only ~40 being known.

The broad middle ground is occupied by the *ordinary chondrites*. They are subdivided into three groups, differing in both their state of oxidation and total iron content. The *H-group* chondrites (H = highest in Fe) contain ~27 weight % Fe, with 18 ± 2 mol % Fe^{2+} in their olivine. The *L* and *LL groups* are progressively lower in total Fe (22 and 20%), but higher in Fe^{2+} (24 ± 2 and 29 ± 2 mole % Fe^{2+} in olivine). With less iron to go around and a larger amount taken up by the silicates, the metal content decreases in the sequence H > L > LL from ~19 to ~3% by weight.

Most people believe that meteorites come from the asteroids, small planets situated between Mars and Jupiter. Other sources have been proposed, however: the moon, by Urey,[5] and comets, by Öpik.[6] Both are distinct possibilities for some of the rarer types of meteorites (calcium-rich achondrites, carbonaceous chondrites), but neither seems satisfactory for the common types, irons and ordinary chondrites.[2]

Chemical Composition of Chondrites

Most of the early work in this field served to establish the major element composition. Although some interesting trends emerged from these data, I shall confine myself largely to the more recent measurements, dealing with trace elements.

The first such measurements were done in the 1930's by the Noddacks[7] and by Goldschmidt[8] using X-ray and optical emission spectroscopy. A real breakthrough came in the 1940's with the advent of neutron activation analysis. A wealth of data has been turned out by Ehmann, Goles, Lovering, Reed, Schmitt, Smales, Wasson, and others. Recent reviews were given by Ehmann,[9] Urey,[10] and Larimer and myself.[11]

One of the first trends established in this work was the primitive composition of chondrites. It led to their acceptance as representative samples of nonvolatile cosmic matter and as the basis of the cosmic abundance curve. There are two reasons for this acceptance: the abundances in chondrites resemble those in the sun,[7,12–14] and they seem to be a smooth function of mass number.[13,14] But the latter rule had its exceptions from the very beginning. In 1947, when Suess[13] attempted to construct a cosmic abundance curve on the basis of Goldschmidt's[8] meteoritic data, he noted that certain elements (Se, Te, Ga, In, Tl, Zn, Cd, Hg, and Re) were underabundant relative to their neighbors by factors of up to 100. Since most of these elements were congeners in the periodic table, he suggested that chemical rather than nuclear factors were responsible for their depletion. Subsequent work by Brown and Goldberg[15] showed some of these measurements (Ga, Re) to be erroneous, and for a while it was believed that all such discrepancies would eventually disappear. Urey noted in a series of classic papers[16–19] that volatile metals (Hg, Cd, etc.) in the earth and meteorites might serve as "cosmothermometers," indicating the temperature that prevailed during their accretion from the primitive solar nebula. A careful analysis of the data then available showed no clear-cut evidence of substantial depletion. Urey therefore concluded that the earth and meteorites accreted at a temperature of ~300°K, low enough to allow quantitative condensation of volatile metals.

But the situation changed in subsequent years. Precision measurements by neutron activation analysis showed that some elements were indeed grossly underabundant. Moreover, Reed, *et al.*,[20] made an im-

(5) (a) H. C. Urey, *J. Geophys. Res.*, **64**, 1721 (1959); (b) H. C. Urey, *Science*, **147**, 1262 (1965); (c) H. C. Urey, *Monthly Notices Roy. Astron. Soc.*, **131**, 199 (1966); (d) H. C. Urey, *Icarus*, **7**, 350 (1967); (e) H. C. Urey, *Naturwissenschaften*, **2**, 49 (1968).

(6) (a) E. J. Öpik, *Mem. Soc. Roy. Sci. Liège*, 575 (1965); (b) E. Öpik, *Advan. Astron. Astrophys.*, **4**, 301 (1965).

(7) (a) I. Noddack and W. Noddack, *Naturwissenschaften*, **18**, 758 (1930); (b) I. Noddack and W. Noddack, *Svensk Kem. Tidskr.*, **46**, 173 (1934).

(8) V. M. Goldschmidt, *Skrifter Norske Videnskaps-Akad. Oslo, I, Mat. Naturv. Kl.*, No. 4 (1937).

(9) W. D. Ehmann, *J. Chem. Educ.*, **38**, 53 (1961).

(10) H. C. Urey, *Rev. Geophys.*, **2**, 1 (1964).

(11) J. W. Larimer and E. Anders, *Geochim. Cosmochim. Acta*, **31**, 1239 (1967).

(12) A. E. Ringwood, *Rev. Geophys.*, **4**, 113 (1966).

(13) (a) H. E. Suess, *Z. Naturforsch.*, **2a**, 311 (1947); (b) H. E. Suess, *ibid.*, **2a**, 604 (1947).

(14) H. E. Suess and H. C. Urey, *Rev. Mod. Phys.*, **28**, 53 (1956).

(15) H. Brown and E. D. Goldberg, *Phys. Rev.*, **76**, 1260 (1949).

(16) H. C. Urey, "The Planets," Yale University Press, New Haven, Conn., 1952.

(17) H. C. Urey, *Geochim. Cosmochim. Acta*, **2**, 269 (1952).

(18) H. C. Urey, "Plenary Lectures," XIII IUPAC Congress, London, 1953, p 188.

(19) H. C. Urey, *Astrophys. J. Suppl.*, **1**, [6], 147 (1954).

(20) G. W. Reed, K. Kigoshi, and A. Turkevich, *Geochim. Cosmochim. Acta*, **20**, 122 (1960).

portant observation: elements which are depleted by factors of 10–1000 in ordinary chondrites (*e.g.*, Hg, Tl, Pb, and Bi) often occur in nearly their predicted, "cosmic" abundances in carbonaceous and enstatite chondrites. Type I carbonaceous chondrites, in particular, tend to match solar composition[12] and are now widely regarded as the best surviving examples of primitive cosmic matter. (Carbonaceous chondrites are divided into three types on the basis of volatile content: C, H, S, and O contents decrease from type I to type III.)

Abundance Patterns. Gradually a pattern emerged:[2,11] 48 elements, mostly in the transition groups of the periodic table (Figure 2), were well-behaved. Their abundances were nearly constant from one class of chondrites to the next and generally matched solar or "cosmic" abundances. But a sizable group of "depleted" elements, some 31 at last count, fell out of line. Their abundances were variable and often fell far below solar or "cosmic" levels.

Let us examine the abundances of depleted elements in *carbonaceous chondrites* (Figure 3). The elements are listed in somewhat unconventional order, based on their depletion pattern in ordinary chondrites. Abundances of the first 26 elements are normalized to those in type I carbonaceous chondrites, which in turn are virtually identical with cosmic abundances. Data for the last five, highly volatile, elements are normalized to cosmic abundances.

It is significant that the abundances decrease consistently from type I to type III.[21] Moreover, they decrease by *constant factors*, as shown by the parallelism of the curves. This parallelism extends even to the last five elements on the chart whose absolute depletion is very much greater. Mean relative abundances for the three meteorite types are 1, 0.55, and 0.32.

Ordinary chondrites show a somewhat different pattern. The points in Figure 4 represent means; error bars indicate the observed range of variation. Manganese and the alkalis (except Cs) show little or no depletion. The next nine elements parallel the type II and III carbonaceous chondrite curves. Deviations from horizontality match in both figures, which suggests that they are due to errors in the type I carbonaceous chondrite data used for normalization. The mean depletion factor is 0.28.

The remaining elements show markedly greater depletion, to a factor of 0.002 for In and Tl. They also tend to be more variable. We shall refer to these elements as "strongly depleted," in contrast to the preceding, "normally depleted," group.

Fractionation Mechanisms. Let us focus our attention on the carbonaceous chondrite pattern, as the simpler of the two. The only property shared by all depleted elements is volatility. Two opportunities exist in the history of meteorites for volatility-dependent fractionations: accretion of meteorite parent bodies from the solar nebula, during which they lost

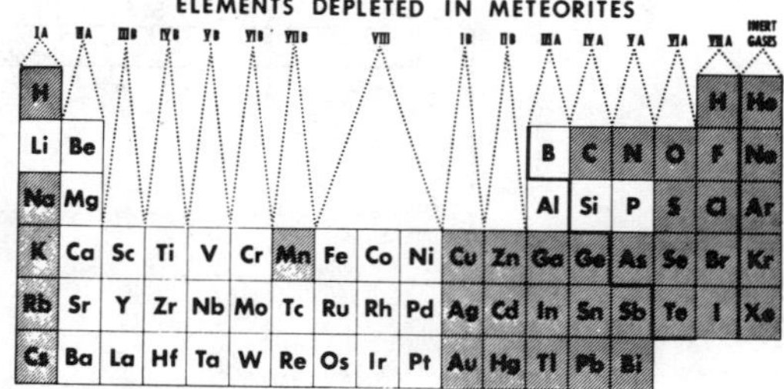

Figure 2. With the exception of Mn, all depleted elements (shaded) are situated in the main groups of the periodic table. Their only common property is volatility.

their complement of gases (H_2, He, etc.), and subsequent reheating in these bodies. The separation of dust from gas during accretion of the planets was the first, and probably largest, chemical fractionation in the solar system.[22] Let us therefore see whether the trace element depletion could have occurred at the same time.

We can at once dismiss the simplest explanation: that the constant depletion factors resulted from partial condensation or partial volatilization of the elements in question. Conditions for fractionation in a cooling solar nebula were discussed by Urey[16–19] and more recently by Lord[23] and by Larimer.[24] Briefly, an element E remains gaseous if its vapor pressure P_e exceeds its partial pressure in the nebular gas; otherwise it condenses.

It can be shown that the condition for partial condensation of some fraction $(1 - \alpha)$ is

$$P_e(T) = 2\alpha P_t(E/H) = \alpha E \cdot \text{const}$$

where E and H are the atomic abundances of element E and hydrogen, and P_t is the total pressure.

If the fraction condensed is to remain constant for a series of elements, the ratio of vapor pressure to cosmic abundance, P_e/E, likewise must remain constant. Cosmic abundances of many trace elements in Figure 3 lie within the narrow range of 0.1–1 atom per 10^6 Si atoms, but their vapor pressures vary by much larger factors. There exists *no single temperature* at which the ratio P_e/E is the same for any 5 of these elements, let alone 26.

The only way to account for the observed pattern is to assume that the meteorites are a mixture of two materials: a high-temperature fraction that lost all its volatiles and a low-temperature fraction that retained them. To explain the observed depletion factors, one can postulate that type I carbonaceous chondrites contain 100% low-temperature fraction, type II, 55%, and type III, 32%.

There exists some circumstantial evidence favoring this model. Broadly speaking, the minerals of carbonaceous chondrites can be divided into two groups: high-

(21) Apparent exceptions occur at Au, Cl, Br, and Bi, but past experience suggests that these may be due to sampling or analytical errors. Some points are based on only one or two measurements.

(22) H. Suess, *Ann. Rev. Astron. Astrophys.*, **3**, 217 (1965).
(23) H. C. Lord, III, *Icarus*, **4**, 279 (1965).
(24) J. W. Larimer, *Geochim. Cosmochim. Acta*, **31**, 1215 (1967).

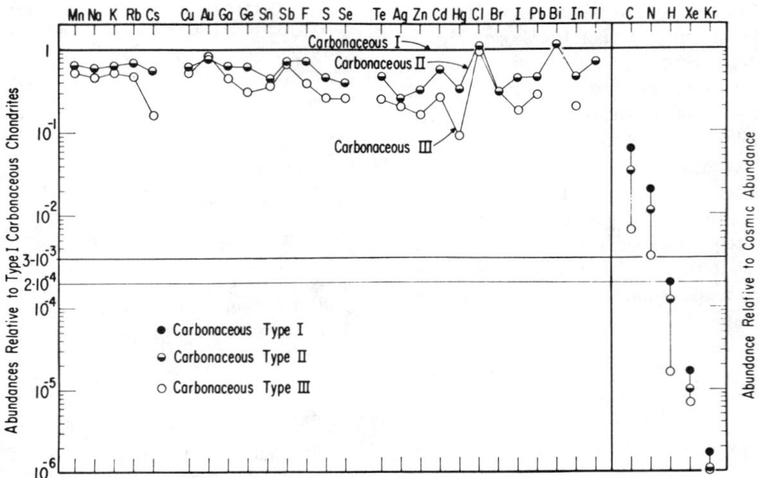

Figure 3. Depleted elements are underabundant in type II and III carbonaceous chondrites by nearly constant factors. Even for the strongly depleted elements C to Kr, relative abundances in types I, II, and III stand in nearly constant ratio (1:0.6:0.3) (data from ref 11 with minor revisions).

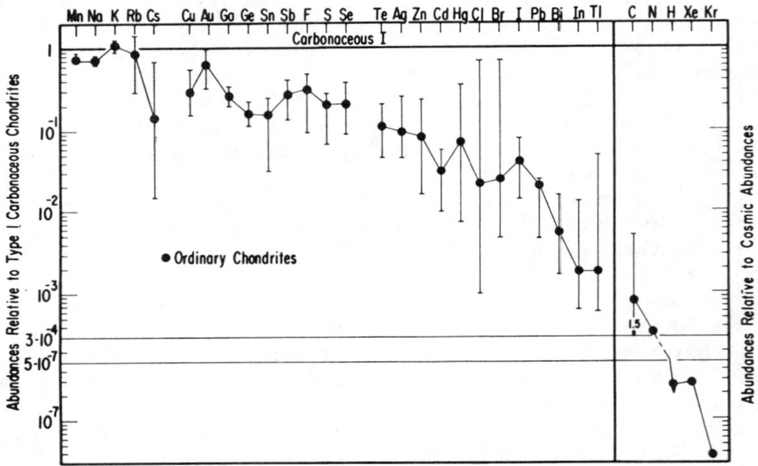

Figure 4. Abundances in ordinary chondrites show a more complex pattern. Mn and alkalis (except for Cs) are not significantly depleted. The next nine elements, Cu to Se, are depleted by a nearly constant factor of ~0.28. Irregularities in this curve segment parallel those in Figure 3, which suggests that the fault may lie with the type I carbonaceous chondrite data used for normalization. The remaining elements show increasingly greater depletion. Error bars indicate the total observed range of variation (data from ref 11 with minor revisions).

temperature (olivine, pyroxene, metal) and low-temperature (hydrated silicates, $MgSO_4$, S, FeS, Fe_3O_4).[2,25] Significantly, the *observed* content of low-temperature minerals in the three types agrees quite well with the *postulated* content of low-temperature fraction, and so does the ratio of matrix to chondrules. Type I consists entirely of a fine-grained matrix of low-temperature minerals; type III consists mainly of chondrules of high-temperature minerals. Furthermore, Schmitt, *et al.*,[26] have shown that the chondrules are depleted in Na and Mn relative to the matrix. This suggests that the low-temperature fraction may be identical with the matrix, and the high-temperature fraction, with the chondrules and metal particles.

The data for ordinary chondrites look less straightforward (Figure 4). The nine "normal" elements from Cu to Se again are depleted by nearly constant factors. This seems to call for a two-component model. But the

(25) E. R. DuFresne and E. Anders, *Geochim. Cosmochim. Acta*, **26**, 1085 (1962).

(26) R. A. Schmitt, R. H. Smith, and G. G. Goles, *J. Geophys. Res.*, **70**, 2419 (1965).

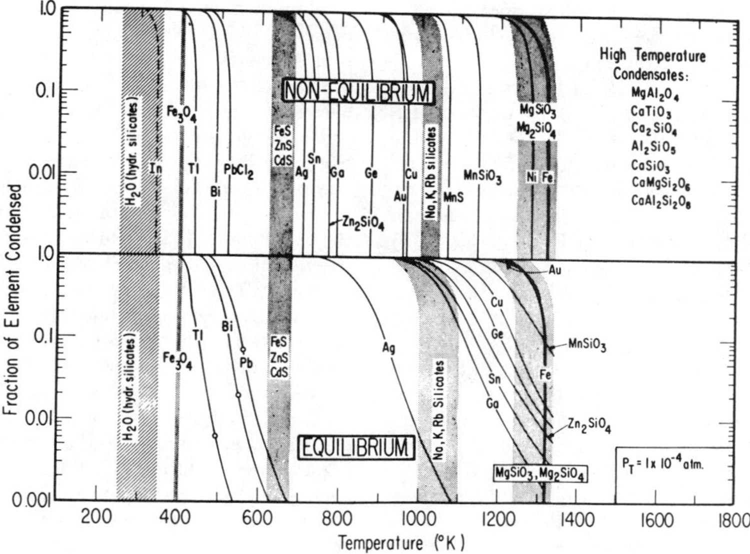

Figure 5. Condensation sequence of a gas of cosmic composition, at a total pressure of 1×10^{-4} atm.[27] "Equilibrium" sequence assumes diffusional equilibrium between grain surfaces and interiors, with formation of solid solutions, while "non-equilibrium" sequence corresponds to decomposition of pure elements or compounds, without interdiffusion. Shaded areas represent condensation or chemical transformation of major constituents. The formation range of hydrated silicates is poorly known. Abundances in ordinary chondrites (circles) suggest an accretion temperature of ~550°K for meteorites of this class.

remaining elements depart from the pattern: Mn, Na, K, and Rb are not depleted at all, while the Te–Tl group shows progressively greater depletion. Both trends can be explained by *ad hoc* assumptions, *e.g.*, a lower condensation temperature of the chondrules which allowed Mn and the alkalis to condense, and a higher accretion temperature of the matrix which inhibited condensation of the more volatile elements. (Here we are tacitly assuming that the strongly depleted elements Te–Tl are indeed more volatile than the rest.)

Condensation Sequence of the Elements. Larimer[24,27] has made a quantitative study of the condensation process in order to clarify some of these problems. The situation becomes more involved if the condensed phase is not the free element, but a compound: oxide, sulfide, silicate, or chloride. Pertinent chemical equilibria must then be considered. A further complication is the formation of solid solutions between a trace element and a major phase, *e.g.*, Mg_2SiO_4, Fe, or FeS. The activity of the trace element is lowered, resulting in a higher condensation temperature. Activity coefficient data were not available for most systems of interest, and Raoult's law was therefore assumed to be valid in cases of complete miscibility. Cases of partial miscibility were more troublesome.[27] For most trace metals of interest, low-temperature solubilities in iron or nickel–iron of cosmic composition (6% Ni) were unavailable and had to be extrapolated from high-temperature data, sometimes on related systems. The activity of the saturated solution was set equal to 1 and the activity at lower concentrations found from Henry's law.

Kinetic factors must also be considered. Formation of a solid solution requires interdiffusion between surface and interior of the grain. Hence it is useful to consider two limiting cases: *nonequilibrium*, where pure elements and compounds condense on each grain in successive layers without interdiffusion, and *equilibrium*, with complete interdiffusion.

Condensation sequences for these two limiting cases are shown in Figure 5. The total pressure is 1×10^{-4} atm, a value appropriate to the center of the asteroid belt.[28]

Let us follow the condensation sequence of a cosmic gas, using the *nonequilibrium* diagram for illustration (Figure 5). Near 2000°K, a few refractory substances condense: platinum metals, silicates, and spinels.[23] At 1340–1240°K, two major phases appear: nickel–iron and magnesium silicates. Together, they account for some 90% of all condensable material. Manganese and the alkalis come next, followed by a series of trace metals, Cu to Ag. At 680°K, metallic iron begins to react with H_2S, giving FeS as the third major phase. Several elements still in the gas phase which tend to form stable sulfides or iron compounds, respectively (Cd, Zn, Se, and Te), may be expected to condense at this stage. The cosmic abundance of Fe is somewhat greater than that of S, so that some metallic iron is left over. The four elements Pb, Bi, Tl, and In are the last to condense, owing to their low heats of vaporization.

(27) J. W. Larimer, to be published.

(28) (a) A. G. W. Cameron, *Icarus*, 1, 13 (1962); (b) A. G. W. Cameron, *ibid.*, 1, 339 (1963).

(The In curve is not well determined, owing to complexities introduced by the gaseous species In_2S.[24] The true In curve may lie closer to the Tl curve.)

At 400°K, another chemical reaction sets in: conversion of the remaining Fe to Fe_3O_4 by reaction with water vapor. Finally, at about 300–350°K, olivine transforms to hydrated silicates. The exact temperature cannot be determined for lack of thermodynamic data. As noted by Urey, an upper limit for magnesium silicates can be obtained from $Mg(OH)_2$ ($T \approx 300$°K,[16]) which holds water more tenaciously than do hydrated magnesium silicates (*e.g.*, talc, $T \approx 250$°K[11,29]). Temperatures for iron-bearing silicates should be somewhat higher.

The *equilibrium* case (Figure 5) is similar to the nonequilibrium case insofar as the major phases are concerned. However, owing to the formation of solid solutions the trace elements condense more gradually and at higher temperatures. The shift in condensation temperature is smallest for those four elements that show limited solubility in iron: Ag, Pb, Bi, and Tl. (In is probably similar, but is omitted for lack of solubility data.) In fact, the last three elements show a marked break in their condensation curves, corresponding to the point where the available nickel–iron has become saturated. Below this point, these elements condense as pure metals, as in Figure 5. Of course, these curves are based on *estimated* low-temperature solubilities and may be in error by 50° or more. A threefold increase in solubility would also eliminate the break.

To decide whether the equilibrium or nonequilibrium case is appropriate to the solar nebula, we must know diffusion coefficients, cooling times, and grain sizes. The last two, in particular, are rather uncertain. Let us therefore proceed empirically and see which diagram best explains the observed fractionation patterns in Figures 3 and 4.

Interpretation of Abundance Patterns. The *high-temperature fraction of carbonaceous chondrites* seems to have lost all its volatiles: Mn, alkalis, and the elements Cu to In. This pattern agrees well with the nonequilibrium sequence for a condensation temperature between 1150 and 1240°K. It does not agree with the equilibrium sequence; here the elements Cu to Ga condense ahead of the alkalis.

The *low-temperature fraction of carbonaceous chondrites* must have accreted at a temperature no higher than ~350°K, judging from the presence of magnetite (≤400°K) and hydrated silicates (approximately ≤350°K). The trace element content does not provide any more restrictive limits. All elements from In on up are present in nearly cosmic proportions, which implies accretion at ≤340 or ≤400°K according to either the nonequilibrium or the equilibrium sequences.

The data for *ordinary chondrites* also seem to fit the condensation sequence, supporting our previous speculations. A lower condensation temperature of the *high-temperature fraction* (970–1000°K according to the nonequilibrium sequence) would indeed account for the nondepletion of Mn, Na, K, and Rb. Again, the equilibrium sequence does not fit the observed pattern: the *depleted* elements Cu, Ge, Zn, Sn, and Ga condense ahead of *undepleted* Na and K.

The *low-temperature fraction* can also be explained along the lines previously suggested. The four most strongly depleted elements, Pb to In, indeed are the last to condense. The equilibrium sequence clearly gives the better match. The nonequilibrium curves slope too steeply and would cause Pb and Bi to condense quantitatively before any Tl had condensed. The accretion temperature of ordinary chondrites may thus be bracketed between rather close limits. It must have been greater than 520°K to prevent substantial condensation of $PbCl_2$ and less than 680°K to allow formation of FeS.

We can try to refine our estimate by plotting actual abundances in Figure 5. Data on bulk meteorites from Figure 4 were multiplied by 1/0.28 to convert them to abundances in the low-temperature fraction. The abundances of Pb, Bi, and Tl, recalculated in this manner, are 7×10^{-2}, 2×10^{-2}, and 6×10^{-3} (circles in Figure 5). As expected, they decrease with increasing volatility. Corresponding accretion temperatures are 560, 550, and 490°K. The last value can be brought into accord with the other two by assuming a higher solubility of Tl in nickel–iron. In any case, it seems that the mean accretion temperature of ordinary chondrites was in the neighborhood of 550°K. Judging from the abundance variation in Figure 4 (factors of 6, 14, and 40 for Pb, Bi, and Tl), the dispersion in accretion temperatures was small, ±30°.

A few other observations must be accounted for. The strongly depleted elements show a remarkable degree of covariance: if a given meteorite is low in one element, it tends to be low in all others as well. One striking example is the correlation between In and Xe (or Ar), discovered by Tandon and Wasson[30] (Figure 6). These obviously dissimilar elements correlate over more than two orders of magnitude. Indeed, such interelement correlations seem to extend to all strongly depleted elements. Larimer and I have plotted abundances of depleted elements in the three most widely studied meteorites.[11] Surprisingly, the strongly depleted elements Tl to C show a consistent abundance trend in the order Beardsley > Richardton > Allegan, with a tenfold spread over-all. It seems that Beardsley was uniformly more successful in picking up volatiles than were Richardton or Allegan.

This difference can be explained by equilibrium factors alone. Beardsley may have accreted at a lower temperature, where condensation of the strongly depleted elements was more complete. A temperature drop of 40° would suffice to raise abundances tenfold. Alternatively, kinetic factors may be invoked. Beardsley may have had a finer grain size, or a longer contact time with the gas phase. Either of these would favor

(29) K. O. Bennington, *J. Geol.*, **64**, 558 (1956).

(30) (a) S. N. Tandon and J. T. Wasson, *Science*, **158**, 259 (1967); (b) *Geochim. Cosmochim. Acta*, in press.

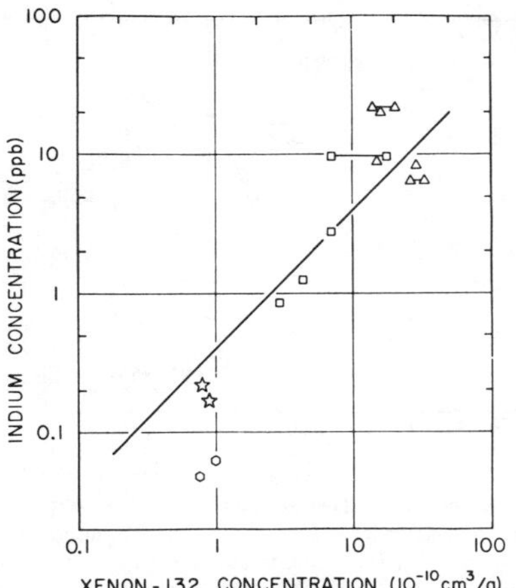

Figure 6. Correlation between In and ^{132}Xe in L-group chondrites (after Tandon and Wasson[30b]). These two elements correlate strongly, in spite of their markedly different volatilities. Horizontal bars connect measurements on two specimens of the same meteorite. Point symbols were chosen so that the number of corners corresponds to "petrologic grade" number,[31] the degree of recrystallization increasing from grades 3 to 6.

more complete equilibration, and hence condensation. It is not clear, however, that the In–Xe correlation in Figure 6 can be explained along these lines. This correlation apparently requires that trapping of a noble gas and condensation of a metal both have similar temperature dependence. More data are obviously needed on the solubility of noble gases in meteoritic minerals.

Origin of the High-Temperature Fraction. Most of our concern thus far has been with the low-temperature fraction. It appears that it precipitated from the gas at high temperatures, collected a succession of volatiles on cooling, and finally became separated from the gas by accretion into larger bodies. Clues to the history of the high-temperature fraction are not very specific, and thus its origin is explained largely by *ad hoc* postulates.

One possibility, first suggested by Wood,[32,33] is that the high-temperature fraction represents that portion of the condensate which passed through the liquid field of the phase diagram. Droplets of silicate and metal would grow by coalescence into millimeter-sized chondrules and metal particles, while material condensing from vapor directly to solid would remain in the form of a micron-sized smoke (=matrix). On cooling, the matrix material with its large surface-to-volume ratio would continue to equilibrate with the gas, unlike the coarse chondrules-plus-metal fraction. Thus the formation of an undepleted and a depleted fraction can be understood.

Wood's model requires pressures of $\geq 10^2$ atm, hardly attainable in a solar nebula except during transient shocks. However, Blander and Katz[34] have pointed out that supercooling can change the situation radically, condensation to liquids being possible at pressures as low as 10^{-4} atm.

Another possibility has been suggested by Whipple[35] and was later elaborated by Cameron.[36] Whipple notes that considerable charge separation can occur in dust-laden, nonconducting gases; hence the solar nebula should have passed through a stage of intense lightning activity. Such lightning discharges would melt any dust agglomerates in the vicinity, changing them into chondrules and metal grains. Even though the time scale of remelting would be very short, volatiles would be largely lost. Thus a depleted, high-temperature fraction would result. Inasmuch as its formation on this model is a purely local event, its amount and formation temperature may be expected to vary from place to place. The alkali depletion in carbonaceous chondrites may well imply an origin at greater distances from the sun, where pressures and volatilization temperatures were lower.

Other Models. One must consider the possibility that the fractionation happened in the solar nebula, as outlined here, but at higher or lower pressure or at a different composition. It turns out that there is not much latitude in these parameters. Pressures outside the range 10^{-2}–10^{-6} atm would shift the condensation curves by a prohibitive amount, leading to various inconsistencies. *Major* deviations from solar composition are also unlikely, although minor fractionations of gas and dust actually seem necessary to account for the highly reduced state of carbonaceous and enstatite chondrites.[37,38] A further alternative is that the fractionation happened *after* accretion, in the meteorite parent bodies. Chondrites have undergone recrystallization to a variable degree: one finds a continuous transition from meteorites with a fine-grained opaque matrix and sharply delineated chondrules to highly recrystallized ones in which the chondrules are barely resolvable from the coarse-grained matrix.[32,39,40] It is striking that the trace element depletion tends to parallel the degree of recrystallization (Figure 6). This suggests a causal relationship. Wood[41] has proposed that trace element depletion is caused by reheating in the parent body ("metamorphism"), while Ringwood[12,42] has suggested volatilization at an earlier, igneous stage. Either model implies that the meteorite parent bodies were open systems, free to lose volatiles.

(31) W. R. Van Schmus and J. A. Wood, *Science*, **31**, 747 (1967).
(32) (a) J. A. Wood, Technical Report No. 10, Smithsonian Institute of Astrophysics Observatory (ASTIA Document No. AD 158364), 1958; (b) J. A. Wood, *Nature*, **194**, 127 (1962).
(33) J. A. Wood, *Icarus*, **2**, 152 (1963).
(34) M. Blander and J. L. Katz, *Geochim. Cosmochim. Acta*, **31**, 1025 (1967).
(35) F. L. Whipple, *Science*, **153**, 54 (1966).
(36) A. G. W. Cameron, *Earth Planetary Sci. Letters*, **1**, 93 (1966).
(37) J. A. Wood, *Geochim. Cosmochim. Acta*, **31**, 2095 (1967).
(38) J. W. Larimer, *ibid.*, **32**, 965 (1968).
(39) J. A. Wood in "The Moon, Meteorites, and Comets," B. M. Middlehurst and G. P. Kuiper, Ed., University of Chicago Press, Chicago, Ill., 1963, Chapter 12, p 337.
(40) (a) R. T. Dodd, Jr., and R. Van Schmus, *J. Geophys. Res.*, **70**, 3801 (1965); (b) R. T. Dodd, Jr., W. R. Van Schmus, and D. M. Koffman, *Geochim. Cosmochim. Acta*, **31**, 921 (1967).
(41) J. A. Wood, *Icarus*, **6**, 1 (1967).
(42) A. E. Ringwood, *Geochim. Cosmochim. Acta*, **30**, 41 (1966).

The same correlation between trace element content and recrystallization would be expected for the condensation model. Meteorites last to accrete would be richest in volatiles, especially if temperatures declined during accretion. Being situated in the outermost layers of the asteroid, these meteorites would be least reheated during metamorphism. Here we are assuming that the body was a closed system during metamorphism, with negligible loss or transport of volatiles.

The available evidence seems to favor the condensation model. Reheating during metamorphism cannot account for the uniform depletion of the "normally depleted" elements, whose volatilities span a wide range. Other objections exist as well.[11]

Discussion

The evidence presented here suggests that most meteorites accreted at a temperature of 520–680°K. This is more than three times the present black-body temperature in the asteroid belt, 170°K. Apparently a powerful, transient heat source was present in the early solar system. One possibility is the primitive sun itself.[11] Hayashi[43] showed several years ago that the sun must have passed through a brief high-luminosity stage before settling down to its present state on the main sequence. Recent work by Cameron[28,44] on the evolution of the solar nebula shows, however, that the sun would be a rather late phenomenon, forming at the expense of the solar nebula during the final stages of its existence. Part of the nebular material would be absorbed by the growing sun, the remainder being expelled from the solar system. Only rather sizable bodies would survive this process, and they must therefore have accreted prior to the sun's formation. High temperatures during accretion hence must have been maintained by some mechanism other than solar heating.

The nebula itself would be heated by the release of gravitational potential energy during contraction. For the asteroid belt, Cameron predicts a temperature of ~500°K, in close agreement with our estimate of 520–680°K.

Earth. What bearing does this conclusion have on the inner planets, especially the Earth? If temperatures were high in the asteroid belt, they must have been high elsewhere in the inner solar system, and this, in turn, should have led to important chemical fractionations during accretion of the planets. Such fractionations may be hard to recognize in the Earth, where large-scale redistribution of elements has occurred and only a thin outer crust is accessible to chemical analysis. There is a possibility, however, that at least the most volatile elements are quantitatively concentrated in the crust, atmosphere, or oceans.[45] Let us therefore calculate nominal "whole-earth" abundances for a number of volatile elements on the assumption that the amount

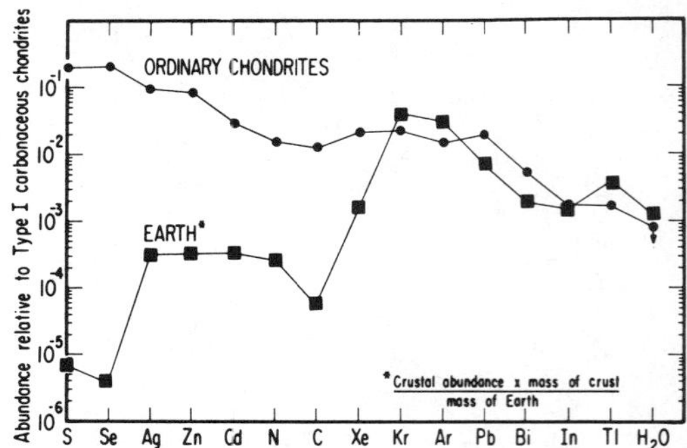

Figure 7. Abundance of volatile elements on Earth and in ordinary chondrites. Terrestrial abundances were computed from Taylor's crustal abundances on the assumption that the amount in the crust represents Earth's entire inventory of a given element (S. R. Taylor, *Geochim. Cosmochim. Acta*, **28**, 1273 (1964)). This assumption is realistic only for highly volatile elements that were largely outgassed from Earth's interior; the values for all others will be too low. It is interesting that the last seven values, Kr to H_2O, are high, and coincide with the values in ordinary chondrites. Apparently Earth, too, accreted near 600°K.

in the crust represents the Earth's entire inventory of this element.

It is instructive to compare these "terrestrial" abundances with the meteoritic ones (Figure 7). For the first eight elements the terrestrial curve is systematically low by several orders of magnitude. This may imply that these elements are largely concentrated in the earth's interior. The remainder of the curve, from Kr to H_2O, virtually coincides with the chondritic curve. These elements are the most volatile ones, for which the postulate of surface concentration is most likely to be correct. It is remarkable that the curves match so well in both slope and position. This suggests that Earth and ordinary chondrites accreted at about the same temperature, around 600°K.

One cannot rule out an alternative possibility: that the bulk of Earth accreted at still higher temperatures, with a small amount of volatiles being brought in at a late stage *via* carbonaceous chondrite-like material. About 3% of such material would be required to account for Earth's Kr and Ar. The whole-earth abundances of the remaining elements in Figure 7 should then also be at the 3% level; the fact that they are lower must be ascribed to retention in Earth's interior. A slight disadvantage of this model is its failure to account for the resemblance of the two curves in the Kr–H_2O region, but whichever model is correct, the bulk of Earth apparently accreted at a rather high temperature, at least ~600°K.

This temperature is much higher than Urey's original estimate, ~300°K. The reason is not hard to find. The abundance data available in 1952–1954, though incomplete, showed no clear-cut depletion of volatile metals. This ruled out high-temperature condensation in its simplest form (without formation of solid solutions) and seemed to suggest low-temperature condensation instead. A decade later the picture had changed.

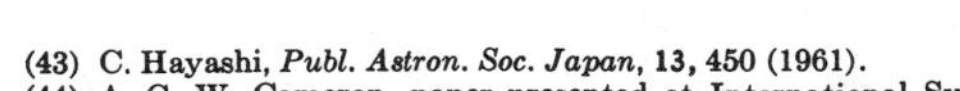

(43) C. Hayashi, *Publ. Astron. Soc. Japan*, **13**, 450 (1961).
(44) A. G. W. Cameron, paper presented at International Symposium on Meteorites, Vienna, Aug 7–13, 1968. To be published by Reidel, Dordrecht, Netherlands.
(45) H. C. Urey, *Proc. Roy. Soc.* (London), **A219**, 281 (1953).

Most of the volatile elements turned out to be depleted, but by small and frequently constant factors. Thus a more complicated, two-fraction model became necessary, involving solid-solution formation. Let us see how well this high-temperature condensation model accounts for the chemical differences among the remaining inner planets.

Inner Planets. It has been known for some time that the inner planets differ markedly in density, from 5.5 g/cm^3 for Mercury to 4.0 g/cm^3 for Mars (both corrected to zero pressure). Noddack[7] and Urey[16] attributed these differences to variations in metal–silicate ratio. Such variations are hard to explain in terms of the low-temperature model, which assumes that the planets accreted from highly oxidized material, containing Fe_3O_4 rather than Fe. The high-temperature model,[32] on the other hand, produces metal and silicate particles from the nebula. Given their differences in density, brittleness, and magnetic susceptibility, it is not hard to imagine how they might be fractionated from each other.[32b]

The situation is similar in regard to the volatile content of inner planets. According to the low-temperature model, only slight differences might be expected, corresponding to the $r^{-1/2}$ variation of temperature with distance from the sun. The high-temperature model provides two additional degrees of freedom. First, temperatures probably varied with time as well as distance. This difference might be enhanced by variations in the effective cutoff times for accretion. Mars, a small planet sweeping out a large volume of space, may have been much less effective than Earth or Venus in collecting the late, volatile-rich condensate prior to its dissipation. Second, the proportion of high- to low-temperature fraction may have varied from planet to planet, just as it seems to have varied from one class of meteorites to the next.

One interesting case in point is Venus. It seems to have less than 10^{-3} Earth's water content, although both planets contain similar amounts of C and N. Their sizes are nearly identical, and there is no obvious way in which Venus could have lost large amounts of water. Libby[46] has recently suggested that most of the water is present in the form of giant ice caps. An alternative possibility is that temperatures in the neighborhood of Venus did not fall below 400°K (the estimated formation temperature for hydrated silicates at 0.7 astronomical unit) until very late, when little accretable material remained in the nebula. Earth would have passed this point several hundred thousand years earlier.[11]

Of course it is a risky undertaking to deduce the formation conditions of a planet from the volatiles in the accessible portions: crust and atmosphere. One must know what fraction of the planet's initial endowment these surface volatiles represent. A major part may still be trapped in the interior or may have escaped from the planet altogether; hence the surface volatiles merely provide a lower limit. Nonetheless, even a lower limit can be quite informative, particularly if an upper limit can be obtained from other evidence. Mueller[47] and Lippincott, *et al.*,[48] have studied this problem in detail, using the present atmospheric composition of Venus to deduce the surface mineralogy and the history of the atmosphere.

Organic Compounds. Finally, this model seems to provide the proper setting for production of organic compounds in meteorites.[49] Studier, *et al.*,[50] have recently shown that all organic compounds reliably identified in meteorites can be made from CO, H_2, and NH_3 by a Fischer–Tropsch type reaction if iron meteorite particles are present to act as catalysts. This reaction produces a number of compounds, including some biologically important ones: normal alkanes and isoalkanes, isoprenoid alkanes, adenine, guanine, polynuclear aromatics, alkylbenzenes, etc. The starting materials for this reaction, CO, H_2, NH_3, and metal particles, would certainly be present in a cooling solar nebula. On cooling, methane should form as the equilibrium product, but laboratory experiments show that high-molecular weight organic compounds form instead as metastable products.[50] Indeed, the existence of fairly nonvolatile carbon compounds in the inner solar system must be postulated on independent grounds: methane, with a condensation temperature of 48°K at 10^{-3} atm, is too volatile to condense in the region of the inner planets.[16,18] The suggestion that carbon was incorporated in Earth and meteorites as organic compounds was first made by Urey in 1953.[18] He showed that carbides and graphite were unlikely to form in a gas of solar composition and suggested that "complex tarry organic compounds" might have formed as metastable intermediates in the reaction of CO or C_2H_2 with H_2. Dayhoff, *et al.*,[51] have shown by thermodynamic calculations that complex aromatic hydrocarbons can indeed form metastably under such conditions, and Studier, *et al.*,[50] have observed such compounds, along with alkanes, in their experiments.

Some Remaining Problems. The two-component model seems to account for all the available evidence. However, quite a few details remain to be filled in. Estimates and guesses must be replaced by accurate measurements. It is possible that such new data may prove incompatible with the model, forcing its revision or abandonment.

Not enough is known about the chemistry involved

(46) W. F. Libby, *Science*, **159,** 1097 (1968).

(47) (a) R. F. Mueller, *ibid.*, **141,** 1046 (1963); (b) *Nature*, **203,** 625 (1964); (c) *Icarus*, **3,** 83 (1964); (d) *ibid.*, **3,** 285 (1964); (e) *ibid.*, **4,** 506 (1965); (f) *J. Chem. Educ.*, **42,** 294 (1965).

(48) E. R. Lippincott, R. V. Eck, M. O. Dayhoff, and C. Sagan, *Astrophys. J.*, **147,** 753 (1967).

(49) A recent review by J. M. Hayes, *Giochim. Cosmochim. Acta*, **31,** 1395 (1967), provides an excellent introduction to this controversial subject.

(50) (a) M. H. Studier, R. Hayatsu, and E. Anders, *ibid.*, **32,** 151 (1968); (b) R. Hayatsu, M. H. Studier, A. Oda, K. Fuse, and E. Anders, *ibid.*, **32,** 175 (1968); (c) M. H. Studier, R. Hayatsu, and E. Anders, Enrico Fermi Institute preprint, EFINS 65-115, 1965.

(51) (a) M. O. Dayhoff, E. R. Lippincott, and R. V. Eck, *Science*, **146,** 1461 (1964); (b) R. V. Eck, E. R. Lippincott, M. O. Dayhoff, and Y. T. Pratt, *ibid.*, **153,** 628 (1966).

in trapping strongly depleted elements. Noble gases were probably dissolved in the crystal lattices of major phases (silicates, FeS, and Fe_3O_4). Solubilities in these phases are unknown, at least in the temperature range of interest. Small amounts of water would be trapped in similar manner. Again, data are lacking.

The problem of carbon condensation has not yet been completely solved. True, aliphatic and aromatic hydrocarbons of 30 or more carbon atoms can be produced by Fischer–Tropsch reactions or pyrolysis of methane,[50a,c,52] but nobody has yet duplicated the aromatic polymer which accounts for the bulk of the carbon in carbonaceous and ordinary chondrites.[49,53] Similarly, the chemical state of condensed nitrogen has not been settled. Carbonaceous chondrites contain about 0.3% N,[54] mostly as heterocyclics and other organic nitrogen compounds,[50b,55] not as NH_4^+.[25] Ordinary chondrites contain a few parts per million of N of unknown chemical state (iron nitride, titanium nitride, organic polymers?). The inner planets presumably acquired their N in the same form, and for this reason alone it would be interesting to learn more about the chemistry involved.

The condensation behavior of several other partially condensed trace elements cannot be accurately determined for lack of pertinent data. Thermodynamic data are lacking for FeSe and FeTe. Solubilities are largely unknown for solid solutions of Pb, Bi, In, Tl, Cd, Zn, Se, and Te in Fe, Fe–Ni alloys, and FeS. The problem is not an easy one from an experimental point of view, diffusion rates near 600°K being rather slow. Moreover, the situation is complicated by the existence of intermetallic compounds, but these systems can certainly be studied with the tools of modern physical chemistry and metallurgy. Perhaps the great cosmochemical interest of these systems will provide the needed incentive.

(52) J. Oró and J. Han, *Science*, **153**, 1393 (1966).

(53) M. C. Bitz and B. Nagy, *Proc. Natl. Acad. Sci. U. S.*, **56**, 1383 (1966).

(54) B. Mason, *Space Sci. Rev.*, **1**, 621 (1963).

(55) R. Hayatsu, *Science*, **146**, 1291 (1964).

Conclusions

If our analysis is valid, the chemical composition of the chondrites was established in the solar nebula and has changed but little in subsequent eons. As first noted by Wood, this has opened up the exciting prospect of having at our disposal samples of the legendary "primordial dust" from which the planets formed. Some of the properties of meteorites were undoubtedly established in the meteorite parent bodies, but these late alterations have not obscured the record of the earlier, far more interesting stage. The meteorites falling on earth represent a fairly wide range of environments, from high to low temperatures and strongly reducing to oxidizing conditions. One of the great remaining challenges to meteoriticists is to determine the place of origin of each kind of meteorite. With this information, it should be possible to derive a fairly detailed picture of chemical and physical conditions in the early solar system and their variation with time.

My special gratitude goes to John W. Larimer, who provided data and much valuable advice. Ole J. Kleppa kindly furnished data and educated guesses on the properties of binary alloys. This work was supported in part by AEC Contract AT(11-1)-382 and NASA Grant NsG-366.

7

Reprinted from *Geochim. et Cosmochim. Acta* **34**:89–103 (1970)

Bismuth contents of chondrites

J. C. LAUL,* D. R. CASE,†, F. SCHMIDT-BLEEK‡ and M. E. LIPSCHUTZ
Department of Chemistry, Purdue University, Lafayette, Indiana 47907

Abstract—A neutron activation technique has been developed for determination of the bismuth concentrations in C-, H-, L- and LL-group chondrites. The technique was tested by duplicate analyses of eight standard rocks and a single synthetic sample. The results obtained for standards AGV-1, BCR-1, DTS-1, G-2, GSP-1 and PCC-1 are the first reported. Our analyses of 14 carbonaceous and 5 equilibrated ordinary chondrite samples have confirmed that Bi is indeed a "strongly-depleted" element in meteorites. The mean Bi abundances in carbonaceous chondrites are 143, 71 and 37 atoms/10^9Si atoms for groups Cl, C2 and C3, respectively, which relative to Cl yield ratios of: C1: C2: C3 = 1·00: 0·50: 0·26—nearly the same as those predicted by a two-component model. In 12 unequilibrated ordinary chondrites (UOC), the Bi contents generally decrease exponentially with increasing equilibration of the silicate phases although there are apparent exceptions to this trend for chondrites having highly unequilibrated silicates. The Bi concentrations in the UOC correlate well with those of carbon and primordial Ar, Kr and Xe determined by others in these same meteorites.

INTRODUCTION

DESPITE the enormous amount of research that has been conducted in recent years on elemental abundances in meteorites, Bi is one of the few elements whose abundance has not been well established. This is particularly surprising in view of the twofold importance of Bi to cosmochemistry: it is the heaviest stable element, hence has special significance in nucleosynthesis; and it is highly volatile, so that its abundance provides clues to the temperature history of meteorites. Thus, the establishment of its content in chondrites is important in understanding the processes that resulted in the formation of the meteoritic parent bodies.

Prior to this study, the Bi contents (or limits) in ten chondrites and two achondrites (Johnstown and Nuevo Laredo) had been measured (EHMANN and HUIZENGA, 1959; REED *et al.*, 1960; REED, 1963). In terms of the chemical–petrologic classification of VAN SCHMUS and WOOD (1967) the chondrites studied previously included replicate samples of: one C1; one C2; seven equilibrated ordinary chondrites—four H5, two L6, and one LL6; and one E4. The results of these studies clearly demonstrate that the extreme Bi contents of chondrites differ by a factor of more than a thousand and, in equilibrated ordinary chondrites at least, the Bi concentrations are quite low—on the order of a few ppb. These previous results also indicate that the Bi contents of chondrites depend on

* In partial fulfillment of requirements for the Ph.D. Degree.
Present address: Enrico Fermi Institute for Nuclear Studies, University of Chicago, Chicago, Illinois.

† Present address: Air Force Nuclear Energy Center, Wright-Patterson Air Force Base, Dayton, Ohio.

‡ Present address: Department of Chemistry, University of Tennessee, Knoxville, Tennessee.

chondritic type and decrease in the order carbonaceous–enstatite–equilibrated ordinary chondrites. In detail, though, some results are puzzling. These include the C2/C1 Bi abundance ratio (Reed *et al.*, 1960) and extremely variable amounts of Bi in the equilibrated ordinary chondrites (Ehmann and Huizenga, 1959; Reed *et al.*, 1960; Reed, 1963).

In view of these facts, we decided to measure the Bi contents in a larger number of chondrites. In choosing the samples to be studied we decided to concentrate upon carbonaceous and unequilibrated ordinary chondrites as examples of "primitive" meteorites, and to include only a few equilibrated ordinary chondrites for comparison. This study was undertaken in conjunction with a broader study of other trace element contents in these same meteorites by an adaptation of neutron activation analysis (Laul, Case, Lipschutz, Pelly, Schmidt-Bleek and Wechter, in preparation). In this paper we report on the Bi contents in 31 samples including 27 different chondrites together with a preliminary interpretation of the data. A more complete discussion will be made in concert with the results of our broader study.

Experimental

In view of the generally low Bi contents in chondrites, neutron activation appeared to be the most suitable method. Our attempts to use previously-described radiochemical purification techniques for Bi proved unsuccessful since we could neither reproducibly separate Po^{210} for α-counting (Ehmann and Huizenga, 1959) nor obtain radiochemically-pure Bi^{210} using the scheme of Reed *et al.* (1960). Accordingly we were forced to devise another procedure which we tested by analyzing standard rocks and a synthetic sample (Table 1). This procedure was then used to determine the Bi contents of the chondrites listed in the Appendix and in Tables 2 and 3. These chondrites were all received as relatively large fragments except for Ivuna [B], which was a powdered sample collected when the specimen that yielded Ivuna [A] was cut in Tanzania (see Appendix for the sources and identification numbers of these samples). All samples were prepared for irradiation as described elsewhere (Laul *et al.*, 1969). In general the sample weights ranged from 0·7–1·4 g except for the Ivuna samples, Clovis #1 and Mezö-Madaras where the weights ranged from 0·3 to 0·5 g.

The monitors were prepared in batches of eight. To simulate the geometry of the samples and to minimize any effects due to self-shielding, monitor "beds" of 100–200 mg of MgO powder (Mallinckrodt reagent grade, previously analyzed by us to assure its low Bi content) were weighed into pre-cleaned quartz irradiation vials. The monitor solutions were prepared freshly for each batch by appropriate dilution of a previously standardized carrier solution (10 mg Bi/ml). A 100 λ aliquot (20 μg) of the monitor solution was pipetted onto the prepared "bed". The monitor vials were dried for 4–6 days at 65°C and sealed. The synthetic sample was prepared by "doping" 1-g of MgO powder with μg to mg quantities of all elements that we studied, including 20 μg of Bi. After preparation, this synthetic sample was treated exactly as the standards and chondrites.

Each irradiation (initially two, later three samples per monitor) was performed in the thermal column of the Argonne CP 5 Reactor at a flux of $\sim 8 \times 10^{12}$ thermal neutrons/cm² sec for 6 days. After irradiation the samples were dissolved in the presence of 100 mg of inactive Bi carrier and were processed as described elsewhere (Laul *et al.*, 1969) to remove specific gamma-ray emitters. The Bi-containing acid–sulfide precipitate was obtained by addition of thioacetamide to the solution remaining after the samples were extracted with diethyl ether and distilled in the presence of $HBr–H_2SO_4$.

Radiochemical purification

After this initial decontamination, the sulfide was processed extensively to assure its radiochemical purity. The basic procedure used was that developed by Prestwood (1963) for the separation of Bi from mixed fission products. This procedure was, however, modified extensively for our purposes:

(a) The acid–sulfide precipitate was dissolved in conc. HCl and the solution boiled to remove H_2S. This solution was diluted to 20 ml with 6 N HCl and leaded onto a 36 ml Dowex AG 1 × 8 (100–200 mesh) anion exchange column. The column was washed successively with 20 ml 6 N HCl, 75 ml 0·5 N HCl and 50 ml H_2O (Ehmann and Huizenga, 1959) and the Bi fraction eluted with 125 ml 1 M H_2SO_4 (Strelow and Bothma, 1967).

(b) To the effluent was added 10 ml 1 M thioacetamide and after digestion at 80°C Bi_2S_3 was precipitated and the supernate discarded.

(c) The Bi_2S_3 was dissolved in 10 ml conc. HCl and the solution evaporated nearly to dryness. BiOCl was precipitated by addition of H_2O and the supernate discarded.

(d) The BiOCl was dissolved in 2 ml conc. HCl and diluted to about 20 ml. After addition of 5 drops of Ag^+ carrier (25 mg/ml), an AgCl scavenge was performed and the filtrate collected.

(e) $Bi(OH)_3$ was precipitated with 5 ml conc. NH_4OH and the supernate discarded.

(f) The $Bi(OH)_3$ was dissolved in 2 ml conc. HCl and 5 drops each of Al, Fe, Cu, As, Se, Mo, Cd, In, Sn, Sb, Te, Ir and Hg holdback carriers (~10 mg/ml) in nitrate or chloride form were added. After evaporation to dryness, the residue was taken up in 25 ml of 6 N HCl and extracted with 25 ml of hexone. The organic layer was discarded. After addition of 1–2 g NaI to the aqueous phase, BiI_3 was extracted into 25 ml of hexone and the aqueous layer discarded. Following one wash with 25 ml 6 N HCl containing 1 g NaI, Bi^{3+} was extracted from the organic layer with 25 ml 6 N HCl containing 1 g $NaNO_2$. The organic phase was discarded.

(g) $Bi(OH)_3$ was precipitated with 5 ml conc. NH_4OH, washed once with H_2O, and both supernate and wash were discarded. The precipitate was dissolved in 3 ml conc. HNO_3, diluted to ~20 ml and 1 g Na_3PO_4 was added. After boiling, the $BiPO_4$ precipitate was washed once with H_2O and the supernate and wash discarded.

(h) The $BiPO_4$ was dissolved in 5 ml conc. HCl and 10 ml of 1 M thioacetamide was added to precipitate Bi_2S_3. The precipitate was washed with 25 ml conc. NH_4OH to remove soluble basic-sulfides and supernate and wash were discarded.

(i) Steps (c)–(g) were repeated except that no hold-back carriers were added in the repetition of step (f).

(j) $BiPO_4$ was dissolved with 2 ml conc. HCl, diluted to 10 ml and $Bi(OH)_3$ was precipitated by addition of 6 N NaOH. The supernate was discarded.

(k) $Bi(OH)_3$ was dissolved in 2 ml conc. HCl and, after dilution, the BiOCl precipitate was washed once with H_2O and the supernate and wash were discarded. The BiOCl precipitate was filtered onto a pre-weighed glass-fiber filter circle, washed with H_2O and ethanol, and dried at 110°C for 10 min. After cooling for 10 min, the precipitate was weighed and mounted for counting.

In nearly all of our experiments, we obtained chemical yields of 20–40%. For PCC-1 [A], AGV-1 [A] and Warrenton, however, we obtained chemical yields on the order of ~5%. Each BiOCl precipitate was counted for 10–15 min with a Ge(Li) detector and 1024-channel analyzer system in an attempt to detect gamma-emitting impurities. Since we found no evidence for such contaminants and the decay of the samples followed the known 5·01 day half-life of Bi^{210g} (Lederer *et al.*, 1967) for at least 5 half-lives, we decided not to recycle all samples to constant specific activity. We were encouraged to make this decision after recycling two early samples (AGV-1 [A] and BCR-1 [A]). After application of the self-absorption correction the specific activities of the recycled samples agreed with the original specific activities to within 1%.

We did not find it necessary to process the flux monitors as exhaustively as the samples. After irradiation, the opened vial was placed in a 40 ml centrifuge tube containing 10 ml 6 N HCl and 50 mg each of Bi^{3+} and Tl^{3+} inert carriers. After heating to dissolve the MgO "bed", the solution and several 6 N HCl washes were placed in a 100 ml volumetric flask and the solution was diluted with H_2O. A 5-ml aliquot of this solution was then added to a centrifuge tube containing 5 ml of Bi^{3+} carrier solution (10 mg/ml). This solution was evaporated nearly to dryness, diluted to 20 ml with 6 N HCl and further processed as in steps (a)–(c), (e), (f, omitting the hold-back carriers), (g), (j) and (k). The chemical yields for the monitors ranged, in general, from 50–80%.

Counting and data analysis

The BiOCl samples (3·08 cm^2 area) were placed onto stainless steel counting discs and covered with mylar film (1·79 mg/cm^2). The samples were counted with a Beckmann–Sharp Low-beta II low background proportional counter. Normally, four different samples were counted, then a background, for 200 min each. However, 500-min counts were taken for low-activity samples with initial activities comparable to the 0·65 cpm typical background count-rate. The initial counting of the Bi samples was made 4–5 days after the end of irradiation and the activities were followed for at least 5 half-lives to assure that they decayed with the proper Bi^{210g} half-life. Only in the cases of G-1 [A] and DTS-1 [A] was there any significant evidence for a contaminant decaying with a half-life of other than 5·01 days. We used an average background, obtained by summing all of the individual background counts accumulated during counting of a sample, to calculate the net Bi^{210g} activity in the sample. These average backgrounds represented 2500–5000 min of counting time and ranged from 0·61 ± 0·02 to 0·71 ± 0·02 cpm. We observed no unusual background fluctuations during our study.

The raw data were processed using a CDC-6500 computer and a decay curve analysis program (CUMMING, 1963) assuming both two-component and one-component systems. In all cases except for G-1 [A] and DTS-1 [A] we used the one-component output after comparative consideration of it and the two-component output for goodness-of-fit and number of discarded points. In general the time-zero net count rates for each output pair agreed to within 3% indicating the radiochemical purity of the samples. For each set of samples in a given irradiation, the net Bi^{210g} count rates at time-zero were corrected for self-absorption relative to the monitor in that irradiation. This empirical correction was seldom as much as 5% and never more than 15%.

RESULTS AND DISCUSSION

The results on the standard rock samples are listed in Table 1. For these, the precision for the duplicate runs seems quite reasonable except for G-1 [A] and PCC-1 [A]. The former appeared to be heavily contaminated with some long-lived radioactivity which may have affected our Bi result considerably and the latter had an extremely low chemical yield. Thus we regard these results with

Table 1. Bismuth contents of U.S.G.S. standard rocks and synthetic sample

Standard	Bismuth concentration (ppb) A	B	mean*
AGV-1	56·1†	56·7	56·4 ± 0·4
BCR-1	46·4	47·0	46·7 ± 0·4
DTS-1	4·8‡	4·7	4·8 ± 0·1
G-1	76·2‡	46·3	46
G-2	35·9	39·3	37·6 ± 2·5
GSP-1	36·4	37·1	36·8 ± 0·5
PCC-1	14·6†	8·0	8
W-1	44·0	43·1	43·5 ± 0·5
"Doped" MgO	19·8 μg §		20·0 μg

* As noted in the text the errors listed are the estimated standard errors of the mean. FLEISCHER (1969) has not recommended any Bi values or magnitudes for standards G-1 and W-1. For the remaining standards FLANAGAN (1969) has listed upper limits of 500 ppb.

† These samples had very low chemical yields (~5%) and the results may be unreliable. We have retained the value for AGV-1 [A] because of its agreement with that of AGV-1 [B].

‡ The decay curves of those samples indicated two-component systems. The contamination was serious only in G-1 [A] whose value is disregarded.

§ From separate experiments on untreated MgO, the Bi contamination in the MgO "bed" was 0·005 μg, which is negligible compared to the amount of Bi added.

considerable reservation. The data from the remaining 6 sample-pairs may be used to estimate the precision of our technique when applied to homogeneous standards. For each pair we calculated the standard error of the mean in the normal fashion from the best estimate of the population variance which was, in turn, obtained from the sample variance after application of Bessel's correction. The estimated relative standard errors of the means ranged from 0·7 to 6·7% and in 5 of the 6 cases were 2% or less.

Table 2. Bismuth contents in carbonaceous chondrites

Meteorite	Type	Bi contents (ppb)	Bi contents (atoms/10^9Si atoms*)
Ivuna [A]	C1	106	134
Ivuna [B]†	C1	126	159
Orgueil [A]	C1	110‡	139
Mean§	C1	114 ± 7	143 ± 10
Cold Bokkeveld	C2	65·4	68·4
Mighei	C2	79·4‡	81·6
Murray [A]	C2	62·1	61·9
Murray [B]	C2	70·7	70·5
Mean§	C2	69 ± 5	71 ± 5
Grosnaja	C3 [V]	68·3	61·1
Vigarano	C3 [V]	51·0	44·0
Mean§	C3 [V]	60 ± 13	53 ± 11
Felix [A]	C3 [O]	38·5	32·4
Felix [B]	C3 [O]	42·8	36·0
Lancé	C3 [O]	48·9	42·0
Ornans	C3 [O]	23·8	20·4
Warrenton	C3 [O]	31‖	26‖
Mean§	C3 [O]	37 ± 5	31 ± 4
Mean§	C3	43 ± 6	37 ± 5

* Si contents taken from MASON (1963).

† This consisted of saw cuttings collected in Tanzania (Appendix).

‡ REED *et al.* (1960) reported Bi concentrations in Orgueil and Mighei of 130 and 180 ppb respectively.

‖ The chemical yield for this sample was quite low (~5%). Despite this, we have included the Bi result for Warrenton in the C3 [O] and C3 means because of its similarity to results obtained from other C3 [O] chondrites.

§ The uncertainty listed for each group mean is the estimated standard error of the mean.

For heterogeneous samples, such as chondrites, the estimated standard errors of the means would be expected to be much larger. Indeed, this is the case since duplicate samples of Ivuna, Murray and Felix yield estimated relative standard errors of the means of 12%, 10% and 23% respectively (Table 2). If we assume that all Cl, C2 and C3 chondrites represent samples of homogeneous populations, these estimated errors are 6, 7 and 14% respectively. However, there is reason to doubt this assumption for C3 chondrites (LAUL *et al.*, in preparation). Thus, at the 95% confidence level the precision of our measurements on heterogeneous samples of homogeneous populations may be estimated at 14%. In the case of the ordinary chondrites our intent was to conduct as broad a survey as possible in the time available. Thus unfortunately we were unable to carry out replicate analyses and cannot estimate the precision of our measurements.

Were there reliable or recommended values for the Bi contents of these standards, we could assess the accuracy of our measurements. Unfortunately this is not the case. The older standards (G-1 and W-1) have no recommended Bi value (FLEISCHER, 1969) while the newer ones have only an upper limit of <500 ppb reported (FLANAGAN, 1969). In all cases the Bi contents measured lie well below this detection limit. The single analysis of the synthetic sample (Table 1) cannot be used to obtain a statistically meaningful estimate of the accuracy of our measurements. However, the agreement of our results with the known Bi content of this sample provides some evidence for our accuracy.

Carbonaceous chondrites

Our values for the Bi concentrations and atomic abundances in the carbonaceous chondrites are listed in Table 2. The Bi concentrations in two of these, Orgueil and Mighei, had been measured previously by REED *et al.* (1960). Their result on Orgueil is somewhat higher than any of our Cl analyses but this is not particularly disturbing. Their Mighei result, on the other hand, is distinctly higher than any of our results on other C2 chondrites or, for that matter, Cl chondrites. This alone causes us to regard the Mighei Bi result of REED *et al.* (1960) with some suspicion. There is also some additional internal evidence which lends considerable support to our results.

From previous studies (EHMANN and HUIZENGA, 1959; REED *et al.*, 1960; REED, 1963) ANDERS (1964) observed that the Bi abundances in ordinary chondrites were 10–1000 times lower than that in Orgueil and concluded that Bi is "strongly-depleted" in ordinary chondrites. Our results are certainly in accord with this conclusion. The Bi abundances that we obtained from the equilibrated ordinary chondrites range from 1 to 12 atoms/10^9 Si atoms (Table 3) corresponding to a "depletion-factor" of 10–100 relative to the Cl results (Table 2). Based on his conclusion that Bi is a "strongly-depleted" element, ANDERS (1964) predicted that the Bi mean abundance ratios (relative to Cl) of the carbonaceous chondrites would be C1:C2:C3 = 1·00:0·55:0·32.

Our results (Table 2) verify this prediction. The mean Bi abundance of the three C1 samples is 143 $\pm$ 10 or 148 $\pm$ 9 atoms/10^9 Si atoms if the Orgueil result of REED *et al.* (1960) is included. The four C2 samples yield a Bi abundance of 71 $\pm$ 5 atoms/10^9 Si atoms. For the C3 chondrites, several mean abundances may be computed. If we assume that these meteorites are of a single group, the Bi abundance of the seven samples studied is 43 $\pm$ 6 atoms/10^9 Si atoms. On the basis of mineralogic and petrographic studies, VAN SCHMUS (1969) has suggested that the C3 chondrites consist of two groups, denoted as Ornans (abbreviated here as C3 [0]) and Vigarano (C3 [V]) subtypes. Of these, we studied five (including a duplicate) and two representatives, respectively, and these yield mean Bi abundances of 31 $\pm$ 4 and 53 $\pm$ 11 atoms/10^9 Si atoms. It is noteworthy that the C3 [V] samples had higher Bi abundances than the C3 [O] samples. While possibly coincidental, these results tend to support VAN SCHMUS' (1969) suggestion and may indicate a chemical compositional difference between the two sub-types. These results then lead to Bi abundance ratios relative to C1 chondrites of C1:C2:C3 = 1·0:0·50:0·26, or C1:C2:C3 [V]:C3 [0] = 1·00:0·50:0·37:0·22, which are quite

Table 3. Analyses of equilibrated and unequilibrated ordinary chondrites

Meteorite*	Type†	PMD in olivine†	Carbon content %	ppb	Bi contents atom/10^6 this work atoms‡
Sharps	H3	37	0·95[a]	88·6	71·5
Bremervörde	H3	15	0·22[b]	24·9	18·9
Prairie Dog Creek	H3	6·9	0·35[b]	7·2	5·6[g]
Clovis #1	H3	5·6	0·22[b]	4·6	3·7
Fayetteville [light]	H	—		7·8	6·1[d,g]
Fayetteville [dark]	H	—		10·9	8·6[d,g]
Plainview	H5	—		15·4[h]	12·1[e]
Krymka	(L)3	45	0·27[b]	17·0	12·1
Mezö-Madaras	(L)3	28	0·46[b]	123	88·8
Khohar	L3	18	0·32[b]	31·0	22·5
Barratta	L4	4·2	0·085[b]	1·58	1·11
Holbrook	L6	—	0·06[c]	2·34[i]	1·68[e]
Bruderheim	L6	—	0·040[b]	1·4	1·0[f]
Ngawi	(LL)3	40	0·39[b]	23·6	16·7
Chainpur	(LL)3	32	0·44[b]	65·0	45·7
Parnallee	(LL)3	19	0·19[b]	11·7	8·1
Hamlet	LL(3,4)	3·5	0·16[b]	6·5	4·5

* Meteorites listed in italics are finds; all others are observed falls.

† All data from VAN SCHMUS and WOOD (1967) except for Fayetteville (HEY, 1966). Parentheses are placed around those classifications that may be uncertain (VAN SCHMUS and WOOD, 1967).

‡ All PMD values and the Si data from DODD *et al.* (1967) unless otherwise noted.

[a]Jarosewich, unpublished analysis listed in DODD *et al.* (1967).

[b]MOORE and LEWIS (1967).

[c]MASON and WIIK (1961).

[d]MASON (1965).

[e]WING (1964).

[f]Average of analyses reported by DUKE *et al.* (1961) and BAADSGAARD *et al.* (1961).

[g]Average Si analysis for group.

[h]Replicate samples of this meteorite have previously yielded Bi concentrations of 3·0, 3·8 and 4·1 ppb (EHMANN and HUIZENGA, 1959).

[i]Replicate samples of this meteorite have previously yielded Bi concentrations of [illegible]·1 ppb (EHMANN and HUIZENGA, 1959) and 1·7, 2·6 and 7·6 ppb (REED *et al.*, 1960).

similar to those predicted by ANDERS (1964, 1968) and LARIMER and ANDERS (1967) on the basis of a two-component model.

Having now established that the Bi contents of the carbonaceous chondrites vary with chondritic type it is appropriate that we consider the question of the "cosmic abundance" of Bi. Since it is generally accepted that C1 chondrites constitute the closest approximation to unaltered cosmic material, it seems likely that the Bi cosmic abundance is $0{\cdot}143 \pm 0{\cdot}010$ atoms/10^6 Si atoms [or $0{\cdot}148 \pm 0{\cdot}009$ atoms/10^6 Si atoms, including the Orgueil value of REED *et al.* (1960)]. This value is virtually identical to that adopted by SUESS and UREY (1956) [based, in turn, on the value adopted by NODDACK and NODDACK (1934)]. The revised Bi abundance value adopted by CAMERON (1968), which is based on the single Orgueil result of REED *et al.* (1960), should therefore be reduced slightly to 0·14–0·15 atoms/10^6 Si atoms in view of our results.

Ordinary chondrites

From the results in Table 3 it appears at first glance that the Bi contents of the equilibrated H-group chondrites may be about 5 times higher than those of the equilibrated L-group chondrites. This difference may well be fortuitous, however, in view of the small number of samples analyzed. The Bi concentration ranges reported previously (EHMANN and HUIZENGA, 1959; REED *et al.*, 1960; REED, 1963) for equilibrated H-, L- and LL-chondrites are about the same and vary in the extreme cases by about a factor of 10^2 (Table 4). Indeed, for at least one L6 chondrite alone (Modoc) the extreme concentrations in 7 replicates vary by about the same factor (EHMANN and HUIZENGA, 1959; REED *et al.*, 1960). Part, if not all, of this variation may be due to sample inhomogeneity although at concentrations of ~1 ppb, experimental difficulties may play a significant role. In view of these variations it would clearly be improper to calculate arithmetic means for the Bi contents in ordinary chondrites. Based on all available data the geometric mean Bi contents in the equilibrated H- and L-groups are 2·5 and 1·3 ppb or 2·0 and 1·0 atoms/10^9 Si atoms respectively (Table 4). In the case of the

Table 4. Bismuth contents in ordinary chondrites

Chondritic group	Equilibrated			Unequilibrated		
	Range* (ppb)	Geometric means ppb	Geometric means (atoms/10^9 Si atoms)	Range* (ppb)	Geometric means (ppb)	Geometric means (atoms/10^9 Si atoms)
H	0·18–15(20)	2·5	2·0	4·6–89(4)	16	13
L	0·07–7·6(13)	1·3	1·0	1·6–120(4)	18	13
LL	0·55–1·0(2)	0·7	0·5	6·5–65(4)	19	14
All	0·07–15(35)	1·9	1·4	1·6–120(12)	18	14

* Data from this study, EHMANN and HUIZENGA (1959), REED *et al.* (1960) and REED (1963). The numbers in parentheses are the number of samples, not meteorites, analyzed of each group.

LL-group, only 2 samples have been studied. Thus, the Bi means for this group (0·7 ppb or 0·5 atoms/10^9 Si atoms) carry little statistical weight. Many further measurements will be required before firm conclusions can be drawn on possible group differences in the Bi contents of equilibrated ordinary chondrites. In this connection, we note that the dark phase of Fayetteville appears to have a Bi concentration 40% higher than that in the light. The only similar chondrite studied previously is Pantar in which the Bi concentration was reported to be 5–6 times higher in the dark phase than in the light (REED, 1963). We will not speculate upon the reasons for these differences until we have further information on the contents of other trace elements in the light and dark phases of Fayetteville.

It is well established that, among the ordinary chondrites, there exists a small number of unequilibrated ordinary chondrites (UOC) in which the silicates are chemically inhomogeneous (DODD and VAN SCHMUS, 1965; DODD *et al.*, 1967). Based on these inhomogeneities in the UOC, DODD *et al.* (1967) have established a semi-quantitative disequilibrium scale expressed as "percent mean deviation" of the Fe^{2+} in the olivine (PMD). Despite considerable efforts there is no consensus for the origin of these UOC. DODD (1969) has reviewed, in detail, the results of those studies pertinent to the origin of the UOC and, indeed, all chondrites, and has critically evaluated the rival theories which attempt to explain their origin. It seems that much further data must be obtained if ever the genesis of these chondrites is to be understood. One important point which is unsettled is whether the relationship of the unequilibrated H-, L- and LL-chondrites to each other is more important than their respective group relationships to the equilibrated chondrites. For much of the discussion which follows we will assume the former relationship to be the more important although both will be considered.

Within this framework then, let us now specifically consider the Bi data of the H-, L- and LL-chondrites (Table 3). In two cases (Sharps, Mezö-Madaras), the Bi concentrations approach those of the Cl chondrites although, even for these two, the abundances are markedly lower. In the remaining UOC the abundances are lower by as much as 10^2, extending to the range of the equilibrated ordinary chondrites. This suggests the possibility that the Bi contents in equilibrated ordinary chondrites are progressively depleted with increasing petrologic type. From our data and those of Ehmann and Huizenga (1959), Reed *et al.* (1960) and Reed (1963), there appears to be no certain trend in this direction. Of course this tentative conclusion is based on relatively few analyses and may be changed when more type 4–6 samples are studied. The geometric mean Bi contents in the unequilibrated H-, L- and LL-chondrites are quite similar (Table 4)—16, 18 and 19 ppb or 13, 13 and 14 atoms/10^9 Si atoms, respectively. Insofar as the meteorites that we studied are representative of the unequilibrated ordinary chondrites these results indicate the lack of a marked dependence of mean Bi content upon chondritic class.

The Bi abundance data (Table 3) are plotted in Fig. 1 vs. the degree of equilibration of the olivine. A striking feature of Fig. 1 is the monotonic decrease in Bi abundance within each group with decreasing PMD (or increasing equilibration), suggesting that whatever process was responsible for equilibration was also reponsible for the decrease in the content of Bi. This general trend might be expected to result from processes similar to those postulated by Dodd *et al.* (1967), i.e. progressive metamorphism, or Larimer and Anders (1967), i.e. accumulation at different temperatures. What is unexpected, however, is the violation of this monotonic trend in the cases of the least equilibrated member of each chondritic group (Fig. 1). If we assume that the remaining chondrites of each group represent discrete evolutionary sequences (the dotted lines in Fig. 1), then the chondrites having the highest PMD values of each group appear to be considerably underabundant in Bi. Alternatively, if we assume a single evolutionary sequence for all UOC (the solid line in Fig. 1), the same conclusion holds. Can it be that these trends result from the omission of Sharps, Krymka and Ngawi in calculating the lines in Fig. 1? If these are included the least-squares line for the UOC as a group has a less steep slope but, even so, both Krymka and Ngawi remain underabundant in Bi. Furthermore, the least-squares line for the H-group UOC can hardly be shifted so as to include Sharps. It seems therefore that our observation is valid and that these three meteorites are distinctly underabundant in Bi. This observation appears equally inconsistent with the genetic models of Dodd *et al.* (1967) or Larimer and Anders (1967). It could be made consistent with either by the adoption of *ad hoc* assumptions the nature of which need not be discussed here.

It seems clear from Fig. 1 that our working hypothesis of an exponential relationship between the Bi contents and degree of equilibration of the UOC satisfies the empirical observations for all UOC except Sharps, Krymka and Ngawi. It must not be assumed however that this simple empirical relationship has any

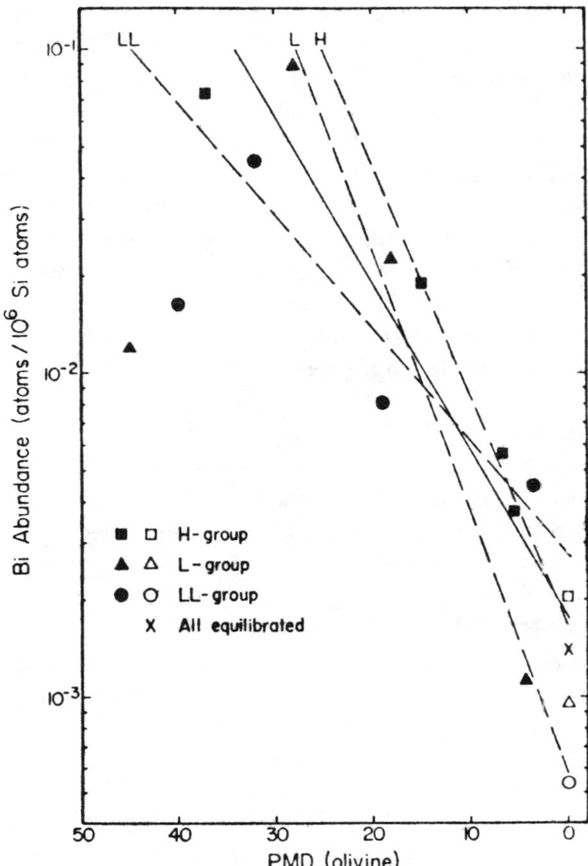

Fig. 1. Bismuth abundances versus the percent mean deviation (PMD) of olivine in ordinary chondrites. The closed symbols indicate individual unequilibrated ordinary chondrite analyses and the open symbols indicate geometric mean Bi abundances in the equilibrated ordinary chondrite groups. The geometric mean abundance for all equilibrated ordinary chondrites is indicated by "X". Except for the most unequilibrated member of each chondritic group, the Bi concentrations appear to decrease exponentially with PMD. Least-squares lines are shown for: (a) each chondritic group separately (dashed lines); (b) the UOC as a group (solid line). For the calculations, the chondrite having the highest PMD of each group was omitted since these meteorites appear to be distinctly underabundant in Bi (see text).

theoretical basis at the moment or can be extrapolated beyond the bounds indicated in Fig. 1. Indeed if such extrapolations are made to the extremes two paradoxical situations arise. The first of these is that extrapolation of the H- and L- group lines to PMD > 30 and the LL-group line to PMD > 50 would indicate Bi abundances in excess of those in C1 chondrites—a very unlikely prospect. In highly unequilibrated ordinary chondrites, the relationship between Bi and disequilibrium must be more complex than is indicated by our working hypothesis and, indeed, we have already noted Bi abundances to be low in such meteorites. The second paradox is that the Bi abundances in members of each group of equilibrated ordinary chondrites (i.e. at zero PMD) would be expected to be quite similar when, in fact, they are not. In order to account for this, it seems obvious that after the PMD parameter ceases to be a measure of equilibration other processes must occur to redistribute Bi or cause its loss in equilibrated ordinary chondrites. Thus, in the remainder of this discussion, where we refer to the "nearly exponential" depletion of Bi with decreasing PMD or increasing equilibration it must be understood that we are describing a phenomenon and not implying any particular process.

With respect to Fig. 1, we may note some speculative but nonetheless interesting observations. When the lines for the H- and L-group UOC are extrapolated to zero PMD (Fig. 1), the respective Bi abundances are within 20% and 40% of the means calculated for equilibrated chondrites of these groups (Table 4). In the case of the LL-group the agreement is not good since the zero-PMD intercept is 6 times higher than the mean Bi abundance in equilibrated LL-group chondrites. The least-squares line for the LL-group UOC is much less steep than those of the H- and L-groups primarily because of the Hamlet datum. Can it be that the PMD datum for this chondrite is in error? If so, then the zero-PMD intercept of a line through the Chainpur and Parnallee points is within 20% of the mean Bi abundance of the equilibrated LL-group. We are reluctant to speculate further upon this since a straight line can always be drawn through 2 points and the equilibrated LL-group mean represents only 2 analyses. We note however that assuming a single evolutionary sequence for all UOC, the zero-PMD intercept is within 30% of the mean Bi abundance for all equilibrated ordinary chondrites. It may be that these seeming agreements are fortuitous and it would be well were the PMD values reexamined (DODD, personal communication).

Returning to less speculative matters it is interesting that carbon shows the same overall pattern as Bi—i.e. nearly exponential, progressive depletion with decreasing PMD, except for the least equilibrated chondrites, in which carbon is underabundant (Table 3; see also Fig. 14 of DODD, 1969). We would therefore expect a mutual correlation between the Bi and C concentrations in the UOC and this is indeed observed (Fig. 2), suggesting the influence of some process in common. It is worthwhile noting that the overall abundance trends for Bi alone (Fig. 1) or C alone (Fig. 14; DODD, 1969) are duplicated when one plots the Bi/C ratios vs. PMD for the UOC.

It seems scarcely possible that the Bi–C correlation (Fig. 2) represents the elements' proportional volatilization in some critical temperature region if one considers the primordial gas contents of the UOC (e.g. HEYMANN and MAZOR, 1968; ZÄHRINGER, 1968). In Fig. 3 we have plotted the primordial Ar^{36} content vs. Bi for the UOC. Despite the scatter in the noble gas data, there appears to be a good correlation between the concentrations of primordial Ar^{36} and Bi. While there seem to be correlations of Bi with primordial Kr^{84} and Xe^{132} as well, these are less striking than the one evident in Fig. 3 because of the smaller variation in the Kr^{84} and Xe^{132} contents of UOC. In connection with these correlations it seems rather surprising that the overall Bi concentrations in the UOC decrease at a far more rapid rate than do the noble gas concentrations (a factor of 10^2 for Bi compared with factors of 10 for Ar^{36} or 3–4 for Kr^{84} and Xe^{132}). It is also noteworthy

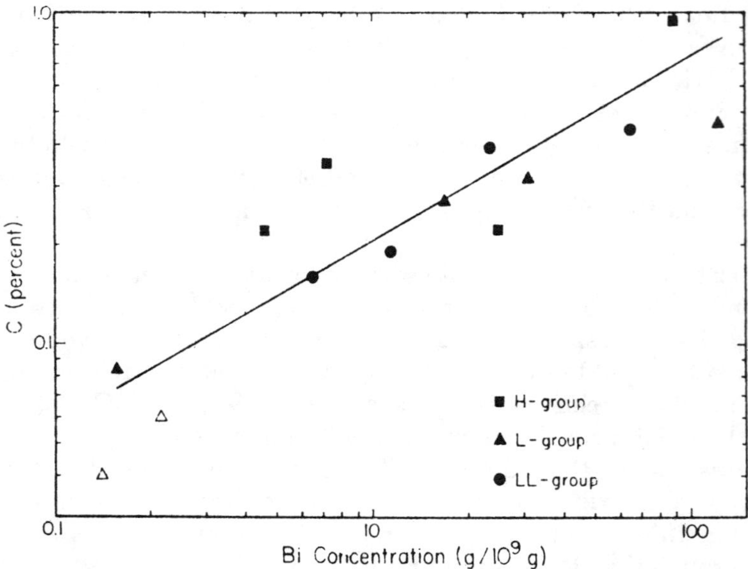

Fig. 2. Correlation between the bismuth and carbon contents of the ordinary chondrites. The solid line is calculated by least-squares analysis of all UOC data. The open symbols indicate data for equilibrated chondrites (L6).

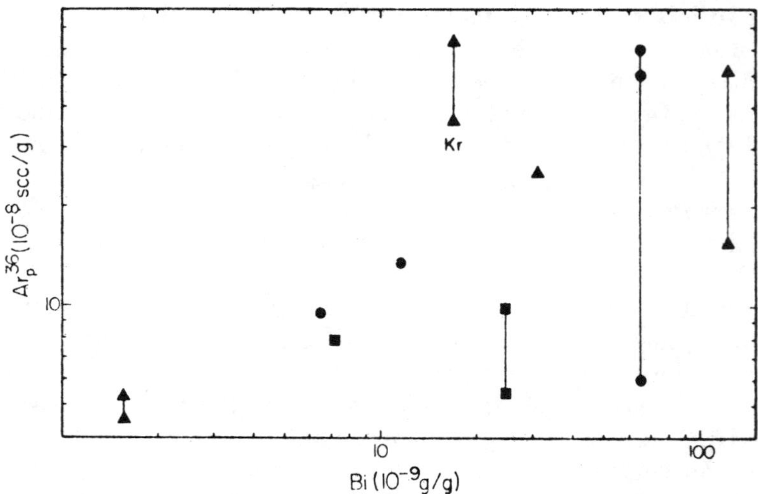

Fig. 3. Primordial Ar^{36} concentrations vs. Bi concentrations in UOC. Despite the scatter in the noble gas data (HEYMANN and MAZOR, 1968; ZÄHRINGER, 1968) the data seem well correlated. When Kr^{84} and Xe^{132} are plotted vs. Bi, similar correlations are observed. Krymka (Kr) seems to be an exception to these correlations in that in all three cases it is underabundant in Bi (or overabundant in primordial noble gases).

that with respect to the Ar^{36}, Kr^{84} and Xe^{132} concentrations, the only chondrite whose Bi content is clearly aberrant (e.g. Fig. 3) is Krymka, the L-group chondrite having the highest PMD.* Yet, the Bi contents of the UOC (including Krymka, Ngawi and Sharps) are consistent with their C contents (Fig. 2). Since both Bi and C can scarcely be more volatile than Ar, Kr and Xe, it seems that the trends evident in Figs. 1–3 cannot be the result of late volatilization losses. Instead these trends must be the result of some primordial process(es), the nature of which seems quite nebulous at this time.

Conclusions

Based on this survey of the Bi contents in 31 samples of carbonaceous and equilibrated and unequilibrated ordinary chondrites, a number of apparently significant observations may be made. These observations are:

(a) Bismuth is confirmed to be a "strongly-depleted" element (Anders, 1964).

(b) The mean Bi abundance ratios in the carbonaceous chondrites relative to that of Cl are practically identical to those predicted by Anders (1964). This, in turn, supports his hypothesis that the carbonaceous chondrites are mixtures of two components, (i.e. chondrules and matrix) which are, respectively, depleted and undepleted with reference to a number of trace elements of various geochemical behavior.

(c) The true cosmic abundance of Bi is 0·14–0·15 atoms/10^6 Si atoms based on our Cl analyses.

(d) The Bi contents of the unequilibrated ordinary chondrites are intermediate between those of the carbonaceous and equilibrated ordinary chondrites. There is a nearly exponential progressive decrease in the Bi contents of unequilibrated ordinary chondrites with the degree of equilibration except for the most unequilibrated H-, L- and LL-chondrites, in which Bi appears distinctly underabundant.

(e) The Bi contents of the unequilibrated ordinary chondrites correlate very well with carbon. The concentrations of the primordial gases Ar^{36}, Kr^{84} and Xe^{132} also correlate with Bi in most of these chondrites. The distinct exception appears to be Krÿmka, the only highly unequilibrated ordinary chondrite for which both Bi and rare gas data are available.

Acknowledgments—We are grateful to R. S. Clarke Jr., R. E. Folinsbee, P. K. Kuroda, C. B. Moore and E. Olsen for the meteorite specimens used in this study (Appendix) and E. Anders and J. T. Wasson for very useful comments on an earlier version of this paper. We thank N.T Porile for the use of low-level beta counters and E. Culp and Dr. I. Pelly for valuable help. Various aspects of this study were supported by grants from the Advanced Research Projects Agency, the National Aeronautics and Space Administration (grant NGL 15-005-021) and the U.S. National Science Foundation (grant GA-1474).

References

Anders E. (1964) Origin, age and composition of meteorites. *Space Sci. Rev.* **3,** 583–714.

Anders E. (1968) Chemical processes in the early solar system, as inferred from meteorites. *Accounts Chem. Res.* **1,** 289–298.

* The Ar^{36}, Kr^{82} and Xe^{132} contents of Sharps and Ngawi, the most unequilibrated H- and LL- group chondrites that we studied, have not been reported. We would expect that when these are measured, both Ngawi and, to a lesser extent, Sharps will also be aberrant with respect to the Bi-noble gas correlations.

Baadsgaard H., Campbell F. A., Folinsbee R. E. and Cumming G. L. (1961) The Bruderheim meteorite. *J. Geophys. Res.* **66,** 3574–3577.

Cameron A. G. W. (1968) A new table of abundances of the elements in the solar system. In *Origin and Distribution of the Elements,* (editor L. H. Ahrens), pp. 125–143. Pergamon.

Cumming J. B. (1963) CLSQ, the Brookhaven decay curve analysis program. U.S.A.E.C. Rep. NAS-NS-3107, (editor G. D. O'Kelley), pp. 25–33.

Dodd R. T. (1969) Metamorphism of the ordinary chondrites: a review. *Geochim. Cosmochim. Acta* **33,** 161–203.

Dodd R. T. and Van Schmus W. R. (1965) Significance of the unequilibrated ordinary chondrites. *J. Geophys. Res.* **70,** 3801–3811.

Dodd R. T., Van Schmus W. R. and Koffman D. M. (1967) A survey of the unequilibrated ordinary chondrites. *Geochim. Cosmochim. Acta* **31,** 921–951.

Duke M., Maynes D. and Brown H. (1961) The petrography and chemical composition of the Bruderheim meteorite. *J. Geophys. Res.* **66,** 3557–3563.

Ehmann W. D. and Huizenga J. R. (1959) Bismuth, thallium and mercury in stone meteorites by activation analysis. *Geochim. Cosmochim. Acta* **17,** 125–135.

Flanagan F. J. (1969) U.S. Geological Survey standards—II. First compilation of data for the new U.S.G.S. rocks. *Geochim. Cosmochim. Acta* **33,** 81–120.

Fleischer M. (1969) U.S. Geological Survey standards—I. Additional data on rocks G-1 and W-1, 1965–1967. *Geochim. Cosmochim. Acta* **33,** 65–79.

Hey M. H. (1966) *Catalogue of Meteorites,* 3rd edition. British Museum.

Heymann D. and Mazor E. (1968) Noble gases in unequilibrated ordinary chondrites. *Geochim. Cosmochim. Acta* **32,** 1–19.

Larimer J. W. and Anders E. (1967) Chemical fractionations in meteorites—II. Abundance patterns and their interpretation. *Geochim. Cosmochim. Acta* **31,** 1239–1270.

Laul J. C., Case D. R., Wechter M., Schmidt-Bleek F. and Lipschutz M. E. (1969) To be published in *J. Radioanal. Chem.*

Lederer C. M., Hollander J. M. and Perlman I. (1967) *Table of Isotopes,* 6th Edition. John Wiley.

Mason B. (1963) The carbonaceous chondrites. *Space Sci. Rev.* **1,** 621–646.

Mason B. (1965) The chemical composition of olivine–bronzite and olivine–hypersthene chondrites. *Amer. Mus. Novitates* 2223.

Mason B. and Wiik H. B. (1961) The Holbrook, Arizona, chondrite. *Geochim. Cosmochim. Acta* **21,** 276–283.

Moore C. B. and Lewis C. F. (1967) Total carbon content of ordinary chondrites. *J. Geophys. Res.* **72,** 6289–6292.

Noddack I. and Noddack W. (1934) Die geochemischen Verteilungkoeffizienten der Elemente. *Svenskr. Kem. Tidskr.* **46,** 173–201.

Prestwood R. J. (1963) Bismuth. *Collected Radiochemical Procedures.* U.S.A.E.C. Rep. LA 1721, (2nd edition) Bi pp. 1–6.

Reed G. W. (1963) Heavy elements in the Pantar meteorite. *J. Geophys. Res.* **68,** 3531–3535.

Reed G. W., Kigoshi K. and Turkevich A. (1960) Determinations of concentrations of heavy elements in meteorites by activation analysis. *Geochim. Cosmochim. Acta* **20,** 122–140.

Strelow F. W. E. and Bothma C. J. C. (1967) Anion exchange and a selectivity scale for elements in sulfuric acid media with a strongly basic resin. *Anal. Chem.* **39,** 595–599.

Suess H. E. and Urey H. C. (1956) Abundances of the elements. *Rev. Mod. Phys.* **28,** 53–74.

Van Schmus W. R. (1969) Mineralogy, petrology and classification of Types 3 and 4 carbonaceous chondrites. *Meteorite Research,* (editoı P. M. Millman), pp. 480–491. D. Reidel.

Van Schmus W. R. and Wood J. A. (1967) A chemical–petrologic classification for the chondritic meteorites. *Geochim. Cosmochim. Acta* **31,** 747–765.

Wing J. (1964) Simultaneous determination of oxygen and silicon in meteorites and rocks by non-destructive activation analysis with fast neutrons. *Anal. Chem.* **36,** 559–564.

Zähringer J. (1968) Rare gases in stony meteorites. *Geochim. Cosmochim. Acta* **32,** 209–237.

Appendix—Sources and Sample Numbers of Meteorites Studied

1. R. S. Clarke Jr.; U.S. National Museum; Washington, D. C. *Clovis* (USNM-2170), *Felix* [B] (USNM-235), *Hamlet* (USNM-3455), *Ivuna* [A] (USNM-2478), *Ivuna* [B] (USNM-2478), *Krymka* (USNM-2488), *Murray* [A] (USNM-2122), *Murray* [B] (USNM-1519), *Ngawi* (USNM-2962), *Prairie Dog Creek* (USNM-1115), *Sharps* (USNM-640).
2. R. E. Folinsbee; University of Alberta; Canada *Bruderheim* (B-93).
3. P. K. Kuroda; University of Arkansas; Fayetteville, Ark. *Fayetteville*.
4. C. B. Moore; Arizona State University; Tempe, Ariz. *Chainpur* (ASU-424-1), *Holbrook* (ASU-452), *Khohar* (ASU-623-1), *Plainview* (ASU-92-423).
5. E. Olsen; Field Museum of Natural History; Chicago, Ill. *Barratta* (Me-1464), *Bremervörde* (Me-1758), *Cold Bokkeveld* (Me-1738), *Felix* [A] (Me-1330), *Grosnaja* (Me-1732), *Lancé* (Me-589), *Mezö-Madaras* (Me-1607), *Mighei* (Me-1456), *Orgueil* [A] (Me-509), *Ornans* (Me-1766), *Parnallee* (Me-437), *Vigarano* (Me-732), *Warrenton* (Me-1720).

8

Reprinted from *Geochim. et Cosmochim. Acta* **36**:897–902 (1972)

Bismuth in stony meteorites and standard rocks

P. M. Santoliquido and W. D. Ehmann
Department of Chemistry, University of Kentucky, Lexington, Kentucky 40506 (U.S.A.)

Introduction

The abundance pattern of Bi in meteorites has interested many investigators. Early work by Reed *et al.* (1958, 1960) and Ehmann and Huizenga (1959) showed that Bi is a severely depleted element whose abundance varies with chondrite class in the order carbonaceous > enstatite > ordinary. More recently, Laul *et al.* (1970) have found Bi abundances within the carbonaceous chondrite classes to be in good agreement with the ratios predicted by Anders (1964, 1968) and Larimer and Anders (1967) on the basis of a two-component model. Additionally, Laul *et al.* (1970) pointed out a progressive decrease in the Bi contents of unequilibrated ordinary chondrites with the degree of equilibration. Keays *et al.* (1971) reported a decrease in Bi content with increasing petrographic grade of L-group chondrites. In the present work, we have attempted to complete the study of the Bi abundance pattern in meteorites by analyzing previously omitted chondrite classes as well as a number of achondrites and separated meteoritic phases.

Experimental

Bismuth was determined by alpha counting of the ^{210}Po daughter activity of the ^{210}Bi produced by thermal neutron irradiation.

Samples were prepared by weighing aliquants of meteorite and rock powders into clean quartz vials, which were then heat-sealed. Sample sizes varied from 100 to 300 mg for meteorites and from 500 to 800 mg for rocks.

To prepare flux monitors, high purity Bi metal was dissolved in concentrated HNO_3, and a standard flux monitor solution prepared on a weight basis by successive dilution with 4 M HNO_3. Aliquants of this standard solution were evaporated onto approximately 200 mg of 'Specpure' SiO_2 contained in cleaned quartz vials.

At the beginning of this work, samples were irradiated at the Air Force Nuclear Engineering Center Reactor, Dayton, Ohio for 100 hr at a flux of 10^{13}n cm^{-2} sec^{-1}. When this reactor terminated operation, samples were sent to the University of Missouri Research Reactor, Columbia, Missouri, where the flux varied from 2 to 5 $\times$ 10^{13}n cm^{-2} sec^{-1}. For the first half of the samples analyzed, the irradiation time was 100 hr; in later runs two successive 100 hr irradiations were used.

At least 20 days were allowed to pass between the end of irradiation and the beginning of chemical work, so that essentially all the ^{210}Bi decayed to ^{210}Po. The irradiated sample was transferred to a Teflon dish and an aliquot of a standard ^{208}Po solution added to serve as an indicator nuclide for the determination of chemical yield. After dissolution with concentrated H_2SO_4 and HF, aqua regia was added to dissolve metal grains. The solution was transferred to a disposable, polypropylene beaker and 1·0 g of hydroxylamine and 0·5 g of sodium citrate added. Sulfur dioxide was bubbled through the solution for several min, and the pH adjusted to 4 with dilute NH_4OH. Polonium was plated out of solution onto a silver disc without applied potential for 4 hr at 65°C. The silver disc was cemented with epoxy resin to a Teflon rod which was affixed to a stirring motor. The back of the disc was covered with epoxy resin so that polonium would plate only on the front surface. After plating, the disc was rinsed with distilled water, dilute NH_4OH and ethanol and allowed to dry. Counting was done with a surface barrier silicon detector.

Table 1. Bismuth in various classes of stony meteorites

Specimen	Class*	Bismuth, ppb (10^{-9} g/g)	Mean	Cond. temp.† (°K)
Murchison	C2	50·1, 44·5, 38·2	44·3	—
Allende	C3	59·6, 58·6, 55·0	57·7	—
Coolidge	C4	3·1	3·1	—
Karoonda	C4	23·3	23·3	—
Abee	E4	78·3, 62·9	70·6	487
Indarch	E4	84·8, 76·9	80·9	486
Atlanta	E5	1·9, 1·2, 1·1, 1·0	1·3	534
St. Marks	E5	15·2, 4·2	9·7	503
Blithfield	E6	8·7, 7·4, 5·8	7·3	509
Hvittis	E6	10·1, 6·5	8·3	508
Bath	H4	15·2, 9·0	12·1	455
Kesen	H4	32·3, 28·2, 21·3	27·3	450
Ochansk	H4	56·2, 39·8	48·0	435
Weston	H4	14·3, 9·1, 6·2	9·9	455
Forest City	H5	9·1, 7·7	8·4	455
Lost City	H5	1·8, 1·7, 1·5	1·7	461
Plainview	H5	7·0, 5·1	6·1	456
Gladstone	H6	5·6, 4·3	5·0	456
Mt. Browne	H6	4·8, 4·4	4·6	456
Cynthiana	L4	7·9, 6·4	7·2	456
McKinney	L4	16·2, 12·5	14·4	454
Arriba	L5	8·3	8·3	456
Crumlin	L5	7·8	7·8	456
Melrose	L5	6·6, 5·4	6·0	456
Bruderheim	L6	1·4	1·4	464
Harleton	L6	3·3	3·3	457
Chainpur	LL3	43·1	43·1	445
Appley Bridge	LL6	1·3, 1·1	1·2	467
Dhurmsala	LL6	4·1	4·1	456
Jelica	LL6	17·2, 12·4	14·8	454
Kapoeta‡	Howardite	1260, 1250	1260	—
Pasamonte	Eucrite	1·1, 1·0	1·1	—
Cumberland Falls	Aubrite	6·1, 4·9	5·5	—
Norton County	Aubrite	0·85, 0·63	0·74	—
Shallowater	Aubrite	2·4, 2·1	2·2	—
North Haig	Ureilite	5·4, 4·3	4·8	—

* Classification according to VAN SCHMUS and WOOD (1967).

† Calculated using Si abundances in the same powders, as reported by VOGT and EHMANN (1965), EHMANN and DURBIN (1968), and EHMANN *et al.* (1970).

‡ Our sample of Kapoeta has yielded unusually high abundances in other trace element studies (see text) and may not be representative of the meteorite as a whole.

Results and Discussion

The results obtained for the whole meteorites, separated meteoritic phases and standard rocks analyzed are given in Tables 1, 2 and 3, respectively. These include

Table 2. Bismuth in separated meteoritic phases and in spherules

Meteorite	Phase	Bismuth (ppb)
Allende	chondrules	23·4
Allende	white phase	20·7
Antofagasta	olivine	3·9, 2·8, 1·8
Canyon Diablo	metal	1·6
Canyon Diablo	spherules	17·3, 13·5
Canyon Diablo	troilite	147, 102
Sardis	troilite	82·0, 79·9, 67·8, 67·3
Springwater	olivine	13·7

Table 3. Bismuth in U.S.G.S. standard rocks

Sample	(Split/position)	Bismuth (ppb)	Mean value (ppb)
Andesite, AGV-1	(110/21)	50·1, 44·8, 36·9	44
Basalt, BCR-1	(1/8)	36·7, 34·7, 34·4	35
Dunite, DTS-1	(10/14)	6·8, 6·1, 4·8	5·9
Granite, G-2	(100/6)	77·5, 67·2, 45·4	63
Granodiorite, GSP-1	(64/10)	35·5, 32·3, 30·3	33
Peridotite, PCC-1	(44/23)	18·6, 13·3, 8·4	13

the first analyses for Bi in the chondrite classes C4, H4, H6, E5, E6, and LL6. The U.S.G.S. standard rocks were previously analyzed by Laul *et al.* (1970) and most of our values are in good general agreement.

General trends

For the carbonaceous chondrites, the results obtained are in harmony with those of Laul *et al.* (1970). The Allende values are within the range reported for the Vigarano subtype of class C3. The Murchison C2 classification is that given by Jarosewich (1971). In Bi content, however, it is right on the average reported by Laul *et al.* (1970) for C3. The C4 chondrites analyzed in this study continue the trend of decreasing Bi abundance with increasing petrographic grade, since the values for Karoonda and especially Coolidge are lower than those of the C3 class.

In the enstatite chondrites, Bi abundance is high for the trace-element-rich group and low for the trace-element-poor group, with a sharp hiatus between the two. Type E5 chondrites have about the same Bi abundance as E6, rather than being intermediate between E4 and E6. The relatively poor precision obtained for the St. Marks meteorite may reflect an inhomogeneous distribution of Bi in this meteorite.

The ordinary chondrites decrease in Bi content with increasing petrographic grade. Since there is a great deal of overlap among the H, L, and LL groups, it would appear that the metal–silicate fractionation has not affected the Bi abundance.

The achondrites yield both the lowest (Norton County) and the highest (Kapoeta) Bi values of this work. Most achondrites are at the same general level as the most metamorphosed ordinary chondrites. Kapoeta is clearly a special case. Other authors have found it to be greatly enriched in Sb (TANNER and EHMANN, 1967) and solar noble gases (ZÄHRINGER and GENTNER, 1960).

The Allende phases (Table 2) show a depletion of Bi in the chondrules and 'white phase' relative to the whole meteorite analyses. It should be noted that the chondrules were merely separated from the matrix with a vibrating needle and were not surface etched. Therefore, very small amounts of matrix material may have been present on the chondrule surfaces. Allende black matrix was not analyzed separately, but it is evident from the other values that Bi must be enriched there. This would lead to another case of enrichment in dark over light phase which would parallel those already noted by Pantar (REED, 1963) and Fayetteville (LAUL *et al.*, 1970), although in contrast to Allende, these meteorites show no appreciable difference in bulk chemistry between their dark and light portions.

Two possible host minerals seem to be indicated. The high levels of Bi found in the separated troilite phases indicate that this is probably the principal host mineral. The enstatite chondrite results suggest that oldhamite could well be an accessory host mineral. The enstatites are the only class of meteorites in which oldhamite (CaS) has been found. The ionic radius of Bi(III) (1·09 Å) is very nearly the same as that of Ca(II) (1·06 Å), so that it would be possible for bismuth to substitute for calcium in the crystal lattice. The one enstatite chondrite to yield a very low bismuth abundance, Atlanta, is the only one in the present study to have been reported (MASON, 1966) as containing no oldhamite.

Implications for meteorite thermal histories

The condensation behavior of Bi has been discussed by LARIMER (1967) and KEAYS *et al.* (1971). In the latter article, condensation temperatures were calculated for some L-group chondrites on the basis of a solar nebula pressure of 10^{-4} atmospheres. Recently it has been brought to our attention (Dr. E. Anders, private communication) that a better assumption is a total pressure of 1×10^{-5} atm. for the ordinary chondrites and of 5×10^{-4} atm. for the enstatite chondrites. Under these conditions, the condensation equations of KEAYS *et al.* (1971) become:

(1) Condensation as pure element, $\log (1 - \alpha) = \frac{-10310}{T} + B'$

where $B' = 22{\cdot}56$ at 1×10^{-5} atm. and $20{\cdot}86$ at 5×10^{-4} atm.

(2) Condensation as alloy, $\log (\alpha/1 - \alpha) = \frac{6617}{T} - B$

where $B = 15{\cdot}85$ at 1×10^{-5} atm. and $14{\cdot}15$ at 5×10^{-4} atm.

In these equations, T is the absolute temperature, and α is the fraction condensed, which is calculated by dividing the experimental Bi abundance for a given sample by the value for 100 per cent condensation. This latter value is calculated from the equation:

(3) Abundance for 100 per cent condensation (ppb) $= \frac{10(0{\cdot}164)(209)(0{\cdot}25)(\%\text{Si})}{28{\cdot}09}$

where 0·164 is the atomic abundance of Bi relative to 10^6 atoms Si (based on the solar system abundance of CAMERON (1968)), 209 and 28·09 are the atomic weights of Bi and Si, %Si is the Si content of each sample in per cent by weight, and 0·25 is the fraction of matrix in the ordinary chondrites. Matrix fractions of 0·5 and 0·7 were used for the trace-element-poor and trace-element-rich enstatite chondrites, respectively.

Table 1 lists condensation temperatures calculated using the above equations. A computer readout of T as a unique function of α with the metastable mode of condensation suppressed was furnished to us by Dr. E. Anders which facilitated these calculations. Our data for the ordinary chondrites yield condensation temperatures which are generally within the broad range of 510^{+80}_{-60}°K given by KEAYS *et al.* (1971) for the L-group chondrites. However, our values exhibit a smaller dispersion and appear to cluster at about 456°K. The smaller dispersion partly reflects the fact that we analyzed no type-3 ordinary chondrites. The lower median temperature is due in part to the different solar nebula pressure used in the calculations. The enstatite chondrites yield higher condensation temperatures of 486°K and 507°K for the trace-element-rich and trace-element-poor groups, respectively, if the high value for the find, Atlanta, is excluded.

Acknowledgements—The authors wish to express their appreciation to the many individuals who provided meteorite specimens for this work. These individuals are specifically acknowledged in SANTOLIQUIDO (1971) while a list of institutional sources and specimen numbers is given in the appendix section of this paper. We are grateful to the personnel of the U.S. Air Force NEC Nuclear Reactor, and the University of Missouri Research Reactor facilities for their assistance in providing irradiations. We also wish to thank Dr. JOHN WASSON and Dr. EDWARD ANDERS for their helpful comments and suggestions which aided in improving the manuscript of this paper. This work was supported in part by U.S. Atomic Energy Commission Contract AT-(40-1)-2670, by a NASA Traineeship to one author (P. M. S.), and by the University of Kentucky Research Foundation.

REFERENCES

ANDERS E. (1964) Origin, age, and composition of meteorites. *Space Sci. Rev.* **3,** 583–714.

ANDERS E. (1968) Chemical processes in the early solar system as inferred from meteorites. *Acc. Chem. Res.* **1,** 289–298.

CAMERON A. G. W. (1968) A new table of abundances of the elements in the solar system. In *Origin and Distribution of the Elements* (editor L. H. Ahrens) pp. 125–143, Pergamon Press, Oxford.

EHMANN W. D. and DURBIN D. R. (1968) Silicon abundances in some meteorites and standard rocks by activation analysis. *Geochim. Cosmochim. Acta* **32,** 461–464.

EHMANN W. D., GILLUM D. E., MORGAN J. W., NADKARNI R. A., REBAGAY T. V., SANTOLIQUIDO P. M. and SHOWALTER D. L. (1970) Chemical analyses of the Murchison and Lost City meteorites. *Meteoritics* **5,** 131–136.

EHMANN W. D. and HUIZENGA J. R. (1959) Bismuth, thallium and mercury in stone meteorites by activation analysis. *Geochim. Cosmochim. Acta* **17,** 125–135.

JAROSEWICH E. (1971) Chemical analysis of the Murchison meteorite. *Meteoritics* **6,** 49–52.

KEAYS R. R., GANAPATHY R. and ANDERS E. (1971) Chemical fractionations in meteorites—IV. Abundances of fourteen trace elements in L-chondrites, implications for cosmothermometry. *Geochim. Cosmochim. Acta* **35,** 337–363.

LARIMER J. W. (1967) Chemical fractionations in meteorites—I. Condensation of the elements. *Geochim. Cosmochim. Acta* **31,** 1215–1238.

LARIMER J. W. and ANDERS E. (1967) Chemical fractionations in meteorites—II. Abundance patterns and their interpretation. *Geochim. Cosmochim. Acta* **31,** 1239–1270.

LAUL J. C., CASE D. R., SCHMIDT-BLEEK F. and LIPSCHUTZ M. E. (1970) Bismuth contents of chondrites. *Geochim. Cosmochim. Acta* **34**, 89–103.

MASON B. (1966) The enstatite chondrites. *Geochim. Cosmochim. Acta* **30**, 23–29.

REED G. W. (1963) Heavy elements in the Pantar meteorite. *J. Geophys. Res.* **68**, 3531–3535.

REED G. W., KIGOSHI K. and TURKEVICH A. (1958) Activation analysis for heavy elements in stone meteorites. Second United Nations International Conference on the Peaceful Uses of Atomic Energy, Paper 953.

REED G. W., KIGOSHI K. and TURKEVICH A. (1960) Determinations of concentrations of heavy elements in meteorites by activation analysis. *Geochim. Cosmochim. Acta* **20**, 122–140.

SANTOLIQUIDO, P. M. (1971) The determination of bismuth in meteorites and rocks by neutron activation analysis. Ph.D. dissertation, University of Kentucky, Lexington, Kentucky.

TANNER J. T. and EHMANN W. D. (1967) The abundance of antimony in meteorites, tektites and rocks by neutron activation analysis. *Geochim. Cosmochim. Acta* **31**, 2007–2026.

VAN SCHMUS W. R. and WOOD J. A. (1967) A chemical petrological classification for the chondritic meteorites. *Geochim. Cosmochim. Acta* **31**, 747–766.

VOGT J. R. and EHMANN W. D. (1965) Silicon abundances in stony meteorites by fast neutron activation analysis. *Geochim. Cosmochim. Acta* **29**, 373–383.

ZÄHRINGER J. and GENTNER W. (1960) Uredelgase in einigen Steinmeteoriten. *Z. Naturforsch.* **15a**, 600–602.

APPENDIX

Specific sources of all the meteorite specimens used in this work are given by SANTOLIQUIDO (1971). Specimens from major museum collections which bear a specific identification number are as follows: *Arizona State University*—Bath (713), Coolidge (397.2x), Weston (238.2); *British Museum of Natural History*—Appley Bridge (1920,41), Crumlin (86115), Dhurmsala (33763), Kapoeta (1946,141), Karoonda (1931,311), Ochansk (1922,157); *Geological Survey of Canada*—Blithfield (0219102); *U.S. National Museum*—Chainpur (1251), Lost City (4848), St. Marks (486); *University of Melbourne, Department of Geology*—Murchison (3565); *Western Australian Museum*—North Haig (12809d).

Part III

ROCK-FORMING PROCESSES

Editors' Comments on Papers 9 Through 12

9 **BROOKS and AHRENS**
Some Observations on the Distribution of Thallium, Cadmium and Bismuth in Silicate Rocks and the Significance of Covalency on Their Degree of Association with Other Elements

10 **MAROWSKY and WEDEPOHL**
General Trends in the Behavior of Cd, Hg, Tl, and Bi in Some Major Rock Forming Processes

11 **GURNEY and AHRENS**
The Bismuth Contents of Some Rare-Earth Minerals, Notably Gadolinite

12 **GREENLAND, GOTTFRIED, and CAMPBELL**
Aspects of the Magmatic Geochemistry of Bismuth

Our understanding of bismuth concentration in rocks is only in its infancy at the time of the editing of this book. Historically, Goldschmidt (1954) and Rankama and Sahama (1950) first suggested the basic theories of bismuth behavior in rock-forming processes. Thus they were able to pave the way for future studies in much the same way Anders and Larimer did for Bi in meteorites. According to the hypotheses of Goldschmidt and Rankama and Sahama, bismuth in rock-forming processes should: (1) substitute for Ca in minerals, (2) be found in minerals of rare earth elements, (3) exhibit chalcophilic and lithophilic tendencies, and (4) be concentrated in the late stages of magma development. The data from which they formulated these properties of Bi were scant, and in some cases their conclusions were based upon "geochemical intuition." Earlier workers (Noddack and Noddack 1934; Preuss 1941) had insufficient data to make any kind of generalizations about the geochemistry of bismuth. Little work has been done on bismuth since the summaries of Goldschmidt, which makes the four papers chosen for this section important. These papers are essentially tests of the previous hypothesis. In addition to the papers of this section, the reader is directed to the 1969 paper by Mintser (Paper 16), which we believe to be one of the most thorough reviews on the geochemistry of Bi in the Russian literature. Mintser's conclusions on the geochemistry of Bi in rock-forming processes are quite compatible with those of the papers in this section.

The problems of studying the bismuth concentrations in rocks had much the same history as the studies of bismuth concentrations in meteorites—that is, lack of data and lack of a good method of analysis. It wasn't until the development of a nonspectrochemical procedure by Brooks et al. (1960) that more data became readily available. In Paper 9, using the new technique and with more abundant data, Brooks and Ahrens examine some of the original hypotheses concerning bismuth in rocks. These hypotheses are further examined by G. Marowsky and K. H. Wedepohl (Paper 10) using data obtained by neutron activation techniques. Apparently, the method of analysis can make a difference as to the absolute concentration of Bi measured in a sample. Comparing Papers 9 and 10, we can see that for similar rock types, higher bismuth concentrations are found using spectrochemical procedures rather than neutron activation techniques. However, trends in concentrations can be of more significance than absolute concentrations in determining the geochemistry of an element. Thus, using the relative concentration values between rocks, Bi was found to be enriched in the crust by the authors of both Papers 9 and 10, as predicted by Goldschmidt, but no evidence was found of Bi substitution for Ca.

Paper 10 is important because it represents the first time sufficient data became available to allow making comparisons of Bi contents among the three major rock types. In general, the average sedimentary rock type was found to contain less Bi than the average igneous rock type, which might imply another source for Bi in sediments other than just rock weathering. We believe more data is necessary to confirm this observation. As might be predicted from the study of the Bi contents of meteorites, there was found to be a depletion of Bi with increasing grade of metamorphism.

In Paper 11 by J. J. Gurney and L. H. Ahrens, the hypothesis of bismuth substitutions for rare earth elements (RE) is examined by studying the Bi contents of the mineral gadolinite. The authors conclude that Bi is a trace constituent of the mineral; therefore, the Bi-RE substitutions is probably real. The absolute concentration of Bi in the mineral was thought to be dependent on the geographic area from which the sample was taken. Although we think that this finding is a significant contribution to our understanding of the geochemistry of Bi, more work still needs to be done on Bi-

RE relationships. Since the data presented in Paper 11 was obtained using optical spectroscopy, it would be particularly interesting to see whether the interpretation would change using results obtained by neutron activation techniques.

Paper 12 by L. P. Greenland, D. Gottfried, and E. Y. Campbell is, as far as we know, the only work dealing specifically with Bi and magmatic processes and is a major contribution to the understanding of the geochemistry of bismuth. Studying three differentiated batholiths, Greenland and his colleagues have been able to put together an excellent description of the behavior of Bi during the magmatic differentiation process. Similar to the prediction by Goldschmidt, they demonstrate that Bi is enriched in residual magmas. The relatively high Bi levels found in some apatites (Clark 1965) have frequently been suggested as evidence for the Bi-Ca association. In Paper 11 we are shown that this association in determining the behavior of Bi in rock-forming processes is minor. A similar conclusion was reached by Marowsky and Wedepohl in Paper 10. Bismuth, accordingly, was not found concentrated in any major rock-forming mineral but was distributed evenly among the minerals. Greenland et al. speculated that Bi exists in inclusions of the minerals in the form of the sulfide. Gurney and Anders also suggest this to be the major means of Bi occurrence in apatites.

REFERENCES

Brooks, R. R., L. H. Ahrens, and S. R. Taylor. 1960. The determination of trace elements in silicate rocks by a combined spectrochemical-anion exchange technique. *Geochim. et Cosmochim. Acta* **18**:162.

Clark, A. H. 1965. The mineralogy and chemistry of the Ylojarvi Cu–W Deposit Southwest Finland: Bismuth bearing apatite. *Bull. Comm. Geol. Finlande* **218**:195–99.

Goldschmidt, V. M. 1954. *Geochemistry.* Oxford University Press.

Noddack, I., and W. Noddack. 1934. The geochemical distribution coefficients of the elements. *Svensk. Kem. Tid.* **46**:173.

Preuss, E. 1941. The determination of volatile elements by fractional distillation. *Z. Angew. Min.* **3**:8.

Rankama, K., and Th. G. Sahama. 1950. *Geochemistry.* Chicago University Press, Chicago.

9

Reprinted from *Geochim. et Cosmochim. Acta* **23**:100–115 (1961)

Some observations on the distribution of thallium, cadmium and bismuth in silicate rocks and the significance of covalency on their degree of association with other elements

R. R. Brooks and L. H. Ahrens
Department of Chemistry, University of Cape Town

Introduction

Brooks *et al.* (1960) have described an anion exchange–spectrochemical procedure which they used to estimate Zn, Cd, Tl, Bi and Sn in nineteen rocks of various types. For some of these elements, geochemical data are still quite scant; Bi is a notable example. Data on Cd are also comparatively few. There is also a distinct need for more data on Tl in basic rocks. Accordingly, a modification of this anion exchange–spectrochemical procedure was used to estimate Cd, Tl and Bi in a further forty-four rocks. The selection includes a high proportion of basic rocks and five deep-seated rocks (eclogites). The results are given in Table 1(a); description of the rock samples is given in Table 1(b).

Cd, Tl and Bi form bonds with oxygen that have varying degrees of covalent character and it is in part the purpose of the present study to investigate the significance of covalency on the degree of association of pairs of elements whose cations carry the same charge and have similar radii.

Discussion

In the discussions which follow, each element will be considered in turn.

Thallium

(i) *Abundance.*. Estimates of the abundance of thallium in the crust and in certain specific rock types are given in Table 2.

The crustal value recommended in the present paper is based on the data of Ishimori and Takashima, those of the large number of determinations recently carried out by Russian workers (see Table 2) and the new data given here (Table 1a). This value (0·7 p.p.m.) is lower than some recent (post—1945) crustal abundance

Table 1(a). The determination of Bi, Cd and Tl in rocks

No.	Type	Concentrations (p.p.m.)								
		Tl			Bi			Cd		
		1	2	Av	1	2	Av	1	2	Av
1	Soevite	—	—	—	0·04	0·03	0·04	0·19	0·11	0·15
2	Syenite	0·08	0·08	0·08	0·02	0·03	0·03	—	—	—
3	Syenite	0·05	0·06	0·06	0·02	0·02	0·02	0·04	0·04	0·04
4	Syenite	0·20	0·23	0·22	0·03	0·03	0·03	—	—	—
5	Foyaite	0·11	0·11	0·11	0·04	0·04	0·04	0·04	0·04	0·04
6	Foyaite	0·06	0·05	0·06	0·03	0·05	0·04	—	—	—
7	Rhyolite	0·05	0·04	0·05	0·03	0·03	0·03	—	—	—
8	Rhyolite	0·34	—	0·34	0·02	—	0·02	0·05	—	0·05
9	Acid dyke	0·12	0·11	0·11	0·04	0·04	0·04	—	—	—
10	Rhyolite	0·98	0·80	0·89	0·20	0·23	0·22	—	—	—
11	Rhyolite	0·20	—	0·20	0·21	—	0·21	0·05	—	0·05
12	Q. Feldspar	0·07	—	0·07	0·04	—	0·04	0·03	—	0·03
13	Dolerite	0·04	—	0·04	0·08	—	0·08	0·12	—	0·12
14	Dolerite	0·08	—	0·08	0·15	—	0·15	0·10	—	0·10
15	Dolerite	0·10	—	0·10	0·80	—	0·80	—	—	—
16	Dolerite	0·06	—	0·06	0·15	—	0·15	0·10	—	0·10
17	Dolerite	0·04	0·02	0·03	—	—	—	0·04	0·04	0·04
18	Dolerite	0·15	0·14	0·15	0·28	0·28	0·28	0·22	0·27	0·25
19	Dolerite	0·04	0·04	0·04	—	—	—	0·04	0·04	0·04
20	Dolerite	0·12	0·08	0·10	0·53	0·42	0·48	0·07	0·07	0·07
21	Dolerite	0·08	0·06	0·07	0·18	0·16	0·17	—	—	—
22	Dolerite	0·03	—	0·03	0·76	—	0·76	0·26	—	0·26
23	Dolerite	—	—	—	0·16	—	0·16	0·01	—	0·01
24	Dolerite	0·02	0·02	0·02	0·01	0·02	0·02	—	—	—
25	Dolerite	0·03	—	0·03	0·13	—	0·13	0·08	—	0·08
26	Dolerite	0·04	0·03	0·04	0·03	0·04	0·04	0·02	—	0·02
27	Dolerite	0·03	0·03	0·03	0·03	0·04	0·04	—	—	—
28	Dolerite	0·03	—	0·03	0·01	—	0·01	—	—	—
29	Dolerite	0·03	0·02	0·03	0·03	0·03	0·03	0·04	0·03	0·03
30	Dolerite	0·40	0·50	0·45	—	—	—	0·02	0·02	0·02
31	Dolerite	0·80	1·40	1·10	0·27	0·19	0·23	0·21	0·27	0·24
32	Eclogite	0·03	0·03	0·03	0·02	0·02	0·02	—	—	—
33	Eclogite	0·54	0·58	0·56	—	—	—	0·02	0·04	0·03
34	Eclogite	0·06	0·06	0·06	0·10	0·10	0·10	0·04	0·04	0·04
35	Eclogite	0·16	0·12	0·14	0·03	0·02	0·03	—	—	—
36	Eclogite	0·07	0·07	0·07	—	—	—	—	—	—
37	Basalt	0·07	—	0·07	0·02	0·02	0·02	0·03	—	0·03
38	Carbonatite	0·24	0·13	0·19	0·08	0·05	0·06	0·17	0·08	0·13
39	Carbonatite	0·03	0·02	0·03	0·08	0·04	0·06	0·46	0·26	0·36
40	Glauconite	0·03	—	0·03	0·13	—	0·13	0·03	—	0·03
41	Crocidolite	0·01	—	0·01	0·01	—	0·01	—	—	—
42	Fe-stone	0·01	—	0·01	—	—	—	—	—	—
43	Biotite	5·11	—	5·11	0·21	—	0·21	—	—	—
44	Shale	0·18	—	0·18	0·10	—	0·10	—	—	—

Duplicates were taken only after the ion-exchange enrichment step and represent one rock sample only. The reproducibility of these duplicates therefore refers to the spectrochemical part of the procedure.

Table 1(b). Description of rocks investigated

No.	Description
1	Soevite — All from Ondurakorume alkali complex. Farm Etaneno. No. 44 Otjiwarongo. S.W.A.
2	Nepheline syenite (coarse) — All from Ondurakorume alkali complex. Farm Etaneno. No. 44 Otjiwarongo. S.W.A.
3	Nepheline syenite (fine) — All from Ondurakorume alkali complex. Farm Etaneno. No. 44 Otjiwarongo. S.W.A.
4	Monomineralic syenite. K.Felspar rock. — All from Ondurakorume alkali complex. Farm Etaneno. No. 44 Otjiwarongo. S.W.A.
5	Foyaite (fine grain) — Members of small intrusive complex in NW margin of Paresis* complex.
6	Foyaite (coarse grain) — Members of small intrusive complex in NW margin of Paresis* complex.
7	Intrusive plug into riebeckite rhyolite of intermediate Series, NW Paresis* complex.
8	Na-rhyolite, lower member of central extrusive complex. Paresis*.
9	Acid dyke rock, intruding W. margin of intermediate volcanic series. W. part of Paresis*.
10	Riebeckite rhyolite flow. W. Paresis*.
11	Rhyolitic dyke rock, intrusive into Damara basement. E. Paresis*.
12	Quartz felspar. Flow underlying agglomerate. S. Paresis*.
13	240 ft — All from upper dolerite sill, main shaft, Jagersfontein diamond mine, O.F.S. S. Africa. Total thickness of sill 425 ft. Upper contact 150 ft below surface Base of sill 575 ft below surface.
14	260 ft
15	280 ft
16	300 ft
17	340 ft
18	360 ft
19	380 ft
20	420 ft
21	460 ft
22	480 ft
23	500 ft
24	500 ft
25	520 ft
26	660 ft — All from 2nd dolerite sill. Total thickness 390 ft. Upper contact 575 ft, lower contact 970 ft.
27	840 ft
28	860 ft
29	920 ft
30	960 ft
31	Chilled upper margin of 1st dolerite sill.
32	Eclogite nodule. Bultfontein mine. Kimberley. S. Africa.
33	Eclogite nodule. Roberts Victor mine. O.F.S. S. Africa.
34	Eclogite nodule. Du Toits Pan mine. Kimberley. S. Africa.
35	Eclogite nodule. No locality.
36	Eclogite nodule. No locality.
37	Amygdaloidal basalt. Intrusive into Damara basement. Paresis*.
38	Micaceous carbonatite. — Same locality as nos. 1–4.
39	Dolomitic carbonatite. — Same locality as nos. 1–4.
40	Glauconite. Lower portion of Boornaat. Glen Allen Mine. Prieska.
41	Crocidolite asbestos. Buisvlei. Orangeview, Prieska. S. Africa.
42	Banded ironstone, Tseloan, Hemingvlei, Vryburg. S. Africa.
43	Biotite. Clifton, Cape Town, S. Africa.
44	Shale. Malmesbury shale, Mouille point, Cape Town. S. Africa.

* The Paresis igneous complex lies between Outjo and Otjiwarongo, South West Africa.

Table 2. Estimated thallium concentrations (in p.p.m.) in the crust, chondrites and in certain igneous rocks

	Crust	Granites + related rocks	Basalt diabase or gabbro	Sedi-mentary rocks	Chondrites	No. of samplex
CLARKE and WASHINGTON (1924)	0·0009	—	—	—	—	EST
NODDACK and NODDACK (1934)	0·1	—	—	—	0·15	?
OTTEMAN (1940)	—	1·5	—	—	—	4
PREUSS (1941)	2·1*	3·0	0·30	2·0	—	4 COMP
RANKAMA (1949)	0·6	—	—	—	—	EST
GOLDSCHMIDT (1954)	0·3	—	—	—	—	EST
SHAW (1957)	1·3	2·5	0·12	0·72		28
AHRENS (1947)	3·0	—	—	—	—	EST
AHRENS (as above, recalculated)	0·9‡	—	—	—	—	EST
CANNEY (1952)	—	—	—	0·36	—	323
ISHIMORI (1955)	0·7*	0·90	0·30	—	—	24
VINOGRADOV (1956)	1·7	—	—	—	—	—
DIOMIN and KHITAROV (1958)	—	0·50	—	—	—	21
VOSRESENSKAYA (1959)	—	0·80	—	—	—	43
KOGARKO (1959)	—	0·50	—	—	—	27
EHMANN and HUIZENGA (1959)	—	—	—	—	0·004	5
MORRIS and KILLICK (1960)	—	1·30†	0·09	—	—	6
This paper	0·7*	0·73	0·11	0·34	—	52

* Estimated on 2:1 acid/basic rock ratio.
† One sample only(G–1).
‡ Based on a Rb crustal abundance of 120 p.p.m. The comparatively high value (3·0 p.p.m.) reported by AHRENS (1947) was based on the assumption that his observations on the constancy of the Rb/Tl ratio, estimated at 90, in 164 specimens of potassium minerals from pegmatites held also for common igneous rocks for which at that time very few Tl but considerably more Rb data were available; assuming a value of 280 p.p.m. for Rb (see references below) in the crust, the Tl abundance (calculated at Rb/Tl = 90) = 3 p.p.m. The crustal abundance for Rb has since been considerably lowered to 120 p.p.m. which if used in place of the earlier crustal value reduces the estimated thallium abundance considerably. As, however, considerable data have become available on the abundance of Tl in igneous rocks, calculations as above serve little purpose now.

estimates and is close to the value recommended by RANKAMA (1949). Recent work by MORRIS and KILLICK (1960) on the determination of Tl in five silicates and in G–1 and W–1 using a neutron activation method show results very similar to those reported by the authors. In particular there is a gratifying agreement between the respective values for the abundances in G–1 and W–1.

This comparison is shown in Table 3 below:

Table 3. The thallium content of G–1 and W–1

	G–1	W–1
BROOKS *et al.* (1960)	1·3	0·1
MORRIS and KILLICK (1960)	1·3	0·17

In general, the estimated abundances given in this paper for granites are very similar to those of Goldschmidt, Ishimori and of most of the several Russian workers, but somewhat less than those of Preuss and of Shaw.

(ii) *The rubidium–thallium association.* The Rb–Tl coherence and the variation of the Rb/Tl ratio has been discussed by Ahrens (1945, 1948); Goldschmidt (1954); Shaw (1952); Tauson and Stavron (1957); Diomin and Khitarov (1958); Zlobin (1958); Borovik-romanova and Sosedko (1960); Taylor and Heier

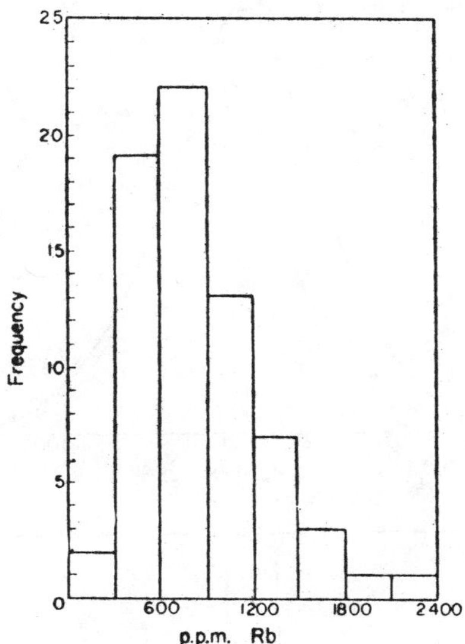

Fig. 1. Frequency distribution diagram for rubidium in 68 biotites from igneous rocks.

(1960), and Zlobin and Lebedev (1960). The principal reasons for this close association are equality of the radii of Rb^+ and Tl^+ and the fact that although the Tl—O bond is evidently more covalent than that of Rb—O both bonds are essentially ionic.

In a discussion of the magnitude of the ratio Rb/Tl in granites, gabbro and basalt and pegmatites, Shaw (1952) concluded that the ratio was at a maximum in granites as indicated in Table 2. Shaw's data were pre-1952. Explanations were offered as to why the Rb/Tl ratio should rise when passing from gabbro and basalt to granite and then fall again in pegmatites. Subsequent data do not support the granite–gabbro + basalt difference.

Table 4 gives some data on Rb, Tl and Rb/Tl ratios in chondrites, granite, gabbro + basalt, alkali rocks, biotite from igneous rocks, pegmatites and lepidolites. Biotites are of interest because though much less abundant than feldspar, they carry the highest Rb and Tl concentrations of igneous rock minerals; maximum concentrations (all silicate minerals including those from pegmatites) for both Rb and Tl are reached in lepidolite and hence this mineral has been included in Table 4.

Some of the data given in Table 4 are shown graphically (Figs. 1 and 2). Fig. 1 is a frequency distribution diagram of Rb in biotites from igneous rocks. The distribution shows positive skewness with a modal (most frequent) Rb concentration of 650 p.p.m. which may be compared with the average of 900 p.p.m. given in Table 1. Fig. 2 compares the Rb/Tl data of AHRENS (1948), based mainly on 167 minerals from various pegmatites, with those of TAYLOR and HEIER (1960) which refer to alkali feldspars from the Pre-Cambrian basement rocks of southern Norway; the match is quite close.

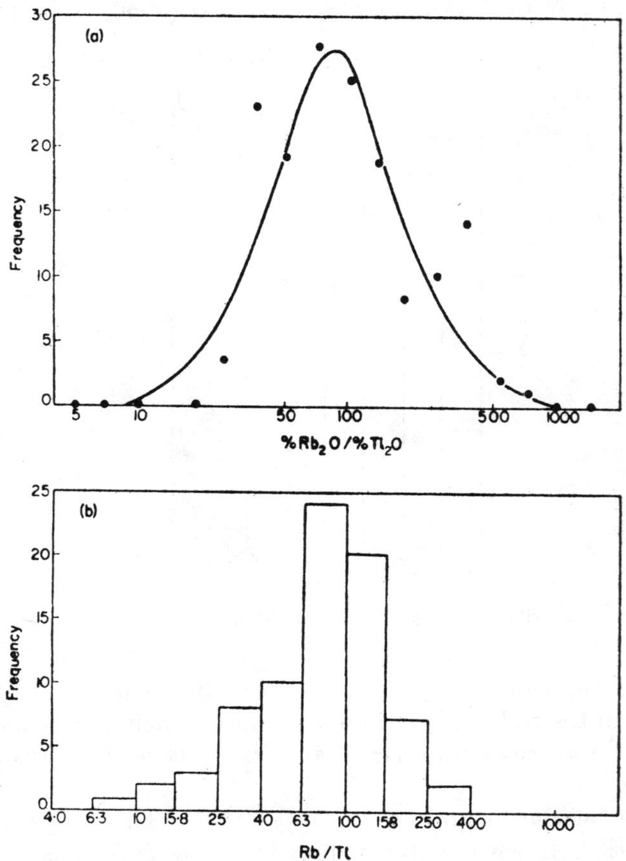

Fig. 2. Frequency distribution diagrams for the rubidium/thallium ratio, (a) in 167 minerals from various pegmatites (AHRENS, 1948). (b) for alkali feldspars from Pre-Cambrian basement rocks of southern Norway (TAYLOR and HEIER, 1960). Correspondence between the two sets of data is quite close.

Lepidolites are often regarded as being formed at a late stage in the sequence of mineralization in a Li-bearing pegmatite and if Rb^+ and Tl^+ enter the structures of these minerals with equal facility—that is to say Rb/Tl (lepidolite) = Rb/Tl (pegmatite emanation or fluid from which lepidolite crystallized) it is of interest to note that the ratio Rb/Tl remains roughly at the same general magnitude (100–300) in the sequence, basalt + gabbro → granite (and also alkali rocks) → pegmatites (as a whole) → pegmatites (very late stage). The available data indicate a slight but distinct fall in the ratio in pegmatites and lepidolites. This could be taken to support TAYLOR and HEIER's (1960) suggestion that the Rb/Tl ratio may fall in a differentiated sequence due to the fact that the Tl—O bond is weaker than that of Rb—O because of its greater covalency. Their observations were made on some Pre-Cambrian feldspars from Norwegian pegmatites; they report a drop in the Rb/Tl ratio from 100 to 20.

The geochemical coherence of the pairs Rb–Tl and K–Rb make an interesting comparison. In the well known K–Rb pair, K^+ and Rb^+ both form dominantly ionic bonds with O^{2-} and radii differ slightly ($K^+ = 1{\cdot}33$ Å and Rb^+ 1·45 Å) whereas in the Rb–Tl pair, radii are identical but the bonds which Tl^+ forms with

Table 4. Rb, Tl and Rb/Tl ratios in certain meteorites, rocks and minerals

	Rb (p.p.m.)	Tl (p.p.m.)	Rb/Tl	
			This paper	Shaw (1952)
Chondrites	4*	0·0004**	10,000	
Basalts + gabbros	30†	0·11††	270	140
Granites	170†	0·75††	200	340
Biotites from granite	900‡	4‡‡	230‡‡	
Alkali rocks	—	—	250§§	
Pegmatites	—	—	115***	90††
Lepidolites	16,500§	120§	140	

* Based on data of Herzog and Pinson (1955) and K/Rb ratio considerations.
† Based mainly on data of Horstman (1956).
‡ Based mainly on sixty-eight mass spectrometer–isotope dilution estimations of Rb in igneous rock-biotite used for geological age investigations (various workers). See Fig. 1.
§§ From Zlobin and Lebedev (1960).
§ From Ahrens (1948) and Taylor (thirteen unpublished analyses). Only these data have been used because both Rb and Tl have been determined in the same specimen.
** Ehmann and Huizenga (1959) and Reed *et al.* (1960).
†† From this paper, see Table 1(A).
‡‡ Shaw (1952), Diomin and Khitarov (1958), Vosresenskaya (1959), Zlobin (1958), and this paper.
*** This value is based on the average Rb/Tl of Ahrens (1948), one of ninety-two given by Taylor and Heier (1960) and one of 103 given by Borovik-Romanova and Sosedko (1960).
††† This is the modal (most frequent) concentration reported by Ahrens (1948)—see Fig. 2 above. The interval in Fig. 2 is given by a power factor (a log scale has been used) and if a linear scale is used the distribution is strongly positively skewed and of a lognormal type; hence the average (130) is distinctly greater than the modal value (90).

O^{2-} though mainly ionic (depending on the ligand) will probably have some covalency (see above). Covalency arises from the fact that the polarizing power of $Tl^+ > Rb^+$, as indicated by their first ionization potentials 6·1 and 4·2 V, respectively (see Ahrens (1953) in this respect).

Although the Rb–Tl association in igneous rocks may be disturbed by local geological processes (presence of sulphide minerals and halogen transport of Tl, both due to covalency) superimposed on general operative processes (see Editor's Note p. 573 of *Geokhimiya*, No. 5, 1958) and perhaps also by some covalency in the Tl—O bond, the data given in Table 4 indicate that aside from the possible greater dispersion of the Rb/Tl ratio in specific rock types, the Rb–Tl coherence is closer than that of K–Rb in certain respects. Thus, whereas the value of the Rb/Tl ratio has about the same magnitude (100–300) in each of the terrestrial materials including lepidolites referred to in Table 4, K/Rb may drop significantly from its igneous rock ratio to lower values in some biotites and late phase granites and various small volume residuals reaching a minimum in lepidolite. On the other hand there is the extraordinary drop (x 1/40) between chondrites and the terrestrial Rb/Tl ratio whereas K/Rb in chondrites is apparently the same as in the crust as a whole (AHRENS *et al.*, 1952). The abundances of Tl and of Bi and Hg in common chondrites may in a sense, however, be anomalous (REED *et al.*, 1960); see discussion below.

Although the presence of covalency in the Tl—O bond has not in general strongly disturbed the Rb–Tl association, a high degree of covalency is probably the reason for little or no coherence in other pairs of univalent elements which form cations of about the same size. For example,

Na–Cu
K,Rb–Ag
K,Rb–Au

Whereas the alkali metal–O bonds are essentially ionic, those of Cu, Ag or Au and oxygen are probably essentially covalent. It will be recalled that the order of

Table 5. Estimated Cd concentrations (p.p.m.) in crust and in silicate rocks

	No. rocks analysed	Crustal abundance*	Acid rocks	Basic rocks	Sediments
PREUSS (1941)	—	0·14	0·2	0·02	0·3
SANDELL and GOLDICH (1943)	8	0·14	0·12	0·19	—
VINCENT and BILEFIELD (1960)	15†	0·31	0·36	0·21	—
This paper	35	0·08	0·06	0·11	n.d.

* Estimated on a 2:1 acid/basic rock ratio.
† Skaergaard rocks.

polarizing power of large singly charged cations, as indicated by the first ionization potential is

$$\begin{array}{ccccccccccccc} \text{Na} & < & \text{Rb} & < & \text{K} & \ll & \text{Tl} & \ll & \text{Ag} & < & \text{Cu} & \ll & \text{Au} \\ (5{\cdot}1) & & (4{\cdot}12) & & (4{\cdot}34) & & (6{\cdot}1) & & (7{\cdot}5) & & (7{\cdot}7) & & (9{\cdot}7)\ \text{V} \end{array}$$

In this respect, see also the discussion by RINGWOOD (1955) who used electronegativity as a guide to the nature of the bond.

The significance of covalency on the degree of association of certain pairs of divalent elements is discussed below.

Cadmium

(i) *Abundance.* Table 5 shows various abundance data for Cd.

Early (pre-1940) abundance estimates of Cd in rocks are those of CLARKE and WASHINGTON (1924) and GOLDSCHMIDT (1937) who give, respectively, 0·1–1 p.p.m. (magmatic rocks) and 0·5 p.p.m. (upper lithosphere).

PREUSS (1941) reported 0·2 p.p.m in acid rocks and 0·02 p.p.m. in basic rocks.

This large difference ($\times$ 10) between acid and basic rocks was not observed by Sandell and Goldich (1943) who working with a colorimetric procedure reported a fairly uniform Cd concentration in individual basic and acid igneous rocks; average for gabbros, dolerite and greenstone = 0·19 p.p.m. and for granite, 0·12 p.p.m. On the basis of these values the crustal abundance would come out at about 0·14 p.p.m.

The values given by Brooks *et al.* (1960) and in this paper tend to be slightly lower than the above, namely 0·06 p.p.m. for acid rocks and 0·11 p.p.m. for basic rocks, with one sample (Kloof Nek dolerite) containing the maximum of 0·45 p.p.m. (Brooks *et al.*, 1960). On the basis of these figures the crustal abundance would come to about 0·08 p.p.m.

When comparing the sets of data the possibility of analytical error—notably systematic error—should be borne in mind. Thus Vincent and Bilefield (1960) reported 0·33 p.p.m. Cd in standard diabase W–1 whereas the value (0·08 p.p.m.) of Brooks *et al.* (1960) is considerably less. It is unfortunately not possible at present to resolve this problem.

A few Cd data on rocks other than those above are considered in the discussion which follows on aspects of the geochemistry of Cd.

(ii) *Possible association of* Cd *with* Ca. Before examining the geochemistry of Cd (Cd^{2+} radius = 0·97 Å) and its possible association with Ca (Ca^{2+} radius = 1·01 Å), it may be instructive briefly to review the principal features of the geochemistry of Zn and its association with Mg. Neumann (1949) has emphasized many interesting features about the geochemistry of Zn whereas Wedepohl (1953) has reviewed its geochemistry and provided much fresh data. It is quite clear that although Zn is to some extent enriched in certain mafic minerals there is no significant Zn–Mg coherence despite charge equality and a similarity of ionic radii (Zn^{2+} = 0·69 Å; Mg^{2+} = 0.65 Å)· This is probably due to two main facts; first, the Zn—O bond is distinctly more covalent in character than that of Mg—O, and second co-ordination of O about Zn in most silicates (other than some with F^- and water) is usually tetrahedral (non-ionic) and not octahedral which would have accorded with the ionic radius ratio rules (see Neumann, 1949). When bonded (i) to O in compounds or minerals containing radicals (those with CO_3^{2-}, NO_3^-, SO_4^{2-}, PO_4^{3-}, etc.) and (ii) in the presence of F^- (for example, ZnF_2) and water, co-ordination is usually ionic, in which case Zn^{2+}–Mg^{2+} replacement is likely to take place fairly freely; this may account for the enrichment of Zn in biotite and amphibole reported by Wedepohl (1953).

The greater covalency of the Zn—O bond compared with that of Mg—O is evidently due in the first place to poorer screening efficiency of the electronic structure of Zn. As a result the electropositive field, and hence polarizing power, associated with $Zn^{2+} > Mg^{2+}$ as indicated by the second ionization potentials of these elements (Table 6); data on Ca, Cd, Sr and Hg are given for the sake of completion.

As in the Mg–Zn pair, so in the Ca–Cd and Sr–Hg pairs are the ionization potentials of the 18-electron elements distinctly greater than the 8-electron elements. (That of Hg is particularly high and in fact greater than those of Zn and Cd despite the fact that Hg is the heaviest element in the Group Zn–Cd–Hg. This "anomaly" is evidently associated with the entry of $4f + 5d$ electrons (Ahrens, 1953). The significance of $4f$ and $5d$ entry on the chemistry and geochemistry of elements of the Au, Hg, Tl . . . . period has been emphasized (Ahrens, 1953). In which case, the bonds between the 18-electron elements (Zn, Cd*, Hg) and oxygen are evidently much more covalent than those between the alkaline earth metals and oxygen. Moreover, the ionization potentials of the Zn–Mg and Cd–Ca

Table 6. Ionic radii, second ionization potentials and ionization potential ratios for Group II elements

	Cation radius (Å)	I_2 (V)	$\frac{I}{I}$
8-electron Mg	0·65	15	1·19
18-electron Zn	0·69	17·9	
8-electron Ca	1·01	11·9	1·42
18-electron Cd	0·97	16·9	
8-electron Sr	1·15	11·0	1·71
18-electron Hg	1·12	18·8	

pairs are such that whether considered as simple differences or as ratios, they indicate that the Cd—O, Ca—O bond difference is likely to be greater than that between Zn—O and Mg—O. Electronegativity data support this conclusion; in this respect, see for example, Ringwood (1955). Accordingly, one may predict on these grounds that any Cd–Ca association is likely to be less than that between Zn and Mg.

On the other hand, whereas co-ordination of O about Zn tends to be four compared with six about Mg, both Cd and Ca (as one would expect) are octahedrally co-ordinated in their simple oxides each of which has the ionic NaCl structure. This could be taken to indicate that in silicates Cd^{2+} might be accommodated more easily in octahedral structure sites than Zn^{2+} and that a Cd^{2+}–Ca^{2+} substitution is more likely therefore than a Zn^{2+}–Mg^{2+} substitutuon; in which case Ca and Cd would be more closely associated than Zn and Mg. Let us now consider the available data.

Sandell and Goldich (1943) suggested that the higher Cd concentration they found in basalt compared with granite was presumably due to substitution of Cd^{2+} for Ca^{2+}. Similarly Goldschmidt (1954) has mentioned the detection of Cd in Norwegian bytownite rock and calcic augite from basalt as due to such substitution. On the other hand no sign of a Ca–Cd association was observed by Vincent and Bilefield (1960) in their work on the Cd content of Skaergaard rocks. They found

* Some data other than those of the free atom or ion which indicate that the Cd–ligand bond is likely to be distinctly more covalent than the Ca–ligand bond are:
(i) Refractive indices, for example those of the fluorides (Ahrens, 1958).
(ii) Melting points of CdO and CaO and of their Cd, Ca compounds (m.p. Ca compounds > Cd compounds).
(iii) Lattice energy calculations.
(iv) Colour of certain Ca and Cd compounds; see for example Morris and Ahrens (1956).
Moreover, the far greater tendency of Cd to form stable complexes with various ligands indicates that the binding power of Cd > Ca and presumably therefore that the polarizing power of $Cd^{2+} > Ca^{2+}$, which is in support of the ionization potential data.

that the maximum Cd concentration was in fact present in transgressive granophyres, acid rocks of late formation and low in calcium. In this respect it may be noted that Preuss (1941) reported a higher concentration of Cd in acid rocks (0·2 p.p.m.) than in basic rocks (0·02 p.p.m.).

The data given by Brooks *et al.* (1960) and in this paper indicate that the Cd content of basic rocks appears to be a little greater than that of acid rocks. But whether this should be ascribed to substitution of Cd^{2+} for Ca^{2+} is another matter. If anything, the data indicate that such a Ca–Cd association is unlikely for this main reason, i.e. the Cd concentration varies enormously ($\times$ 25) in specific rock types (say, basalt + dolerite) even in a single suite of rocks as at Jagersfontein where no sign of a trend is observable, and such large and seemingly sporadic changes of Cd concentration could hardly be due to its association with Ca, the concentration variation of which is far smaller. Cd is a distinctly chalcophile element and perhaps the presence of sulphides is one cause for some of the large Cd concentration changes: in this respect, it may be noted that there is sometimes a distinct correlation between a high Cd concentration and a high concentration of Bi and Tl, both chalcophile elements, in Jagersfontein dolerites (see Table 1a).

It will be concluded, somewhat tentatively therefore, that a Ca–Cd association in common igneous rocks either does not exist or is very feeble and that in any case such possible coherence is less than in the pair Zn–Mg. If the observations of Vincent and Bilefield (1960) on the Skaergaard rocks hold for other differentiated suites they would support the statement by Rankama and Sahama (1950) that: "Cd together with other elements remains largely in the residual melts and solutions throughout the main stage of crystallization and the pegmatitic stage" and that "owing to its weaker oxyphile nature, cadmium becomes nearly quantitatively enriched in hydrothermal rocks and minerals and its highest concentrations are met in formations belonging to the lowermost temperatures." The observations also provide support to a rule enunciated by Ringwood (1955) on the consequences of covalency in a metal-oxygen bond on the behaviour of a metal during processes of differentiation.

Contrary to the suggestion by Goldschmidt (1954) that apatite is very likely to be a host mineral for Cd in magmatic rocks—a suggestion which on crystallochemical grounds seems very plausible indeed because of the presence of PO_4^{3-} and some F^- (see above)—Vincent and Bilefield (1960) did not find a distinct enrichment of Cd in apatite from ferrogabbro. A value of 0·15 p.p.m. is reported and further determinations by the authors of this paper on three calcium-rich minerals corroborate this finding, and the Cd contents of wollastonite, Damara marble and actinolite were found to be comparatively low (0·05–0·1 p.p.m.). The highest Cd concentrations have been observed in two out of three carbonatites (0·8 p.p.m. Brooks *et al.*, 1960; and 0·36 p.p.m., this paper) analysed and it is presumably in these rocks only that the Cd enrichment is due to substitution of Ca^{2+} by Cd^{2+}.

Bismuth

(i) *Abundance.* Table 7 shows abundance data for Bi.

Very few data are available on the abundance of this element. Noddack and

NODDACK (1934) give a value of 0·07 p.p.m. based apparently on the frequency of occurrence of Bi minerals whereas GOLDSCHMIDT (1954) gives an estimate of > 0·2 p.p.m. which is based on the proportions of As, Sb and Bi in ferruginous bauxites and sedimentary iron ores. PREUSS (1941) gives a value of 2, nil and 0·6 p.p.m., respectively for acid, basic and sedimentary rocks. BROOKS *et al.* (1960) gave an estimate of 0·2 p.p.m. for the crust and this magnitude is supported by the data given in the present paper which together with the previous results indicate a crustal abundance of 0·17 p.p.m. for Bi.

Table 7. Estimated Bi concentrations (p.p.m.) in crust, chondrites and rocks

	No. rocks analysed	Acid rocks	Basic rocks	Sed. rocks	Chondrites	Crustal abundance
NODDACK and NODDACK (1934)	—	—	—	—	0·02	0·07
PREUSS (1941)	4 comp.	2·0	—	0·6	—	—
GOLDSCHMIDT (1954)	Est.	—	—	—	—	0·2
EHMANN and HUIZENGA (1959)	5	—	—	—	0·0022	—
This paper	50	0·18	0·15	0·11	—	0·17*

* Based on a 2/1 acid–basic rock ratio.

(ii) *Sundry geochemical features.* Several geochemical features, some based on observation and others on crystallochemical speculation have been discussed by GOLDSCHMIDT (1954). Reference is made to the possibility of a Bi^{3+}–Ca^{2+} substitution, particularly in early minerals of magnetic rocks. In this respect GOLDSCHMIDT (1954) several times mentions (p. 480) a Bi^{3+}–Ca^{2+} substitution in apatite, not only as a suggestion, but seemingly as an observation also; unfortunately neither original data nor original sources are given. GOLDSCHMIDT infers from "experience in mineralogy and economic geology" that Bi is concentrated in granitic magmas and in their residual solution and in this latter respect particular mention is made of the concentration of Bi in rare-earth minerals from pegmatites where Bi^{3+} evidently replaces R^{3+} rare earth cations. Several of the above observations will be taken up in the discussion which follows.

If the common igneous rock types are considered, little, if any, Bi–Ca substitution and association are apparent. Thus the Bi content of acid and alkaline rocks (both relatively low in Ca), see also data BROOKS *et al.* (1960), is similar in magnitude to the basic rocks (comparatively rich in Ca). Furthermore, no enrichment of Bi was observed in wollastonite, actinolite, apatite or marble. Traces of Bi (0·06, 0·06 and 0·06 p.p.m.) in three specimens of carbonatite (this paper and BROOKS *et al.*, 1960) may be due to Bi^{3+}–Ca^{2+} substitution, but the concentration level does not represent an enrichment.

We have not estimated Bi in rare-earth minerals but Bi enrichment in monazite for example is well known and presumably Bi^{3+}–RE^{3+} substitution is the cause.

The spread of Bi concentration in rocks of the same type (for example, Jagersfontein dolerite) is very considerable and ranges from < 0·01 p.p.m. to 0·80 p.p.m., a factor of > 80. This could be due in part at least to the chalcophilic properties of Bi and to a significant degree of covalency in the Bi^{III}—O bond. BROOKS *et al.* (1960) have drawn attention to a large difference in Bi content in some granites.

(iii) *Apparent very high degree of enrichment of* Bi *and* Tl *in the earth's crust.* It has been recognized for a considerable time that certain elements (Li, K, Rb and

Table 8. Some selected values of the crustal enrichment index (arranged in order of increasing enrichment)

Element	Abundances in p.p.m. unless indicated as %		Enrichment index
	crust	Earth as a whole	
Ni	35	1·34%	0·0026
Pd	0·004	1·54	0·0028
Au	0·002	0·2	0·01
Co	20	800	0·025
Cr	100	2500	0·04
Ge	1·1	9	0·12
Mg	2·1%	14·4%	0·15
Fe	5·0%	25·1%	0·20
Zn	40	180	0·22
Cd	0·15	0·5	0·30
Cu	55	90	0·61
Si	27·7%	17·8%	1·55
Ca	3·63%	1·39%	2·6
V	120	39	3·0
Sc	20	6	3·3
Na	2·83%	0·67%	4·2
Zr	156	33	5·0
Al	8·13%	1·32%	6·0
Nd + La	42	4·17	10
K	2·59%	0·085%	30
Rb	120	4	30
Sr	450	10	45
Bi	0·2	0·002	100
Ba	1000	6	167
Tl	*1·3	0·004	320
	†0·6	0·004	150

* SHAW's estimate.
† This paper.

Cs for example) are apparently strongly enriched in the earth's crust; others (Mg for example) are concentrated in the mantle and others (the siderophile elements Au, Pd and Ni for example) are presumably strongly enriched in the core. If it is assumed that the earth as a whole originated from cosmic substance with a chemical composition equivalent or similar to that of chondrites, it is possible to estimate the degree to which an element may have been enriched in the crust. Given for example that the abundance of potassium in the crust is 2·6 per cent compared with a value of 0·085 per cent in chondrites, crustal enrichment may be expressed in the form of the ratio,

$$\frac{\text{abundance in crust}}{\text{abundance in earth as a whole}}$$

which for potassium = 30. It is possible to express crustal enrichment in other forms, on a more absolute basis* for example, by taking into account the respective masses of earth ($5{\cdot}976 \times 10^{27}$ g) and of crust ($0{\cdot}024 \times 10^{27}$ g = 0·4 per cent of earth) but for the purpose here the simple ratio above will be used.

Some selected values of the abundance ratios (metals only) are given in Table 8.

It is clear from Table 8 that provided the analytical data are reasonably accurate, Bi and Tl are evidently enriched to a very high degree indeed in the crust. The tendency for Tl^+ to substitute for K^+ in potassium minerals is well established, and hence because of crustal enrichment of potassium, enrichment of Tl (and of Rb) would be expected. An enrichment of Tl which is apparently far greater than Rb is, however, surprising; in fact, the reverse might have been expected because Tl is to some extent chalcophilic and if sulphides are present in the mantle, they could contain Tl but no K + Rb. The crustal enrichment of Bi seems also surprisingly high in this respect. Cd on the other hand is actually depleted in the earth's crust and the index is similar in magnitude to that of Zn (Table 8).

If the index is 250 or greater an element must be concentrated entirely in the crust, as the mass ratio crust/whole earth is 250.

Investigations by REED *et al.* (1960) (see also EHMANN and HUIZENGA, 1959) have recently indicated that unlike most elements, the concentrations of the heavy metals Hg, Tl, Pb and Bi are much higher in carbonaceous chondrites than in the common chondrites; for Tl the concentration ratio (carbonaceous chondrite)/(ordinary chondrite) is 200 and for Bi it is 50. If the rare carbonaceous chondrites rather than the common chondrites are more accurately representative of primitive cosmic substance (see RINGWOOD, 1960, for example) out of which the earth originated, the above comments about high crustal enrichment of Tl and Bi do not apply; this possibility and its implications will, however, not be considered further here.

The possible significance of (i) any high enrichment and (ii) variation of concentration in the principal structural components of the earth will be discussed in detail elsewhere; particular attention will be given to the use of abundance data for setting up limiting compositions*

Acknowledgements—The authors are indebted to Mr. KEY for providing the carbonatites and other specimens, to De Beers Consolidated Mines Ltd. for the eclogites, to Mr. G. SIEDNER for the specimens from the Paresis complex, to Mr. J. GENIS for the Prieska rocks, to Mr. J. F. TRUSWELL for the dolerite suite of rocks and to Dr. S. R. TAYLOR for his help in collecting the specimens and for helpful comments.

REFERENCES

AHRENS L. H. (1945) The geochemical relationship between Tl and Rb in minerals of igneous origin. *Trans. Geol. Soc. S. Africa.* **48,** 207.

* *Note added in proof*: A few calculations and discussions of this sort have recently been made by GAST (P. W. GAST (1960) Limitations on the composition of the upper mantle. *J. Geophys. Res.* **65,** 1287–1297).

AHRENS L. H. (1947) The abundance of Tl in the earth's crust. *Science* **106,** 8.

AHRENS L. H. (1948) The unique association of Tl and Rb in minerals. *J. Geol.* **56,** 578.

AHRENS L. H. (1953) The use of ionisation potentials. *Geochimica et Cosmochim. Acta* **3,** 1.

AHRENS L. H. (1958) Variations of refractive index with ionisation potentials in some isostructural crystals. *Miner. Mag.* **31,** 929.

AHRENS L. H., PINSON W. H. and KEARNS M. M. (1952) Association of Rb and K and their abundance on common igneous rocks. *Geochim. et Cosmochim. Acta* **2,** 229.

BOROVIK-ROMANOVA T. F. and SOSEDKO A. F. (1960) The geochemical ratio between the Tl and Rb content in pegmatites and minerals from the Kola peninsula. *Geokhimiya* **1,** 31.

BROOKS R. R., AHRENS L. H. and TAYLOR S. R. (1960) The determination of trace elements in silicate rocks by a combined spectrochemical-anion exchange technique. *Geochim. et Cosmochim. Acta* **18,** 162.

CANNEY F. C. (1952) Some aspects of the geochemistry of K, Rb, Cs and Tl in sediments. *Bull. Geol. Soc. Amer.* **63,** 1238.

CLARKE F. W. and WASHINGTON H. S. (1924) The composition of the earth's crust. *U.S. Geol. Survey* No. 127.

DIOMIN A. M. and KHITAROV D. N. (1958) The geochemistry of K, Rb and Tl applied to problems of petrology. *Geokhimiya* **6,** 570.

EHMANN W. D. and HUIZENGA J. R. (1959) The determination of Bi, Tl and Cd in stone meteorites by neutron activation analysis. *Geochim. et Cosmochim. Acta* **17,** 1–2.

GOLDSCHMIDT V. M. (1937) The geochemical distribution of the elements. *Skr. Norsk. Vidensk. Akad. Mat. Kl.* No. 4.

GOLDSCHMIDT V. M. (1954) *Geochemistry*. Oxford University Press.

HERZOG L. F. and PINSON W. H. (1955) Sr and Rb content of granite G–1 and diabase W–1. *Geochim. et Cosmochim. Acta* **8,** 295.

HORSTMANN E. L. (1956) The distribution of Li, Rb and Cs in igneous rocks and sedimentary rocks. *Geochim. et Cosmochim. Acta* **12,** 1.

ISHIMORI T. and TAKASHIMA Y. (1955) *Mem. Fac. Sci. Kyushu* **65,** 74.

KOGARKO L. N. (1959) The distribution of alkaline elements and thallium in granitoids of the Turgoyak massif (middle Ural). *Geokhimiya* **5,** 455.

MORRIS D. F. and AHRENS L. H. (1956) Ionisation potentials and the chemical binding and structures of simple inorganic crystalline compounds—1. *J. Inorg. Nucl. Chem.* **3,** 263.

MORRIS D. F. and KILLICK R. A. (1960) The silver and thallium content of rocks. *Geochim. et Cosmochim. Acta* **19,** 139–140.

NEUMANN A. E. (1949) Notes on the mineralogy and geochemistry of Zn. *Miner. Mag.* **28,** 205.

NODDACK I. and NODDACK W. (1934) The geochemical distribution coefficients of the elements. *Svensk. Kem. Tid.* **46,** 173.

OTTEMAN J. (1940) Investigations into the distribution of trace elements in Harz minerals. *Z. Angew. Min.* **3,** 142.

PREUSS E. (1941) The determination of volatile elements by fractional distillation. *Z. Angew. Min.* **3,** 8.

RANKAMA K. A. (1949) A note on the geochemistry of Tl. *J. Geol.* **57,** 608.

RANKAMA K. and SAHAMA TH. G. (1950) *Geochemistry.* Chicago University Press, Chicago.

REED G. W., KIGOSHI K. and TURKEVICH A. (1960) Concentrations of some heavy elements in meteorites by activation analysis. *Geochim. et Cosmochim. Acta* **20,** 122–140.

RINGWOOD A. E. (1955) The principles governing trace element distribution during magmatic crystallisation. *Geochim. et Cosmochim. Acta* **7,** 189.

RINGWOOD A. E. (1961). In preparation.

SANDELL E. B. and GOLDICH S. S. (1943) The rare metallic constituents of some American igneous rocks. *J. Geol.* **51,** 99.

SHAW D. M. (1952) The geochemistry of Tl. *Geochim. et Cosmochim. Acta* **2,** 118.

SHAW D. M. (1957) The geochemistry of Ga, In and Tl. *Progress in Physics and Chemistry of the Earth.* Vol. 2. Pergamon Press, London.

TAUSON L. V. and STAVRON O. D. (1957) On the geochemistry of Rb in granitoids. *Geokhimiya* No. 8, 699.

TAYLOR S. R. and HEIER K. S. (1960) The petrological significance of trace element variations in alkali feldspars. *Proceedings of the Twenty-first International Geological Congress, Copenhagen.*

VINCENT E. A. and BILEFIELD L. I. (1960) Cd in rocks and minerals from the Skaergaard intrusion, East Greenland. *Geochim. et Cosmochim. Acta* **19,** 6.

VINOGRADOV A. P. (1956) The regularity of distribution of the chemical elements in the earth's crust. *Geokhimya* **1,** 6.

VOSRESENSKAYA N. T. (1959) On the geochemistry of thallium and rubidium in igneous rocks. *Geokhimiya* **6,** 495.

WEDEPOHL K. H. (1953) The geochemistry of zinc. *Geochim. et Cosmochim. Acta* **3,** 93.

ZLOBIN B. I. (1958) The geochemistry of Tl in alkaline rocks. *Geokhimiya* **5,**

ZLOBIN B. I. and LEBEDEV V. I. (1960) The geochemical limits of Li, Na, K, Rb and Tl in the alkaline magma and their petrological importance. *Geokhimiya* **2,** 87.

10

Reprinted from *Geochim. et Cosmochim. Acta* **35**:1255–1267 (1971)

General trends in the behavior of Cd, Hg, Tl and Bi in some major rock forming processes

G. MAROWSKY* and K. H. WEDEPOHL
Geochemisches Institut der Universität Göttingen, Germany

1. INTRODUCTION

THE ELEMENTS under investigation have in common: low concentrations in abundant rock types (in the ppb-range) and, except for thallium, volatility in natural thermal processes.

The main purpose of this investigation is an analytical survey to obtain averages for Cd, Hg, Tl and Bi in the sedimentary cover and the magmatic masses of the continental earth's crust for a gross comparison of these units. A test of the behavior of the respective elements in a gradient of regional metamorphism has been included.

An up-to-date, critical review of literature data and general geochemical information has been published in the *Handbook of Geochemistry*, edited by the latter author of this article. The pertinent chapters in this book are:

Cd 48 B-M, 0: WAKITA and SCHMITT (1970)
Hg 80 B-M : TUNELL (1970)
Tl 81 B-N : DE ALBUQUERQUE and SHAW (1971; in press)
Bi 83 B-M, 0: AHRENS and ERLANK (1969).

In addition the two publications of LAUL *et al.* (1970a, b) should be mentioned as the most recent papers on neutron activation investigations on Bi and Tl in standard rocks.

Because of the very low concentrations of Cd, Hg and Bi in natural materials, neutron activation methods are generally superior to standard chemical procedures and those of optical spectrometry.

2. EXPERIMENTAL PROCEDURES

Only a brief outline of the sample preparation, irradiation, radiochemical separations and radiometric procedure is presented here, as details have been published

* Present address: Max-Planck-Institut für biophysikalische Chemie/Laser-Physik 34 Göttingen-Nikolausberg, am Fassberg, Germany.

elsewhere (MAROWSKY, 1971a). A batch of five powdered rock samples of 1–2 g each together with a flux monitor consisting of 1 μg Cd, Hg, Tl and Bi were irradiated for 3 days at a thermal neutron flux of about 6×10^{13} ncm^{-2} sec^{-1} in the FRJ-1 reactor of the Kernforschungsanlage Jülich (Germany). A flow chart of the radiochemical separations employed is presented in Fig. 1. After dissolution of

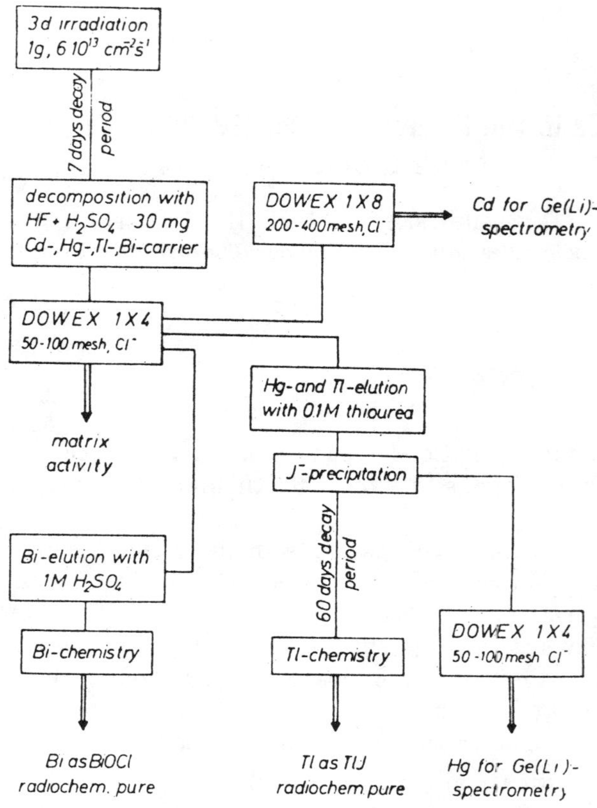

Fig. 1. Flow sheet of the analytical procedures (according to MAROWSKY, 1971a).

the samples in the presence of inactive carriers for each of the 4 elements and suitable holdback carriers for some interfering nuclides the bulk activity of the matrix was removed in an anion exchange step. The nuclear data of the isotopes used for analytical determination are summarized according to LEDERER *et al.*, 1967:

Element	Isotope	Half life	Energy
Cd	Cd-115	53·5 h	E_γ = 0·0530 MeV
Hg	Hg-203	47·0 d	E_γ = 0·279 MeV
Tl	Tl-204	3·8 a	$E_{\beta\text{-max}}$ = 0·766 MeV
Bi	Bi-210	5·0 d	$E_{\beta\text{-max}}$ = 1·116 MeV

The steps of the radiochemical procedure were selected to give a high chemical yield. In nearly all experiments chemical yields varied from 40 to 80 per cent. Cadmium and mercury were purified as far as necessary for γ-spectrometry (with a

28 cm^3 Li-drifted Ge-detector and 4000 channel analyzer). The β-emitters Tl-204 and Bi-210 were radiochemically purified before low level counting with a methane flow counter equipped with anticoincidence shielding (background 5 cpm). The decay of the Bi-samples followed the known 5·0 day half-life of this nuclide. The purity of the final precipitates was checked by measuring standard aluminium absorption curves and analyzing various samples for at least 10 hours with a Ge(Li) detector to find out a possible radiochemical contamination.

To avoid the tedious Tl-radiochemistry, part of the Tl determinations was carried out by evaluating the Hg-X-rays from the 2·1 per cent electron capture of Tl-204. (In the tables of analytical data these Tl values are marked as Tl*.) On the average the γ-spectrometric Tl values are lower than the results of the β-counting method. Several samples were recycled until constant specific activity was achieved. However, when submitted to β- and γ-counting still remained a difference in the counting rates. Perhaps the increased β-counting rate is due to the sum of various very small background activities that could not be identified. For more details of the Tl determination with semiconductor detectors the reader is referred to MAROWSKY (1971b).

During routine analysis the following minimum concentrations could be detected: 0·1 ppb of Tl and Hg, and 1 ppb of Cd and Bi. The problem of limits of detection for these four elements is discussed in detail by MAROWSKY (1970). For concentrations an order of magnitude higher than the limit of detection, the relative standard deviation given by counting statistics was always less than 3 per cent. Taking into consideration the additional errors arising from the various gravimetric steps, a precision of approximately 10 per cent can be assumed. Duplicate determinations—as shown in Table 1a—agreed to within 5–10 per cent of the amount present.

For evaluation of accuracy previous data on U.S.G.S. rock standards by neutron activation analyses and mass spectrometry are listed in Table 1b.

3. Abundance in Common Magmatic Rocks

The abundance of minor elements in magmatic rocks is controlled by the sources and processes of magma formation. According to recent experimental

Table 1a. Compilation of selected duplicate determinations of the present authors

Sample	Laboratory No.	Weight concentration in ppb Cd	Hg	Tl	Bi
W-1	12·5	151	234	108	53
	15·3	155	216	122	47
Arithmetic mean:		153	225	115	50
AGV-1	6·3	—	4	1650	42
	13·3	—	6	1610	64
Arithmetic mean:		—	5	1630	53
PCC-1	12·2	—	3·0	0·7	—
	13·4	—	5·4	0·8	—
Arithmetic mean:		—	4·2	0·75	—
Shale (composite 36 sample)	6·4	218	—	1138	442
	17·7	224	—	1302	418
Arithmetic average:		221	—	1220	430

Table 1b. Comparison of results on standard rocks (ppb) (references in brackets)

Rock	Cd (N/R) (2)	Cd (N/R) (3, 4)	Hg (N/R) (1, 2, 5)	Hg (N/R) (6)	Tl (MS) (1, 2)	Tl (N/R) (7, 9)
G-1		60(3)	340(1)			
			245(2)	70 ± 8	1300(1, 2)	1060 + 160(7)
						1300(9)
G-2			39(5)	29 ± 4		890 ± 130(7)
GSP-1			21(5)	41 ± 6		1300 ± 200(7)
AGV-1			4(5)	16 ± 2		342 ± 50(7)
W-1	270	330(3)	170(1)	94 ± 10	140(1)	102 ± 15(7)
			110(2)		190(2)	170(9)
BCR-1		160(4)	7(5)	4 ± 0·3		330 ± 90(7)
PCC-1			4(5)	3·6 ± 0·5		0·75 ± 0·1(7)

Rock	Bi (MS) (1)	Bi (N/R) (8)	Cd (this investigation)	Hg (this investigation)	Tl (this investigation)	Bi (this investigation)
G-1	100	46	22	59	1420	83
					1080†	
G-2		37·6 ± 2·5	n.d.	38	1200	50
GSP-1		36·8 ± 0·5	50	106	1220	18
					1290†	
AGV-1		56·4 ± 0·4	115	5	1630	53
W-1	250	43·5 ± 0·5	153	225	121	50
					116	
BCR-1		46·7 ± 0·4	n.d.	8	350	50
PCC-1		8	n.d.	4·7	0·75	n.d.
					0·7†	

N/R: Neutron activation; MS: Mass spectrometry; †: Ge(Li)-γ-spectrometry; n.d.: not determined.
1. FLEISCHER (1965); 2. FLEISCHER (1969); 3. VINCENT and BILEFIELD (1960); 4. REY *et al.* in: WAKITA and SCHMITT (1970); 5. FLANAGAN (1969); 6. LAUL *et al.* (1970a); 7. LAUL *et al.* (1970c); 8. LAUL *et al.* (1970b); 9. MORRIS and KILLICK (1960).

evidence, processes of partial melting of ultramafic rocks of the mantle and metamorphic rocks of the continental crust contribute to a large extent to magma generation. The crystal chemical behavior of the elements in rock-forming minerals and partition between solid phases and melts control the elemental abundances in magmatic rocks. Because of the low concentrations of Cd, Hg, Bi and Tl replacement (camouflage) of major elements with related crystal chemical properties can control the tendency of the minor elements to accumulate in specific minerals. Cd^{2+} has close similarities to Zn and even to Fe^{2+} and therefore favors the structures of biotite, amphibolite, pyroxene and magnetite in rocks. It cannot be excluded that Cd might mainly occur as CdS. Tl^{+} with its similarities to K^{+} is accumulated in rock-forming mica and potassium feldspar (in basaltic rocks magnetite can contain relatively high concentrations). A major proportion of mercury in rocks probably occurs as HgS. The information on Bi in rock-forming minerals is too meagre to make definite conclusions about its crystal chemical behavior in magmatic processes. BROOKS (1961) arrived at some crystal chemical relations between Bi, Tl and Pb from a correlation of abundances in major rock types.

Our analytical data on magmatic rocks are compiled in Table 2. If an element has a tendency to light as well as to dark rock-forming minerals there is not much difference in the overall abundance of this element between mafic and granitic rocks as in the case of Cd. If an element has no tendency to major minerals the picture can be similar. The concentration of specific anions etc. may control the abundance of Hg and Bi (?). Tl^{+} with its similarities to K^{+} is as expected higher in granitic than in mafic rocks.

Table 2

Laboratory No.	Rock	Locality	Weight concentration (ppb) Cd	Hg	Tl	Tl*	Bi
14.5	Peridotite nodule from olivine alkali basalt	Baunsberg near Kassel (Germany)	29	195	7	7	1
15.4	Peridotite nodule from olivine alkali basalt	Hirzstein near Kassel (Germany)	23	4	3·4	n.d.	6·6
12.2/13.4	Peridotite, PCC-1	Cazadero, Sonoma County, Calif. (USA)	n.d.	4·7	0·75	0·7	n.d.
5.3	Basalt, BCR-1	Columbia River, Bridal Veil. Wash. Oreg. (USA)	n.d.	8	350	n.d.	50
12.5/15.3	Tholeiitic basalt, W-1	Centerville, Fairfax County, Virginia (USA)	153	225	121	116	50
10.2/17.2	Olivine tholeiitic basalt	Mauna Loa, Honokua Flow, Hawaii (USA)	78	12	11	n.d.	9
14.3	Tholeiitic basalt (composite)	Northern Hessia (Germany) (4 localities)	113	14	88	73	10
10.4/17.4	Tholeiitic basalt	Kottenberg, Ziegenhain (Germany)	86	7	31	22	10
14.4	Olivine alkali basalt (composite)	Northern Hessia, Lower Saxony (6 localities) (Germany)	n.d.	19	49	41	14
10.3/17.3	Olivine alkali basalt	Grefenburg, Göttingen (Germany)	128	11	27	n.d.	43
11.5	Olivine alkali basalt	Hoher Hagen, Göttingen (Germany)	190	16	59	40	25
13.5	Olivine nephelinite	Westberg, Hofgeismar (Germany)	n.d.	43	17	n.d.	3
14.2	Olivine nephelinite	Reissberg, Fritzlar (Germany)	n.d.	103	50	32	11
15.2	Gabbro (composite)	Odenwald, Harz, Saxony Silesia (11 localities) (Germany, Poland)	130	133	109	126	36
6.3/13.3	Andesite, AGV-1	Guano Valley, Lake County Oregon (USA)	115	5	1630	n.d.	53
13.5	Granodiorite GSP-1	Silver Plume, Colorado (USA)	50	106	1220	1290	18
14.6	Granite (composite YCR5, CGII)	Western USA (50 localities)	107	122	690	650	80
11.2	Granite, G-1	Westerly, Bradford, Rhode Island (USA)	22	59	1420	1080	83
5.1	Granite, G-2	Westerly, Bradford, Rhode Island (USA)	n.d.	38	1200	n.d.	50
10.6/17.6	Granite	Wurmberg (Pr. 5), Harz (Germany)	33	60	980	830	90
10.5/17.5	Granite (composite of 14)	Odenwald, Schwarzwald, Fichtelgebirge, Erzgebirge (Germany)	59	(220)†	1560	1300	(930)†

n.d.: not determined.
* Ge-γ-spectrometry.
† The composite contains one greisen sample. Therefore data on mercury and bismuth in this composite should not be used for averages.

If one compares our data for rock types with those recently published in the literature as listed in Table 3 the following trends can be observed. In most cases data obtained by neutron activation are lower than those obtained by spectrographic and chemical procedures such as listed by WAKITA and SCHMITT (1970), TUNELL (1970), DE ALBUQUERQUE and SHAW (1971), AHRENS and ERLANK (1969). To demonstrate this with literature data two sets of values for Hg and Bi, designated as I and II, are given in Table 3. Our averages for Bi are in reasonable agreement with those computed from the few data on standard rocks obtained by neutron activation published by LAUL *et al.* (1970). Our information favors the assumption of accumulation of Bi in granitic rocks relative to basaltic-gabbroic types. Neutron activation analyses for thallium in granitic rocks (including the standard samples) indicate no systematic difference from values obtained from other techniques.

Table 3

		Weight concentration (ppb)					
		Cd	Hg		Tl	Bi	
			I	II		I	II
A	Granitic rocks: average computed from references*	225	64	30†	1700	180	40
B	Granitic rocks: average, this publication	75–100	77		1100†	65†	
			I	II		I	II
A	Basaltic-gabbroic rocks: average computed from references*	170	90	12	180	150	45
B	Basaltic-gabbroic rocks: average, this publication	130†	35–70†		80†	31†	
				II			II
A	Ultramafic rocks: average computed from references*	n.d.		4	50		6
B	Ultramafic rocks: average, this publication	26	?		4	5	

* References: Cd: WAKITA and SCHMITT (1970).
Hg: column I: SAUKOV (1946), STOCK and CUCUEL (1934) by chemical and spectrographic methods.
column II: EHMANN and LOVERING (1967) by neutron activation.
Tl: DE ALBUQUERQUE and SHAW (1971).
Bi: column I: BROOKS (1960), BROOKS and AHRENS (1961) by chemical and spectrographic methods.
column II: LAUL *et al.* (1970) by neutron activation.
† Used by the present authors as the average concentration in a rock type.

However, because of the one order of magnitude lower thallium concentrations in basaltic rocks systematically lower results can be expected from neutron activation. Our data on mercury in abundant magmatic rocks in a few samples are not as low as the neutron activation data published by EHMANN and LOVERING (1967). Samples with very low Hg concentrations can be easily contaminated in conventional storing and laboratory treatment. From this point of view lower values have more probability, and we have tended to give them higher statistical weight. On the basis of our cadmium determinations, the mean for granitic rocks is lower than one computed from the references listed by WAKITA and SCHMITT (1970).

If the close crystal chemical relation between Zn (see WEDEPOHL 1972) and Cd can be used as an indicator of the distribution of Cd in magmatic rocks, this element is expected to be of slightly higher abundance in basaltic than in granitic rocks. For a comparison of magmatic and sedimentary rocks we will use those values of Table 3 which are marked by an asterisk for reasons mentioned in this section. We still need additional data to confirm our conclusions about the distribution of Cd, Hg, Tl and Bi in magmatic rocks.

4. ABUNDANCE IN COMMON SEDIMENTARY ROCKS

Pioneering analytical research on the abundance of readily volatile minor elements in sediments was conducted by PREUSS (1940) using emission spectrography after preconcentration by vaporization. In several cases we have used the same samples Preuss analyzed. Our data are listed in Table 4. Except for mercury and bismuth in the shale composite and bismuth in the graywacke composite there is reasonable agreement between both sets of data. Since the analysis by PREUSS of the

Table 4

Lab. No.	Rock	Locality	Weight concentration (ppb) Cd	Hg	Tl	Tl*	Bi
6.4/17.7	Shale (composite)	36 samples, Germany, Spain, Switzerland, France, Paleozoic age	220	(1590)†	1220	870	430
16.3/18.3	Shale (marl)	2 samples, Upper Triassic, Basle (Switzerland)	450	180	420	n.d.	260
11.4	Pelagic clay	Atlantic, 19°54′N 64°48′W, Lamont A 172 No 14/19	440	53	1000	890	480
11.6	Pelagic clay	Atlantic 22°17′N 55°07′W Lamont A 160 No 5	430	11	800	790	300
12.3	Pelagic clay	Pacific, 14°55′N 124°12′W Scripps 50 BP 16	290	150	800	710	610
12.4	Pelagic clay	Pacific, 20°49′N 125°15′W Scripps Y 2P 101.25	460	68	960	940	710
6.2	Graywacke (composite)	17 samples, Rhine area, Harz etc. Paleozoic age (Germany)	n.d.	220	470	n.d.	100
9.3	Graywacke	Harz Mountains, Kulm system (Germany)	n.d.	1440	1300	n.d.	390
9.4	Graywacke	Harz Mountains, Tanner system, Germany	n.d.	120	360	330	n.d.
6.5	Sandstone, red (composite)	23 samples, Germans, Lower Triassic	n.d.	290	360	n.d.	180
7.4	Quartz sand	Dörentrup (Germany) Tertiary	n.d.	8	88	n.d.	n.d.
9.5	Bituminous Marl "Kupferschiefer"	Mansfeld (Germany) Permian, (composite of mine laboratory)	n.d.	9400	7300	7400	150
9.6	Limestone (composite)	32 samples, Germany Devonian (10% clay residue)	n.d.	290	140	130	200
15.5	Limestone (composite)	16 samples, NW Germany Cretaceous (24% clay residue)	91	29	230	n.d.	54

n.d.: not determined.

* Ge-γ-spectrometry.

† A potential contamination of this sample with time cannot be excluded because PREUSS (1941) and STOCK and CUCUEL (1934) report 300 and 500 ppb Hg respectively for the same sample.

shale composite 30 years ago this sample has probably been contaminated by mercury. The averages for abundant sediment types from our data and from those reported in the recent literature are compiled in Table 5. The data which appear to be the most reliable to the present authors have been marked with an asterisk.

From crystal chemical reasons the Cd, Hg, Tl and Bi concentrations in limestones are expected to be controlled by their clay fraction. In our case the Devonian limestone composite has the lower clay proportion but the higher concentration of bituminous compounds relative to the Cretaceous limestone composite. Bituminous sediments most effectively accumulate Hg, Tl and Bi as evidenced by the Kupferschiefer marl (GOLDSCHMIDT, 1954, mentions a sample of the same bed containing 100 ppm Bi). JOENSUU (1970) reports 1000–25,000 ppb Hg in Black Sea sediments formed under unoxic waters. The composite of 36 European Paleozoic shales contains 1·9%C compared with a value of 0·6%C, which can be taken as the gross average for shales. BOSTRÖM and FISHER (1969) report on the local accumulation of mercury in pelagic sediments from the crest of the East Pacific Rise. From the evidence of Table 4 is to be expected that pelagic clays contain slightly higher Bi concentrations than do nearshore deposited clays with comparably low proportions of organic carbon.

Table 5

		Weight concentration (ppb) Cd	Hg	Tl	Bi
A	Shale, clay: average computed from references*	740	400†	550	1000
B	Shale, clay: average this publication	300†	(1400)	1000†	430†
B	Graywacke average: this publication	n.d.	280	510	120
A	Limestone: average computed from references*	n.d.	40†	n.d.	n.d.
B	Limestone: average, this publication	90†	200	170†	150†

* References: Cd: Wakita and Schmitt (1970).
Hg: Stock and Cucuel (1934), Preuss (1941), Heide and Böhm (1958).
Tl: de Albuquerque and Shaw (1971).
Bi: Preuss (1941), Brooks *et al.* (1960).
† Used by the present authors as the average concentration in a rock type.

5. Comparison of Cd, Hg, Tl and Bi Contents of Magmatic and Sedimentary Rocks

For a great number of chemical elements averages for magmatic rocks of the upper continental crust and averages for sediments are about the same. Ratios of abundances between these two units are close to one (range 0·7–1·5) for at least 55 elements (Wedepohl, 1968). That means that with respect to these elements the total mass of magmatic rocks is converted into sediments by weathering. The averages are computed by considering the areal distribution of major rock types as discussed by Wedepohl (1969). For the elements under consideration here we have weighted granitic and basaltic-gabbroic rocks in the proportion 85:15 and shales-clays and limestones in the proportion 80:20. The latter weighting does not consider sandstones. No doubt in the total column of geologic ages graywackes are the most abundant type of sandstones. A higher proportion of graywackes to shales will lower the gross sedimentary averages of Cd, Hg, Tl and Bi to a certain extent. As will be demonstrated in section 6, metamorphic rocks are close to magmatic rocks in their concentrations of the elements of our investigation. So decomposed metamorphic rocks as well as magmatic rocks contribute relatively small amounts of Cd, Hg and Bi to sediments as detrital fractions. Because the average abundance of an element cannot be much higher in magmatic rocks than in sedimentary rocks (if the element is not extensively accumulated in seawater), the lower thallium average for granitic rocks of Table 3 seems to be more realistic than the higher one.

Table 6 contains tentative averages computed from the recommended data of Tables 3 and 5 and their ratios:

Table 6

		Weight concentration (ppb) Cd	Hg	Tl	Bi
A	Tentative average abundance in magmatic rocks of the upper continental crust	110	30	970	60
B	Tentative average abundance in sedimentary rocks	260	330	830	375
	Ratio B/A	2·4	12	0·87	6

This table clearly demonstrates an accumulation of Hg, Bi and Cd in the sedimentary environment beyond the amounts which are contributed from rock weathering. The only explanation we have for this effect is the accumulation from the degassing of the earth (excess volatile elements as H, O, C, N, Cl, S, B etc.). Whereas H_2O, Cl and N from the degassing mantle and lower crust are mainly accumulated in oceans and atmosphere the major proportion of "excess" C, S, B as well as Hg, Br, Sb, Se, Bi, As, I and Cd are stored in sediments. The cycles of reworking and sedimentation of surface materials act as sinks.

6. Abundance of Cd, Hg, Tl and Bi in Rocks from a Sequence of Increasing Grade of Regional Metamorphism

Rocks from one stratigraphic unit which has undergone regional metamorphism with temperatures up to about 550°C and pressures up to about 6 kbar during Tertiary time have been sampled in the Swiss Alps. This material is either that collected by A. Schneider (S) under the guidance of M. Frey or represents parts of Frey's original samples (MF) from his detailed investigation (Frey, 1969).

Results of our analyses are listed in Table 7. Samples 16.3 and 18.3 have not suffered any thermal metamorphism, sample 18.4 only a slight amount. The unmetamorphosed shales contain 1 Md illite, and mixed layer minerals. These serve as a good indicator of the absence of thermal treatment. Samples 18.5, 16.5 and 18.6 represent rocks of the greenschist facies. The highest grade of metamorphism attained in the area under investigation is that of the kyanite–staurolite–garnet schists in the amphibolite facies (samples 16.6 and 18.2). The large problem in

Table 7. Shales and metamorphic rocks from a profile in the Swiss Alps

Lab. No.	Rock	Locality	Weight concentration (ppb) Cd	Hg	Tl	Tl*	Bi
16.3	Shale, marly (red) S 187	Basel (Frick quarry) 3 m above dolomite, Upper Triassic (Keuper)	730	350	520	n.d.	370
18.3	Shale, marly (light) S 188	Basel (Frick quarry) 2 m below edge, Upper Triassic (Keuper)	170	3	330	n.d.	150
18.4	Shale (red) S 191	Quarten, South of Walen Lake, Creek, (sample locality as MF 7)	49	6	1200	n.d.	220
16.4	Phyllite (containing pyrophyllite) S 196	Klausenpass (sample locality as MF 67)	29	5	n.d.	1400	370
18.5	Phyllite (low in chloritoid) S 199	Garvera near Disentis Vorderrhein-Valley	32	2	1300	n.d.	310
16.5	Phyllite (containing chloritoid) S 202	Garvera near Disentis (sample locality as MF 131)	18	7	1500	1500	80
18.6	Biotite schist S 205	Lukmanier Lake, East rim	5	1	2000	1800	44
18.7	Kyanite staurolite paragonite chloritoide schist S 206	Casaccia, close to Lukmanier road (sample locality as MF 161)	17	4	220	200	38
16.6	Garnet staurolite paragonite schist MF 315	Bronico, close to Lukmanier road	20	9	390	250	14
18.2	Garnet staurolite hornbl. kyanite biotite schist MF 307	Frodalera, close to Lukmanier road	26	14	580	n.d.	(190)†

n.d.: not determined.
* Ge-γ-spectrometry. † Lower value (25 ppb) in a semiquantitative check by flameless atomic absorption.

investigations of chemical systems in a p–t-gradient is the proof of a premetamorphic homogeneity of a sediment layer selected for the study. It is often difficult to trace the test layer through a highly metamorphosed rock unit. Information on the specific test layer is available in M. Frey's (1969) report on his lithostratigraphic, petrographic and chemical investigations and in several references he quotes. There is only limited homogeneity of the test layer to be expected because of the changing proportions of at least dolomite (scatter of Ca, Mg) and mica (scatter of K). Another problem is a potential change of the regional environment of sediment deposition of the test layer as somewhat indicated by an appreciable increase of sodium and aluminium with metamorphic grade. These restrictions are of minor importance because of the magnitude of the effect as demonstrated by this investigation.

The effect of rising temperature and pressure is to cause a tremendous loss of Hg, Cd and Bi from the original sedimentary rocks. The loss of mercury probably begins at a very low grade of metamorphism. For this assumption we need the additional information of Table 4 of a general higher abundance of mercury in sediments, higher than 3–14 ppm. With respect to mercury the available data are meagre for final conclusions. More than half of the original cadmium concentration is already lost before greenschist facies temperatures are attained. Bismuth stays in its typical sedimentary abundance range until after the climax of the greenschist facies has been passed. Thallium is probably not lost from the rock unit during metamorphism. But this possibility cannot be totally excluded. The rocks from the zone of highest thermal alteration are not lower in thallium than the unaltered shales from the Basel area. The scattering of the thallium concentrations around the shale average of about 1000 ppb Tl seems to be normal for phyllites, schists and gneisses as shown by Table 81-M-1 in the report of de Albuquerque and Shaw (1971). The scattering of our thallium values is partially related to some heterogeneity of the test layer (evidenced by a correlated change of potassium).

Another investigation on the behavior of mercury in a metamorphic gradient (test area NE Vermont) has been published by Jovanovic and Reed (1968). They observed a considerable mercury variation within one metamorphic zone, contrary to our observation of systematically low abundances. However, they did not consider zones of low metamorphic grade. In their study the total mercury content even in sillimanite schists was found to vary from 2·5 to 2535 ppb. They also found that the fraction of mercury which will not be lost by heating samples up to 450°C is only in the range of a tenth to a hundredth of the total mercury (representing an average of about 10 ppb Hg). Because all samples are from rock units which have suffered a regional metamorphism in a temperature range above 450°C (even their biotite schists, contrary to their assumption) most of the mercury of their samples must be of secondary origin. Unfortunately they have sampled in an obviously highly contaminated area.

7. Mobilization of Cd, Hg and Bi Under Thermal Treatment: Conclusions

The vaporization of an element from a solid rock depends on the vapor pressure of its compounds, on the nature of the gas phase in equilibrium with the rock and on temperature. If the partial pressure of the element in the gas phase is lower than the vapor pressure of its compound it will be vaporized from the solid rock. The dissociation temperature of the compound in some instances can also control the vaporization. For example the elements under investigation might occur as native compounds (Hg), sulfides (HgS, Bi_2S_3, CdS) or bonded to oxygen in silicates or oxides. Some of these elements have a high vapor pressure under certain natural conditions.

Ordinary chondrites are highly depleted in In, Tl, Hg, Bi, Pb, Cd and several other elements. The abundance ratio of carbonaceous chondrites to ordinary chondrites as taken from the references mentioned in the introduction is: 175 for Tl; 170 for Hg; 60 for Bi and 8 for Cd. This and other differences between types of chondrites can be explained by high vapor pressures of these elements (or their compounds) in a cosmic gas and by an origin of fractional condensation in a thermal gradient (LARIMER, 1967; LARIMER and ANDERS, 1967).

The obvious difference between vaporization in the cosmos and in crustal rocks is the composition of the gas into which readily volatile elements could be vaporized. In addition there is a much higher abundance of potassium in silicates of crustal rocks in which thallium is fixed. The behavior of cosmic gases is characterized by the high abundance of hydrogen, whereas oxygen (and H_2O, CO_2, etc.) controls the important properties of terrestrial surface gases. LARIMER (1967) has computed condensation temperatures of compounds and elements from cosmic gases. Under these conditions mercury does not condense at higher temperatures than about −80°C. Under conditions at the earth's surface one has to expect that the vapor pressures of HgS, Hg, Bi_2S_3 (Bi) and CdS (Cd) can control the vaporization of the respective elements. KRAUSKOPF (1957) computed that at 600°C the dissociation of HgS, Bi_2S_3 and CdS in a gas with 10^{+3} atm H_2O and 10^{-6} atm sulfur (S_2) vapor pressure forms appreciable Hg-, Bi- and Cd-partial pressures (range ppm to ppb Hg, Bi, Cd relative to H_2O vapor). Because of changing gas compositions a simple correlation between the temperature of metamorphism and decrease of Hg, Bi and Cd in metamorphic rocks cannot be expected. However Table 7 does demonstrate some general trend. Almost all of the mercury which can be vaporized from sedimentary rocks is assumed as being below 200°C. From our data vaporization of a major proportion of cadmium is expected to occur around 200°C (sample 18.4), whereas bismuth stays up to a temperature range of 300–400°C. (For discussion of equilibrium temperatures of metamorphism in the sampling area see FREY, 1969.) Zinc is not lost from the samples through regional metamorphism. Except sample 18.5 the range of zinc concentrations is 33–49 ppm with 37 ppm Zn in the marly shale (WEDEPOHL, 1972). Thallium is most probably prevented from major gas transport by fixation in potassium minerals. This general picture is confirmed by the abundance of Hg, Bi, Cd and Tl in granites relative to sediments of similar chemical composition. Granites, having been formed above 600°C, contain Hg, Bi and Cd in concentration similar to highly metamorphic rocks and not as in sediments.

From our observations we must conclude that the major proportion of Hg, Bi and Cd of sedimentary rocks occurs in the form of sulfides and is not incorporated in silicate minerals or oxides.

Acknowledgements—We gratefully acknowledge the irradiations and laboratory facilities, which were kindly placed at our disposal by the Institut für Radiochemie (KFA Jülich). A grant of the Deutsche Forschungsgemeinschaft covered the traveling and minor material costs. Low level β-counting was carried out at the Zentrales Isotopenlaboratorium der Universität Göttingen. Miss K. HORSTMANN participated in a part of the tedious radiochemical procedures. We thank Dr. M. FREY of Berne for his samples.

REFERENCES

AHRENS L. H. and ERLANK A. J. (1969) Bismuth. In *Handbook of Geochemistry* II, (exec. editor K. H. Wedepohl), Chap. 83. Springer.

BOSTRÖM K. and FISHER D. E. (1969) Distribution of mercury in East Pacific sediments. *Geochim. Cosmochim. Acta* **33,** 743–745.

BROOKS R. R. (1960) The use of ion exchange enrichment in the determination of trace elements in sea water. *Analyst* **85,** 745.

BROOKS R. R. (1961) Apparent geochemical association of bismuth and thallium. *Nature* **189,** 910–911.

BROOKS R. R., AHRENS L. H. and TAYLOR S. R. (1961) The determination of trace elements in silicate rocks by a combined spectrochemical-anion exchange technique. *Geochim. Cosmochim. Acta* **18,** 162–175.

BROOKS R. R. and AHRENS L. H. (1961) Some observations on the distribution of Tl, Cd and Bi in silicate rocks and the significance of covalency on their degree of association with other elements. *Geochim. Cosmochim. Acta* **23,** 100–115.

DE ALBUQUERQUE A. R. and SHAW D. M. (1971) Thallium. In *Handbook of Geochemistry* II, (exec. editor K. H. Wedepohl), Chap. 81. Springer. In press.

EHMANN W. D. and LOVERING J. F. (1967) The abundance of mercury in meteorites and rocks by neutron activation analysis. *Geochim. Cosmochim. Acta* **31,** 357–376.

FLANAGAN F. J. (1969) U.S. Geological Survey standards. II. First compilation of data for the new U.S.G.S. rocks. *Geochim. Cosmochim. Acta* **33,** 81–120.

FLEISCHER M. (1965) Summary of new data on rock samples G-1 and W-1, 1962–1965. *Geochim. Cosmochim. Acta* **29,** 1263–1283.

FLEISCHER M. (1969) U.S. Geological Survey standards. I. Additional data on rocks G-1 and W-1, 1965–1967. *Geochim. Cosmochim. Acta* **33,** 65–79.

FREY M. (1969) Die Metamorphose des Keupers vom Tafeljura bis zum Lukmanier-Gebiet. *Beitr. Geol. Karte d.Schweiz, Neue Folge* **137.** Lieferung.

GOLDSCHMIDT V. M. (1954) *Geochemistry* (editor A. Muir). Clarendon Press.

HEIDE F. and BÖHM G. (1958) Zur Geochemie des Quecksilbers. *Chem. Erde* **19,** 198–204.

JOENSUU O. I. (1970) Features of the marine geochemistry of mercury. *GSA Abstracts with Programs* **2,** 7, 588.

JOVANOVIC S. and REED G. W. (1968) Hg in metamorphic rocks. *Geochim. Cosmochim. Acta* **32,** 341–346.

KRAUSKOPF K. B. (1957) The heavy metal content of magmatic vapor at 600°C. *Econ. Geol.* **52,** 786–807.

LARIMER J. W. (1967) Chemical fractionation in meteorites. I. Condensation of the elements. *Geochim. Cosmochim. Acta* **31,** 1215–1238.

LARIMER J. W. and ANDERS E. (1967) Chemical fractionation in meteorites. II. Abundance patterns and their interpretation. *Geochim. Cosmochim. Acta* **31,** 1239–1270.

LAUL J. C., CASE D. R., WECHTER M., SCHMIDT-BLEEK F. and LIPSCHUTZ M. E. (1970a) An activation analysis technique for determining groups of trace elements in rocks and chondrites. *J. Radioanal. Chem.* **4,** 241–264.

LAUL J. C., CASE D. R., SCHMIDT-BLEEK F. and LIPSCHUTZ M. E. (1970b) Bismuth contents of chondrites. *Geochim. Cosmochim. Acta* **34,** 89–103.

LAUL J. C., PELLY I. and LIPSCHUTZ M. E. (1970c) Thallium contents of chondrites. *Geochim. Cosmochim. Acta* **34,** 909–920.

LEDERER C. M., HOLLANDER J. M. and PERLMAN I. (1967) *Table of Isotopes.* John Wiley.

MAROWSKY G. (1970) Sensitivity ranges in neutron activation analysis of some rare trace elements (Cd, Hg, Tl, Bi) in silicate samples. *Proc. NATO Advanced Study Institute on Activation Analysis in Geochemistry and Cosmochemistry*, Kjeller, Norway, 1970.

MAROWSKY G. (1971a) Aktivierungsanalytische Bestimmung von Cadmium, Quecksilber, Thallium and Wismut in Gesteinen. *Z. Anal. Chem.* **253,** 276–271.

MAROWSKY G. (1971b) Aktivierungsanalytische Bestimmung von Thallium in Gesteinen—ein Vergleich mehrerer Meßmethoden. *J. Radioanal. Chem.* **8,** 27–32.

MORRIS D. F. C. and KILLICK R. A. (1960) The determination of silver and thallium in rocks by neutron activation analysis. *Talanta* **4,** 51–60.

PREUSS E. (1940) Beiträge zur spektralanalytischen Methodik II. Bestimmung von Zn, Cd, Hg, In, Tl, Ge, Sn, Pb, *Z. angew. Mineral.* **3,** 8–20.

SAUKOV A. A. (1946) Geokhimiya rtuti (Geochemistry of mercury). *Tr. Inst. Geol. Nauk Akad. Nauk SSSR* 78 *Miner.-Geokhim. Ser.* **17,** 1–129.

STOCK A. and CUCUEL F. (1934) Die Verbreitung des Quecksilbers. *Naturwiss.* 22–24, 390–393.

TUNELL G. (1970) Mercury. In *Handbook of Geochemistry* II, (exec. editor K. H. Wedepohl), Chap. 8. Springer.

VINCENT E. A. and BILEFIELD L. I. (1960) Cadmium in rocks and minerals from the Skaergaard intrusion, East Greenland. *Geochim. Cosmochim. Acta* **19,** 63–69.

WAKITA H. and SCHMITT R. A. (1970) Cadmium. In *Handbook of Geochemistry* II (exec. editor K. H. Wedepohl), Chap. 48. Springer.

WEDEPOHL K. H. (1968) Chemical fractionation in the sedimentary environment. In *Origin and Distribution of the Elements* (editor L. H. Ahrens). Pergamon.

WEDEPOHL K. H. (1969) Composition and abundance of common sedimentary rocks. In *Handbook of Geochemistry* I (exec. editor K. H. Wedepohl), Chap. 8. Springer.

WEDEPOHL K. H. (1972) Zinc. In *Handbook of Geochemistry* II (exec. editor K. H. Wedepohl), Chap. 30. Springer. in preparation.

11

Reprinted from *Geochim. et Cosmochim. Acta* **33**:417–420 (1969)

The bismuth contents of some rare-earth minerals, notably gadolinite

J. J. GURNEY and L. H. AHRENS
Department of Geochemistry, University of Cape Town

Abstract—Bi has been estimated spectrochemically in 29 specimens of gadolinite; it varies from 12 to 262 ppm and averages at 63 ppm. The presence of distinct traces of Bi in gadolinite is evidently due to the ability of Bi^{3+} to replace rare earth cations in the gadolinite structure. Some approximate observations have also been made on the Bi concentration in other rare earth minerals and in apatite.

INTRODUCTION

THE STABLE oxidation state of the rare Group V element Bi (crustal abundance 0·1–0·2 ppm, according to TAYLOR, 1964) is three in geological environments. Although Bi^{3+} is strongly chalcophile, it also displays lithophile tendencies, two of which are of particular interest:

(i) A possible Bi^{3+}–Ca^{2+} substitution in either the common mineral, plagioclase, or in rarer types such as apatite; and

(ii) A possible Bi^{3+}–RE^{3+} substitution in rare earth minerals.

Our main concern in this paper will be on (ii), with particular emphasis on gadolinites, though some observations on the Bi concentration in apatite will also be discussed. Comments bearing on a possible Bi^{3+}–Ca^{2+} substitution in plagioclase and the significance that this could have on a Bi–Ca relationship in common igneous rocks, have been made by BROOKS *et al.* (1960) and BROOKS and AHRENS (1961).

THE Bi^{3+}–RARE EARTH ASSOCIATION

Crystal–chemical considerations

The possibility of significant Bi^{3+}–RE^{3+} substitution in rare earth minerals depends mainly on the general crystal chemistry of Bi^{3+} and rare earth compounds, which in turn depends on the sizes of the cations in question and the nature of the bonds in these compounds, aspects which have been emphasized by GOLDSCHMIDT (1954).

The estimated radius of Bi^{3+} (1·15 Å, according to AHRENS and KABLE, 1968) is similar in magnitude to the radii of the trebly charged rare earth cations (0·9–1·18 Å, according to AHRENS and KABLE, 1968). Whereas, however, bonding in rare earth compounds is essentially ionic, a fairly high degree of co-valency may obtain in Bi^{3+} compounds. Nevertheless, the structures of many Bi^{3+} compounds are ionic (that is, co-ordination conforms with the ionic radius-ratio rules) and it is not surprising that some of these compounds are isostructural, or are at least structurally similar to those of the rare earths. It is of interest that Bi^{3+} forms pucherite ($BiVO_4$) which is structurally similar to xenotime (YPO_4); also that Bi^{3+} can form a silicate, eulytite ($Bi_4Si_3O_{12}$) for example. Substitution in the gadolinite structure has been discussed by KUDRINA and KUDRIN (1961) and by ITO (1965, 1967).

Previous work

In a study devoted mainly to the geochemistry of Re, I. and W. NODDACK (1931) used an X-ray fluorescent method to estimate Bi in four gadolinite specimens and some other rare earth minerals. The estimated Bi content ranged from <1 ppm to 20 ppm in the gadolinite specimens and was less in the other minerals. No other data are known to us.

Specimens used in the present study

Locality details, where available, for the 29 gadolinite specimens appear in Table 1 under "Results and Discussion". Most gadolinite specimens come from pegmatites either from South

Table 1. Estimated Bi concentration in 29 gadolinite specimens

Sample No.	Locality	Bi (ppm)
AAC 3	± 15 m. NNE Augrabies Falls, S.W.A.	46
4	Bokseputs Reserve, N.W. Cape	86
5	Ondermatjie, Warmbad District, S.W.A.	60
7	Vuursteenkop, Warmbad District, S.W.A.	63
8	Steierkraal, Kenhardt District, Cape	114
41	Steierkraal, Kenhardt District, Cape	101
14	Farm Vrede, Kenhardt District, Cape	51
G.S.3	Styr Kraal, Kenhardt District, Cape	37
AAC 24	Kakamas South, S.W.A.	15
26	Kakamas South, S.W.A.	24
27	Kakamas South, S.W.A.	31
29	Kakamas South, S.W.A.	36
G.S.2	Kakamas South, S.W.A.	31
AAC 17	Keikarras, Kakamas District, S.W.A.	22
19	Bassonsdrif, Kakamas District, S.W.A.	56
30	Farm Warmbad, W. of Kakamas	40
33	Farm Jerusalem, 80 m. E. of Warmbad	18
39	Riemswasmaak, Gordonia, Cape	12
G.S.1	Bokwasmaak Bantu Res., Gordonia	84
G.J.B.8	Japie Pegmatite, Gordonia, Cape	73
G.S.4	Daberas, District Kenhardt, Cape	100
AAC 13	Coolegong, W. Australia	262
B.9	W. Australia	167
AAC 1	Unknown	49
2	Unknown	53
6	Unknown	77
9	Unknown	43
10	Unknown	49
12	Unknown	67
	Ave	63

West Africa or from the Cape Province, South Africa; two are from Western Australia. Six specimens are from unknown localities. The absence of locality details for these specimens is unfortunate but they have been included, nevertheless, because Bi data on gadolinites are sparse and the inclusion of these specimens is helpful for obtaining a better overall idea of the distribution of Bi in gadolinites.

In addition to the gadolinites, 16 other rare earth minerals, together with 13 specimens of apatite were analyzed. Locality details are not given as the Bi data on these minerals are approximate and the work on them is of a reconnaissance nature.

Analytical Procedure

Bi was detected and determined by an optical spectro-chemical method in which bismuth's most sensitive line, Bi 3067·7 Å, was used. This line is susceptible to interference from two sources; Fe line interference (Fe 3067·2 Å and Fe 3068·2 Å) and OH band interference. Provided dispersion is large and Fe line emission not excessively intense, Fe line interference is negligible. In the present work, an Ebert grating spectrograph (linear dispersion 5 Å/mm) was used and under the conditions of excitation, no Fe line interference was encountered.

Interference from an OH band component has been examined in detail by GURNEY *et al.* (1965) who observed that such interference could be removed provided the sample was arced in an argon–oxygen atmosphere. In the present work a 4:1 argon–oxygen mixture was used. 10% NaF was added to each sample as a buffer-depressant. Samples were heated at 120°C for 4 hr and then arced in the above atmosphere for 60 sec at 4 A; other details of procedure are given by GURNEY *et al.* (1965). The Bi detection limit is <1 ppm in the rare earth minerals but is somewhat greater (~2 ppm) in apatite because some OH band interference was encountered.

In order to reduce systematic error which is likely to arise if synthetic standards are used, Bi was determined in two gadolinite samples, GS 4 (100 ppm) and GS 2 (31 ppm), by an X-ray fluorescent (XRF) method and these samples were used for the preparation of the working curve for optical spectro-chemical analysis. In the XRF method, varying proportions of Bi (as Spec. Pure Bi_2O_3) were added to GS 4 and GS 2 to give four addition standards; each was mixed with lithium tetraborate and fused. The resultant disc was crushed and pressed into a pellet according to the method described by WELDAY *et al.* (1964). A Mo excitation tube and LiF 200 analyzing crystal were used and also pulse height selection.

As an additional but rough check, pressed pellets of all the powdered minerals were prepared. The bismuth concentration in these samples was estimated using a peak to background method, with GS 4 and GS 2 as standards. The results agreed satisfactorily with those obtained by the optical spectrograph.

No internal standard was used in the spectrochemical procedure. Based on 29 duplicate measurements, the calculated coefficient of variation was 12 per cent

RESULTS AND DISCUSSION

The estimated Bi concentrations in the 29 gadolinite specimens available for our study are given in Table 1. Data and discussions on other minerals follow the discussion on the gadolinites.

Bi was detected and estimated in each of the 29 specimens and it may be concluded therefore that Bi is a characteristic trace constituent of this yttrium rich mineral.

The Table 1 values tend in general to be greater than those reported by I. and W. NODDACK (1931) and average at 63 ppm; the range is considerable, 12–262 ppm. According to the available data, Bi may vary to some extent with geographic locality. Thus, for example, the two specimens from Western Australia contain the highest amounts (167 and 262 ppm). The two specimens from Steierkraal, Kenhardt District, Cape, contain 101 and 114 ppm, distinctly greater than the average, whereas Bi in five specimens from Kakamas South, South West Africa, is characteristically low (15–36 ppm).

Details on the estimated Bi content in additional rare earth minerals which were investigated are discussed below. Because of the effect that matrix has on line intensity in an optical spectrochemical method of analysis, the data on these minerals are only approximate.

Fergusonite—5 specimens

Bi was detected in each specimen. The estimated range is 20–40 ppm.

Allanite—6 specimens

Five of the six allanite specimens contained detectable Bi. The level is relatively low and is estimated as 5–10 ppm.

Monazite—3 specimens

Bi was detected in each specimen; the level appears to be slightly lower than in allanite, and is estimated as 2–5 ppm.

Euxenite—2 specimens

Both specimens contain approximately 40 ppm Bi.

One each of the following minerals was analyzed; fluocerite, yttrofluorite and samarskite. Bismuth was detected in all 3 and the highest concentration (approximately 100 ppm) was found in the samarskite specimen.

Apatite

In addition to the above, a qualitative investigation was carried out on the bismuth content of apatites from igneous and metamorphic rocks. In a discussion on the geochemistry of Bi, Goldschmidt (1954, p. 480) comments on the possibility of acceptance of Bi into the apatite structure, and writes, "Bismuth enters rather regularly into the apatites of igneous rocks, the trivalent Bi ion being captured because of its similar radius to Ca^{2+}." Quantitative data were not given nor is reference made to the source of information.

Because of the presence of distinct traces of Bi in Y + RE rich minerals from pegmatites, one may infer that Bi should be present in Y + RE accessory minerals from granitic rocks; some Bi may also be present in accessory apatite.

Clark (1965) analyzed 15 apatite specimens from one locality, the hydrothermal copper-tungsten deposit of Ylöjärvi, South West Finland; Bi was detected in most specimens, reaching a maximum of 1150 ppm in one. Clark suggests that Bi^{3+} is a structural constituent of apatite and substitutes for Ca^{2+}.

In the present work, Bi was detected in five of the thirteen specimens which were examined. Native Bi and Bi minerals occur at Ylöjärvi as well as at the three localities from which we have examined apatites and found them to contain Bi. Cruft (1966) has observed minute inclusions, including ore minerals, in apatite and the possibility arises therefore that some Bi may be present in sulphide inclusions rather than as a constituent of the apatite structure.

References

Ahrens L. H. and Kable E. (1968) Cationic radii. In preparation.

Brooks R. R. and Ahrens L. H. (1961) Some observations on the distribution of Tl, Cd, and Bi in silicate rocks and the significance of covalency on their degree of association with other elements. *Geochim. Cosmochim. Acta* **23,** 100–115.

Brooks R. R., Ahrens L. H. and Taylor S. R. (1960) The determination of trace elements in silicate rocks by a combined spectrochemical–anion exchange technique. *Geochim. Cosmochim. Acta* **18,** 162–175.

Clark A. H. (1965) The mineralogy and chemistry of the Ylöjärvi Cu–W deposit South West Finland: bismuth bearing apatite. *Bull. Comm. Géol. Finlande* **218,** 195–199.

Cruft E. F. (1966) Minor elements in igneous and metamorphic apatite. *Geochim. Cosmochim. Acta* **30,** 375–398.

Goldschmidt V. M. (1954) *Geochemistry.* Clarendon Press.

Gurney J. J., Oosthuizen C. O. and Willis J. P. (1965) OH band interference on the bismuth line at 3067 A. *Appl. Spectrosc.* **19,** 162.

Ito J. (1965) The synthesis of gadolinite. *Proc. Japan Acad.* **41,** 404–407.

Ito J. (1967) Synthesis of calciogadolinite. *Amer. Mineral.* **52,** 1523–1527.

Kudrina M. A. and Kudrin V. S. (1961) Gadolinite from the alkalic pegmatites of Siberia in Ginsburg, A.I. ed. New data on rare element mineralogy. *Consult. Bur., N.Y.* 84–88.

Noddack I. and Noddack W. (1931) Die Geochemie des Rheniums. *Z. Phys. Chem.* **154,** 207.

Taylor S. R. (1964) Abundances of chemical elements in the continental crust: a new table. *Geochim. Cosmochim. Acta* **28,** 1273–1286.

Welday E. E., Baird A. K., McIntyre D. B. and Madlem K. W. (1964) Silicate sample preparation for light-element analyses by X-ray spectrography. *Amer. Mineral.* **49,** 889–903.

12

Reprinted from *Geochim. et Cosmochim. Acta* **37**:283–295 (1973)

Aspects of the magmatic geochemistry of bismuth*

L. P. Greenland, D. Gottfried, and E. Y. Campbell
U.S. Geological Survey, Washington, D.C. 20242, U.S.A.

Introduction

Very little is known of the geochemistry of bismuth in igneous rocks. Such data as exist, primarily from Brooks *et al.* (1960) and Brooks and Ahrens (1961), have been summarized by Ahrens and Erlank (1969). Subsequently, Marowsky and Wedepohl (1971) have reported bismuth contents of some magmatic, metamorphic and sedimentary rocks. The common occurrence of bismuth in sulfide minerals, notably galena, (Ahrens and Erlank, 1969), suggests its chalcophilic nature. The ionic radius of Bi^{3+} (1·10 Å) is similar to that of Ca^{2+} (1·08 Å) based on the ionic radii given by Whittaker and Muntus (1970) and relatively high bismuth contents have been observed in apatite (Clark, 1965) and in rare earth minerals (Gurney and Ahrens, 1969). Available analyses indicate a general abundance level of 0·0X–0·X ppm Bi in igneous rocks, but are insufficient to delineate the mineralogic residence and behavior of bismuth during magmatic differentiation processes. Inasmuch as bismuth is often found in association with precious metals, knowledge of the abundance of bismuth in igneous rocks from different geographic localities provides background numbers relevant in geochemical exploration for ore deposits.

This paper presents analytical results for a differentiated tholeiitic dolerite suite, two calc-alkaline batholiths, and mineral separates from one of the batholiths. These data illustrate the mineralogic distribution and geochemical variation of bismuth in magmatic processes.

Analytical Methods

Minerals were separated and purified by ordinary heavy liquid and magnetic techniques. Impurities in the final fractions, estimated by microscopic observation, amounted to less than 2 per cent.

Bismuth was determined by a substoichiometric radioisotope dilution technique which has been described in detail elsewhere (Greenland and Campbell, 1972). In brief, 100 mg samples were decomposed in the presence of Bi^{207} tracer with HF–$HClO_4$, and Bi extracted as the iodide to methyl isobutyl ketone. After washing with acid–iodide, Bi was returned to the aqueous phase and reacted with a known, substoichiometric, amount of EDTA. Excess Bi was extracted as the iodide, and Bi^{207} in the EDTA complex in the aqueous phase was counted to determine the

* Publication authorized by the Director, U.S. Geological Survey.

specific activity. The abundance of Bi in the original sample was then calculated by standard techniques (e.g. Ruzikca and Stary, 1968). Standards, both solutions and rocks, were carried through the entire procedure with each batch of samples. All samples were analyzed at least in duplicate; the values reported are averages of these replicates.

Analyses of the U.S. Geological Survey standard rocks (Table 1) are in good agreement with the neutron activation analyses reported by Laul *et al.* (1970). The analytical precision for duplicate determinations, as estimated from the replicate analyses, is ±10 per cent.

Table 1. Comparison of bismuth determinations in U.S. Geological Survey standard rocks

Rock	Bismuth content (ppb) This work	Neutron activation*
G-1, granite	52	46
W-1, diabase	52	44
GSP-1, granodiorite	37	37
AGV-1, andesite	56	56
G-2, granite	41	38
BCR-1, basalt	50	47
DTS-1, dunite	5	4·8
PCC-1, peridotite	6	8

* Laul *et al.* (1970).

Results and Discussion

Great Lake, Tasmania, tholeiitic dolerite

The Great Lake sheet is a differentiated tholeiitic dolerite believed to be representative of the widespread tholeiites of Jurassic age in Tasmania and Antarctica. The petrology and major element composition of several drill cores from this dolerite have been studied by McDougall (1964), and numerous trace element determinations and the Sr^{87}/Sr^{86} ratio have been reported for the 5123 core (Heier *et al.*, 1965; Greenland and Lovering, 1966; Gottfried *et al.*, 1968; Greenland, 1971). The dolerites of the mafic lower zone consist of orthopyroxene, clinopyroxene, plagioclase and poorly crystallized mesostasis; the effect of magmatic differentiation is to increase the abundance of plagioclase and mesostasis at the expense of pyroxene. In dolerites of the silicic central zone, orthopyroxene disappears and clinopyroxene continues to decrease in relative abundance. Granophyres are formed by a large relative increase in mesostasis.

Bismuth content in these rocks (Table 2) gradually increases from the mafic through the silicic dolerites, then decreases through the granophyres. This variation is shown more clearly in Fig. 1. The increasing abundance of bismuth with differentiation implies that bismuth was concentrated in the residual melts until the appearance of a bismuth-rich mineral in the granophyres. This is consistent with the relatively low abundance of bismuth found in plagioclase and hypersthene in the Southern California batholith, to be discussed later.

Bismuth is well known to be relatively enriched in various sulfide minerals (e.g. Malakhov, 1968; Rose, 1967), but there is no evidence for the chalcophilic nature of bismuth in the Great Lake sheet. The 200–600 ft portion of the 5123 core is enriched in copper (Greenland and Lovering, 1966), gold (Rowe, 1969), and Ir (Greenland, 1971), presumably due to crystallization of sulfides. However, bismuth,

Table 2. Abundance of bismuth in rocks of the Great Lake, Tasmania, tholeiitic dolerite.

Depth in 5123 drill core (ft)	Bismuth concentration (ppb)		Average bismuth (ppb)
10	24	Granophyres	34
50	36		
75	42		
150	46	Central zone silicic dolerites	38
250	39		
300	36		
350	32		
450	40		
550	38		
600	46		
700	28		
800	28		
910	33	Lower zone mafic dolerites	28
1000	20		
1150	28		
1200	40		
1250	19		
1360	20		
1440	30		
1500	48		
1540	16		
1640	24		
lower contact drill core 5084	40		

like silver (GREENLAND and FONES, 1971), is not markedly enriched in this zone and it appears that Cu-bearing sulfides have played little part in the variation of bismuth.

The chilled contact of the 5084 core contains more bismuth than most of the 5123 core samples. Although it cannot be proven that bismuth was not lost from the magma during crystallization, the well-known deuteric alteration of the 5084 contact (McDOUGALL, 1964) and the concomitant enrichment of this sample in many elements relative to associated Tasmanian dolerites (e.g. GREENLAND and LOVERING, 1966; GOTTFRIED *et al.*, 1968) makes it probable that this sample does not reflect the initial bismuth content of the magma.

The variation of bismuth in the Great Lake sheet differs markedly from that observed in the Karroo dolerites (from a diamond mine) by BROOKS and AHRENS (1961). They report four samples from one still varying from 10 to 40 ppb Bi, similar to the Great Lake abundance; twelve samples from another sill, however, range from 20 to 800 ppb Bi. In neither sill is there any discernible effect of magmatic differentiation: the variation of bismuth appears to be random. The lack of a systematic trend in the Karroo dolerites probably reflects postmagmatic alteration or differences in analytical procedures.

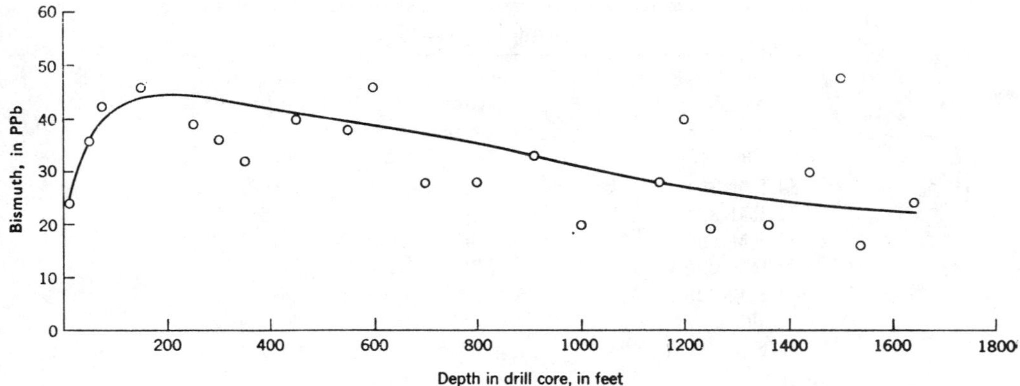

Fig. 1. Bismuth content versus depth in the 5123 drill core from the Great Lake, Tasmania, dolerite.

Southern California batholith

The Southern California batholith occupies the core of the Peninsular Range from southern Baja California northwestward to Riverside, California. The entire mass underlies an area of 40,000 square miles and thus constitutes a quantitatively significant portion of the Earth's crust. Extensive field and laboratory studies of the batholith have been reported by LARSEN (1948), LARSEN and DRAISEN (1950), LARSEN *et al.* (1958), GOTTFRIED and DINNEN (1965), TOWELL *et al.* (1965) and many others. LARSEN'S (1948) interpretation of these rocks as the product of differentiation of a gabbroic magma at depth followed by successive intrusions remains widely accepted.

Bismuth contents of the Southern California batholith rocks are given in Table 3. These data, with some exceptions, show that bismuth increases in abundance from gabbros to tonalites, then decreases in the granodiorites and returns to the original gabbro abundance in the quartz monzonites. This trend is shown more clearly in Fig. 2 where bismuth content is plotted against the Larsen index. The two tonalites relatively low in bismuth (BCl-5, BCl-12) are from a single stock in Baja California and probably represent a geographical, rather than petrochemical, variation of bismuth. Both the granodiorite aplite dike (64GSC-3) and the granite (RG-1) represent late residual liquids: the high abundance of bismuth in these rocks shows that bismuth has been concentrated in residual melts and thus the decreasing abundance of bismuth in the rocks from tonalite to quartz monzonite is due to the depletion* of a mineral phase that concentrates bismuth and does not reflect the content of bismuth in the magma.

Three rocks from the Southern California batholith were selected to determine the mineralogic residence of bismuth. Material balance calculations for a gabbro, a tonalite and a granodiorite are given in Table 4. Plagioclase, hypersthene and magnetite are about equal contributors to the total bismuth content of the gabbro; however, the somewhat high mineral summation and the low hornblende/magnetite

* Assuming a constant crystal/liquid partition coefficient, an increasing Bi content in the magma implies a higher Bi content in the mineral. Therefore a decreasing Bi content in the total rock implies a depletion of the Bi-concentrating mineral in the rock.

Table 3. Bismuth abundance in rocks of the Southern California batholith

Rock	Bi(ppb)	Larsen rock index*
Gabbros		
SLRM-354	15	−16·0
SLRM-334	10	−6·1
SLRM-299	55	−11·6
SLRM-229A	25	−1·5
SLRM-229B	34	−2·2
BL60-5	62	1·9
LTS-1	15	−7·5
Tonalites		
BL60-1	50	9·5
BL60-6	63	1·5
BL60-2	64	8·6
EL38-28	74	8·4
EL38-134	47	5·5
SLR-138	49	9·0
SLR-1016	68	0·4
GSC66-24	45	13·4
BC1-12	21	10·0
BC1-5	18	12·8
Granodiorites		
64GSC-2	44	24·5
64GSC-3 (aplite)	130	29·0
64GSC-4	57	21·6
S-14	20	25·7
S-2	42	23·2
S-9	44	20·6
S-8	130	21·2
S-10	16	28·8
S-6	30	20·9
BL60-7	48	22·0
BL60-8	60	16·9
Quartz monzonites		
EL38-167	18	25·5
EL38-265	22	27·7
LTS-2	30	26·1
Muscovite-granite RG-1	430	—

* Index = $1/3\ SiO_2 + K_2O - FeO - MgO - CaO$

distribution ratio (discussed later) suggests that the bismuth content of the magnetite has been over-estimated, perhaps due to inclusions, and thus only plagioclase and hypersthene are important contributors. In the tonalite, quartz plus plagioclase and hornblende are important contributors, but the low bismuth content based on the mineral summation implies that bismuth is largely concentrated in a minor, unknown, phase. Most of the bismuth in the quartz monzonite is derived from quartz and K-feldspar with significant additions from plagioclase and hornblende.

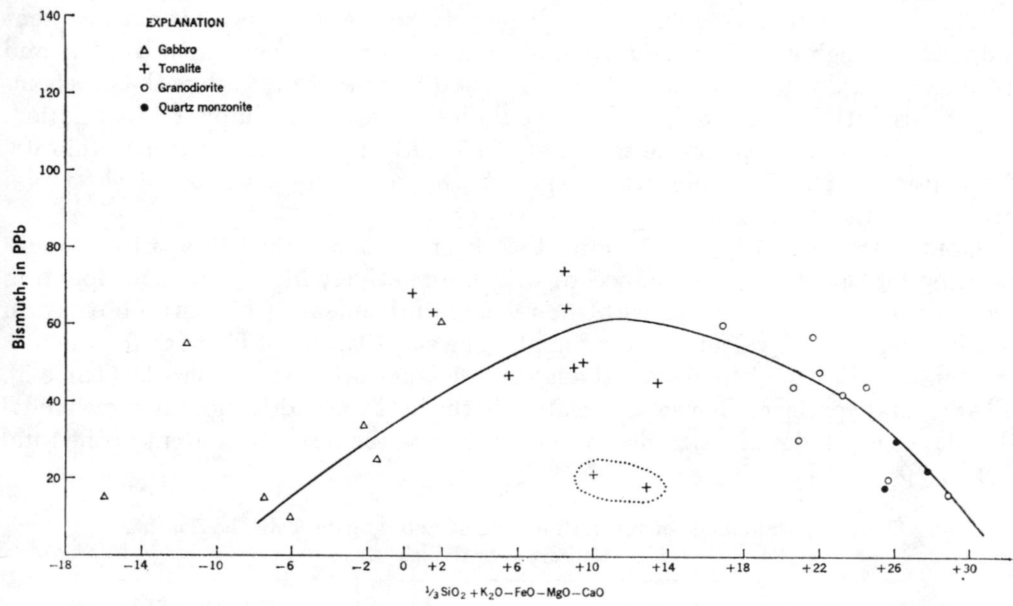

Fig. 2. Bismuth content, as a function of the Larsen index, of the rocks of the Southern California batholith. The two circled tonalites are from Baja California.

Table 4. Mineralogic distribution of bismuth in rocks of the Southern California batholith

Rock	Mineral	Modal abundance (%)	Bi (ppb)	Contribution of Bi(ppb) to rock
Gabbro (SLRM-334)	Plagioclase	65	7·4	4·8
	Hypersthene	28	15	4·2
	Hornblende	1	26	0·3
	Magnetite	2	170	3·4
				Total minerals = 12·7
				Total rock = 10·0
Tonalite (BL60-2)	Quartz + plagioclase	75	23	17
	Hornblende	7	200	14
	Biotite	18	10	1·8
	Apatite	0·02	1000	0·2
				Total minerals = 33·0
				Total rock = 64·0
Quartz monzonite (EL38-167)	Plagioclase	34	7	2·4
	Quartz	33	21	6·9
	K-feldspar	32	20	6·4
	Biotite	2	29	0·6
	Hornblende	1	160	1·6
	Magnetite	0·5	22	0·1
	Hypersthene	0·5	47	0·2
				Total minerals = 18·3
				Total rock = 18

The relatively high contribution of plagioclase to the total rock bismuth-content is due to the high modal abundance of plagioclase. In agreement with BROOKS and AHRENS (1961), plagioclase is relatively depleted in bismuth and thus the geochemistry of bismuth is more complex than would be implied by a simple Bi–Ca relation. The observation of appreciable amounts of bismuth in potassium minerals (biotite, K-feldspar) and in iron minerals (magnetite, hypersthene) also do not support a strong Bi–Ca association.

GOLDSCHMIDT (1964, p. 480) notes that bismuth is enriched in apatite by substituting for calcium; this is in accord with the relatively high abundance found in apatite from the tonalite and with the relative enrichment of bismuth observed in hornblende, another calcium-rich mineral. The substitution of bismuth for calcium was confirmed by analyzing several apatites and sphenes from these rocks (Table 5). All are greatly enriched in bismuth relative to the host rock, although their low modal abundance precludes any significant contribution to the bismuth content of the total rock.

Table 5. Abundance of bismuth in sphene and apatite from the Southern California batholith

Sample no.	Rock type	Mineral	Bismuth (ppb)
SLR-1016	Tonalite	Apatite	1900
Val-1	Tonalite	Apatite	1600
EL3-134	Tonalite	Apatite	7500
BL60-2	Tonalite	Apatite	1000
S-8	Tonalite	Sphene	15000
S-9	Tonalite	Sphene	12000
66-24	Tonalite	Sphene	2600
G-3	Tonalite	Sphene	1400
S-12	Granodiorite	Sphene	12000
S-13	Granodiorite	Sphene	5900

The previous imputation of a Bi–Ca association was due to the high bismuth content of apatite (GOLDSCHMIDT, 1964; CLARK, 1965) and gadolinite (GURNEY and AHRENS, 1969). Our sphene and apatite data support these results, but the mineral analyses discussed above show the generalization to be unwarranted. It is notable that uranium, like bismuth, is known to be enriched in apatite, sphene and rare earth minerals but not in the major rock forming Ca-bearing minerals.

It is possible that bismuth does not occur in the structure of these minerals but is present as inclusions of a minor phase. We have analyzed a number of coexisting hornblende–biotite–magnetite separates to study this possibility (Table 6). We previously inferred that residual magmas in the Southern California batholith were enriched in bismuth, and thus individual minerals, in equilibrium with successive melts, should also be enriched in bismuth. The data in Table 6 partially confirm this expectation: hornblende, biotite and magnetite contain increasingly more bismuth through the rock sequence gabbro–tonalite–granodiorite. Enrichment of the granodiorite mafic minerals in bismuth relative to those of the tonalites, the reverse of the total rock content, is consistent with the relative modal abundances of these minerals and with the enrichment of bismuth in residual magmas; however, in the

Table 6. Abundance and distribution coefficient of bismuth in coexisting hornblende, biotite and magnetite from the Southern California batholith

Rock	Bi(ppb)			Distribution coefficient		
	Hornblende	Biotite	Magnetite	hnb/bio	hnb/mag	mag/bio
Gabbro						
BL60-5	140	44	66	3·2	2·1	1·5
SLRM 334	26	—	170	—	0·15	—
LTS-1	45	—	13	—	3·5	—
Tonalite						
BL60-6	180	32	130	5·6	1·4	4·1
BL60-2	200	10	—	20	—	—
SLR-1016	180	14	—	13	—	—
GSC66-24	190	26	—	7·3	—	—
SLR-138	100	47	210	2·1	0·48	4·5
BC1-12	42	10	—	4·2	—	—
BC1-5	36	22	—	1·6	—	—
Granodiorite						
S-9	580	60	280	9·7	2·1	4·7
S-14	220	15	1100	15	0·20	73
S-8	1100	140	810	7·9	1·4	5·8
S-6	330	490	—	0·67	—	—
BL60-7	490	54	340	9·1	1·4	6·3
BL60-8	230	160	—	1·4	—	—
Quartz monzonite						
EL38-167	160	29	22	5·5	7·3	0·76
EL38-265	540	120	—	4·5		—
LTS-2	150	36	—	4·2	—	—

quartz monzonites these minerals are relatively depleted in bismuth. The approximate equilibrium of the minerals with the inferred liquid composition suggests that bismuth occurs in the mineral structure. On the other hand, co-crystallization of a minor phase appearing as inclusions in these minerals might produce a similar result and, indeed, the great scatter of the individual points and the decreasing bismuth content of the quartz monzonites favor this interpretation.

The data of Table 6 permit a further analysis of the problem. If the individual minerals are in equilibrium with a magma, they must be in equlibrium with each other. Equilibrium of the minerals can be demonstrated by a constant ratio of the bismuth content of hornblende to that of biotite, etc. (Kretz, 1961; McIntire, 1963), assuming little temperature and compositional dependence of the distribution coefficient. Such a relationship for manganese, scandium and cobalt has been shown previously for coexisting biotite and hornblende in these rocks (Greenland *et al.*, 1968, 1971; Tilling *et al.*, 1969). The ratios of bismuth content of the coexisting minerals, calculated in Table 6, show far too much scatter for the minerals to be in equilibrium, but neither are the ratios completely random as would be expected from non-equilibrium conditions: in general, hornblende > magnetite > biotite. It appears, then, that the data of Table 6 show that bismuth occurs both within and without the structure of these minerals; it is probable that much of the bismuth is present as inclusions of a minor phase. The composition of this phase must be purely speculative at this point; in view of the known chalcophilic tendency of bismuth, sulfide minerals are an obvious possibility.

These data support our position on the putative Bi–Ca association. While the generally higher bismuth content of hornblende may be due to the Ca lattice position, the occurrence of bismuth in the biotite and magnetite structures show that Ca sites are not necessary, and perhaps even unimportant, for the incorporation of bismuth.

Idaho batholith

The Idaho batholith is part of the great chain of Mesozoic batholiths extending along the western margins of Mexico, the United States and Canada; it underlies an area of about 16,000 square miles and occupies the main part of the mountains of central Idaho. The chemistry and petrology have been studied and compared with those of the Southern California batholith by LARSEN and SCHMIDT (1958). The rocks of the two batholiths are chemically similar with the major difference that gabbro is less abundant and quartz-monzonite more abundant in the Idaho batholith. LARSEN and GOTTFRIED (1961) have reported uranium analyses of the rocks studied here.

Bismuth content of the Idaho batholith rocks is given in Table 7. The Idaho tonalites and granodiorites are similar to those of the Southern California batholith in the abundance and variation of bismuth, with the granodiorites containing less bismuth than the tonalites. Two of the three quartz monzonites, however, are an order of magnitude richer in bismuth than quartz monzonites of the Southern California batholith.

The porphyritic tonalite gneiss (L-113), which has a relatively high bismuth content (110 ppb), also has an unusually high uranium content (13 ppm; LARSEN and GOTTFRIED, 1961). This rock is characterized by large K-feldspar crystals in a tonalitic matrix and occurs in a relatively small mass (near Yankee Fork, below Stanley) associated with older schists. According to LARSEN and SCHMIDT (1958) this rock may have originated by the introduction of granitic material into schists. It is possible that the relatively high K_2O, U and Bi contents of this sample reflect the introduction of these elements by highly fractionationed residual solutions. The granodiorite (L-112) with the highest bismuth content (77 ppb) is from a mass described by LARSEN and SCHMIDT (1958) as 'fine' granodiorite. This sample was collected about a mile from the bismuth-rich porphyritic tonalite gneiss. The high bismuth content of the latter may have resulted from its introduction by residual solutions derived from the granodiorite magma. Alternatively, the relatively high bismuth contents of these samples may reflect additions from subsequent ore-bearing solutions that formed the silver–gold deposits of the nearby Yankee Fork mining district (UMPLEBY, 1913).

LARSEN and GOTTFRIED (1961) note that the muscovite-bearing quartz monzonites contain less uranium and thorium than the ordinary quartz monzonites, a similar fractionation to that found here for bismuth. The parallelism of uranium and bismuth also occurs within the rock types: both bismuth and uranium are low in the L84 quartz monzonite relative to the other two; in the muscovite-bearing quartz

Table 7. Bismuth abundance in rocks of the Idaho batholith

Rock	Bi (ppb)
Tonalites	
L63A	63
L65	30
L81	54
L227	86
Granodiorites	
L259	28
L112	77
L70	36
L288	21
L295	27
Quartz monzonites	
L207	220
L219	190
L84	45
Muscovite–quartz monzonites	
L272	53
L274	11
L15	28
L253	16
L209	24
L263	68
Porphyritic tonalite gneiss	
L113	110

Note: Location of the samples within the batholith is given in LARSEN and GOTTFRIED (1961)

monzonites, the two richest in bismuth are richest in uranium and L84 is lowest in both elements. The fractionation within the rock types is probably due to magmatic differentiation as both elements are concentrated in residual liquids. The large difference in bismuth and uranium contents between these two closely associated rock types cannot be simply attributed to crystallization differentiation and suggest the possibility that separation and loss of these elements from the muscovite–quartz monzonite magma occurred during crystallization.

Table 8 presents a comparison of the average bismuth content of the Idaho batholith and the Southern California batholith. In view of the small number of samples and the variability within rock types, no great significance can be claimed for these values. Nevertheless, these data suggest an overall similarity of the distribution and abundance of bismuth in the two batholiths; the only important difference between these batholiths is the high abundance of bismuth in two of the three Idaho quartz monzonites which yields a somewhat higher average for the Idaho batholith as a whole.

Table 8. Average bismuth content of the Idaho and Southern California batholiths

Batholith	Rock type	Number of samples	Fraction of batholith*	Average Bi (ppb)
	Tonalite	4	0·12	58
Idaho	Granodiorite	5	0·53	38
	Quartz monzonite	3	0·26	152
	Muscovite–quartz monzonite	6	0·09	33
	Weighted average			70
	Gabbro	7	0·07	31
Southern	Tonalite	10	0·63	50
California	Granodiorite	11	0·28	56
	Quartz monzonite	3	0·02	23
	Weighted average			50

* Data from LARSEN (1948) and LARSEN and SCHMIDT (1958).

CONCLUSIONS

The three suites studied here all demonstrate an enrichment of bismuth in residual magmas during the course of differentiation. It follows from this observation that most of the bismuth in not concentrated in the rock-forming minerals. Material balances show that bismuth is distributed fairly uniformly among the major minerals, with quartz and plagioclase the most important contributors to the total content because of their high modal abundance. However, the decreasing bismuth content in some of the later rocks in spite of the inferred higher liquid abundance of bismuth, the low material balance summation of one of the tonalites, and the lack of equilibrium in coexisting hornblende–biotite–magnetite assemblages suggest that bismuth is largely concentrated in a minor mineral phase (possibly sulfides) and therefore that the concentration of bismuth observed in the major minerals is at least partly due to inclusions.

Evidence for the substitution of bismuth for calcium, expected from the similarity of ionic radii, was seen in the great enrichment of bismuth in sphene and apatite. The high content of bismuth in hornblende, though much less than in sphene and apatite, may be due to the same effect. On the other hand, non-calcium bearing minerals contain more bismuth than does plagioclase, and it appears that any Bi–Ca relationship is of little significance to the magmatic geochemistry of bismuth.

The data reported here are not adequate to infer a crustal abundance of bismuth. Nevertheless, the three suites represent a large fraction of the Earth's crust and the weighted averages, 32, 50 and 70 ppb Bi for the Great Lake dolerite, the Southern California batholith and the Idaho batholith, respectively, suggest that TAYLOR'S (1964) estimate of 170 ppb Bi for the continental crust is too high. We would suggest, tentatively, a value of 50 ppb Bi for the continental crust. MAROWSKY and WEDEPOHL (1971), on the basis of 19 analyses, derive a value of 60 ppb Bi for magmatic rocks of the continental crust in rather surprisingly close agreement with our estimate.

Acknowledgements—We are indebted to R. KEAYS, F. FLANAGAN, M. FLEISCHER and D. J. SWAINE for a critical reading of the manuscript.

REFERENCES

AHRENS L. H. and ERLANK A. J. (1969) Bismuth. In *Handbook of Geochemistry*, (editor K. H. Wedepohl), Vol. II-1, Chap. 28, 25 pp. Springer-Verlag, Berlin.

BROOKS R. R. and AHRENS L. H. (1961) Some observations on the distribution of thallium, cadium, and bismuth in silicate rocks and the significance of covalency on their degree of association with other elements. *Geochim. Cosmochim. Acta* **23,** 100–115.

BROOKS R. R., AHRENS L. H. and TAYLOR S. R. (1960) The determination of trace elements in silicate rocks by a combined spectrochemical–anion exchange technique. *Geochim. Cosmochim. Acta* **18,** 162–175.

CLARK A. H. (1965) The mineralogy and geochemistry of the Ylöjarvi Cu–W deposit, southwest Finland; bismuth-bearing apatite. *Finland, Comm. Geol. Bull.* **218** (*Geol. Soc. Finlande, Compt. Rend.* **37**), 195–199.

GOLDSCHMIDT V. M. (1954) *Geochemistry*, 730 pp. Clarendon Press.

GOTTFRIED D. and DINNIN J. I. (1965) Distribution of tantalum in some rocks and coexisting minerals of the Southern California batholith. *U.S. Geol. Surv. Prof. Paper* 525-B, B96-B100.

GOTTFRIED D., GREENLAND L. P. and CAMPBELL E. Y. (1968) Variation of Nb–Ta, Zr–Hf, Th–U and K–Cs in two diabase–granophyre suites. *Geochim. Cosmochim. Acta* **32,** 925–947.

GREENLAND L. P. (1971) Variation of iridium in a differentiated tholeiitic dolerite. *Geochim. Cosmochim. Acta* **35,** 319–322.

GREENLAND L. P. and CAMPBELL E. Y. (1972) Determination of nanogram amounts of bismuth in rocks by substoichiometric isotope dilution analysis. *Anal. Chim. Acta* **60,** 159–165.

GREENLAND L. P. and FONES R. (1971) Geochemical behavior of silver in a differentiated tholetiitic dolerite sheet. *Neues Jahrb. Mineral. Monatsch.* **9,** 393–398.

GREENLAND L. P., GOTTFRIED D. and TILLING R. I. (1968) Distribution of manganese between coexisting biotite and hornblende in plutonic rocks. *Geochim. Cosmochim. Acta* **32,** 1149–1163.

GREENLAND L. P. and LOVERING J. F. (1966) Fractionation of fluorine, chlorine, and other trace elements during differentiation of a tholeiitic magma. *Geochim. Cosmochim. Acta* **30,** 963–982.

GREENLAND L. P., TILLING R. I. and GOTTFRIED D. (1971) Distribution of cobalt between coexisting biotite and hornblende in igneous rocks. *Neues Jahrb. Mineral. Monatsch* **1,** 33–42.

GURNEY J. J. and AHRENS L. H. (1969) The bismuth contents of some rare-earth minerals, notably gadolinite. *Geochim. Cosmochim. Acta* **33,** 417–420.

HEIER K. S., COMPSTON W. and MCDOUGALL I. (1965) Thorium and uranium concentrations, and the isotopic composition of strontium in the differentiated Tasmania dolerites. *Geochim. Cosmochim. Acta* **29,** 643–659.

KRETZ R. (1961) Some applications of thermodynamics to coexisting minerals of variable composition; examples, orthopyroxene–clinopyroxene and orthopyroxene–garnet. *J. Geol.* **69,** 361–387.

LARSEN E. S. JR. (1948) Batholith and associated rocks of Corona, Elsinore, and San Luis Rey quadrangles, southern California. *Geol. Soc. Amer. Mem.* **29,** 182 pp.

LARSEN E. S., JR. and DRAISEN W. M. (1950) Composition of the minerals in the rocks of the Southern California batholith. Internat. Geol. Cong., 18th, Great Britain, 1948, Rept., pt. 2, 66–79.

LARSEN E. S., JR. and GOTTFRIED D. (1961) Distribution of uranium in rocks and minerals of Mesozoic batholiths in Western United States. *U.S. Geol. Surv. Bull.* **1070-C,** p. 63–103.

LARSEN E. S., JR., GOTTFRIED D., JAFFE H. W. and WARING C. L. (1958) Lead alpha ages of the Mesozoic batholiths of Western North America. *U.S. Geol. Surv. Bull.* **1070-B,** p. 35–62.

LARSEN E. S., JR. and SHMIDT R. G. (1958) A reconnaissance of the Idaho batholith. *U.S. Geol. Surv. Bull.* **1070-A,** 33 pp.

LAUL J. C., CASE D. R., F. SCHMIDT-BLEEK and M. E. LIPSCHUTZ (1970) Bismuth contents of chondrites. *Geochim. Cosmochim. Acta* **34,** 89–103.

MALAKHOV A. A. (1968) Bismuth and antimony as indicators of some conditions of ore formation. *Geochem. Intern.* **5,** 1055–1068.

MAROWSKY G. and WEDEPOHL K. H. (1971) General trends in the behavior of Cd, Hg, Tl and Bi in some major rock forming processes. *Geochim. Cosmochim. Acta* **35,** 1255–1267.

MCDOUGALL I. (1964) Differentiation of the Great Lake dolerite sheet, Tasmania. *Geol. Soc. Aust. J.* **11,** 107–132.

McIntire W. L. (1963) Trace element partion coefficients—a review of theory and applications to geology. *Geochim. Cosmochim. Acta* **27,** 1209–1264.

Rose A. W. (1967) Trace elements in sulfide minerals from the Central district, New Mexico and and the Bingham district, Utah. *Geochim. Cosmochim. Acta* **31,** 547–585.

Rowe J. J. (1969) Fractionation of gold in a differentiated tholeiitic dolerite. *Chem. Geol.* **4,** 421–427.

Ruzikca J. and Stary J. (1968) *Substoichiometry in Radiochemical Analysis*, 150 pp. Pergamon Press.

Taylor S. R. (1964) Abundances of chemical elements in the continental crust: a new table. *Geochim. Cosmochim. Acta* **28,** 1273–1286.

Tilling R. I., Greenland L. P. and Gottfried D. (1969) Distribution of scandium between coexisting biotite and hornblende in igneous rocks. *Geol. Soc. Amer. Bull.* **80,** 651–668.

Towell D. G., Winchester J. W. and Spirn R. V. (1965) Rare-earth distribution in some rocks and associated minerals of the batholith of Southern California. *J. Geophys. Res.* **70,** 3485–3496.

Umpleby J. B. (1913) Some ore deposits in northwestern Custer County, Idaho. *U.S. Geol. Survey Bull.* **539,** 104 pp.

Whittaker E. J. W. and Muntus R. (1970) Ionic radii for use in geochemistry. *Geochim. Cosmochim. Acta* **34,** 945–956.

Part IV

PHASE EQUILIBRIUM

Editors' Comments on Papers 13, 14, and 15

13 **VAN HOOK**
The Ternary System Ag_2S-Bi_2S_3-PbS

14 **GODOVIKOV et al.**
Experimental Study of the PbS-Bi_2S_3 *System*

15 **CRAIG**
Phase Relations and Mineral Assemblages in the Ag-Bi-Pb-S System

Information describing phase equlibria for bismuth under various physical and chemical conditions is widely scattered throughout the literature. The system PbS-Bi_2S_3 was first studied experimentally by Graham et al. (1953). Three studies have brought this material together and at the same time have added new and valuable contributions to our knowledge of middle- and high- temperature phase equilibrium relations within the bismuth-lead-silver sulfide system. Much additional phase equilibria data for bismuth, its oxides, sulfides, carbonates, and so forth can be found in the well-known publication by Levin et al. (1964) and its annual supplements.

The first of the three studies discussed here (Paper 13) is that of Van Hook who has summarized the pertinent data from the literature preceding his study. His experimental work has confirmed the results of earlier studies of the Bi-S system. He showed that Bi_2S_3 melts congruently with Ag_2S and PbS at the pressure of the system. Van Hook gives a list of minerals reported for the system Ag_2S-Bi_2S_3–PbS. Data for other synthetic phases such as $\alpha AgBiS_2$, $Ag_2S \cdot 3Bi_2S_3$, $(Ag_2S)_I$ are also presented. This work was the first detailed experimental investigation of the solubilities and phase relations in this system. The study established the solubility relations of silver sulfide and bismuth sulfide in galena and galena-type solid solutions.

Phase equilibria in the system Bi-S are complicated by relatively high vapor pressures generated at melting temperatures, especially for Bi_2S_3 and for sulfur-rich compositions. For this reason, the system was not studied in earlier investigations. Again, given the importance of this system, Van Hook's study provided the initial breakthrough by providing an understanding of the reaction products and minerals produced with solutions of the type noted above.

He also demonstrated that the sulfur vapor pressure in equilibrium with bismuth sulfide-sulfur mixtures is not negligible at high temperatures. The data suggested that two liquids are more likely than one above 715°C for sulfur-rich compositions of this type.

Van Hook showed that the inversion in matildite (at a temperature of 195°C) in $\alpha AgBiS_2$ is not greatly changed by excess Ag_2S or Bi_2S_3. Furthermore, the inversion in Ag_2S-rich solutions and $AgBiS_2$ requires several days to go to completion. With excess Bi_2S_3, the inversion requires several weeks. In the system $AgBiS_2$-PbS, solid solution is complete at high temperatures, with both the solidus and liquidus temperatures rising continuously from the melting point of $AgBiS_2$ (801°C$\pm$5°) to that of PbS (1115°C$\pm$5°). Information for the ternary system Ag_2S-Bi_2S_3-PbS is summarized in Figure 12 of the article.

The experimental results of studies in this system indicate extensive ternary solubility in the galena-matildite structure. These results also showed that schapbacite and schirmerite are not distinct bismuth mineral species. They are, instead, members of a continuous series of solid solutions and are unstable below the solvus. Van Hook presents a good summary of experimental studies of trace elements in galena and shows that galena that is high in Bi from "high temperature" deposits is almost invariably high in silver. This results from the high amounts of Bi_2S_3 held in the galena phase in the binary system PbS-Bi_2S_3. Of considerable importance was the clear demonstration that when bismuth is present, the solubility of silver in galena is extensive even at low temperatures. A primary problem with Van Hook's study is that for many of the equilibrium curves presented, few data points were used and in many instances an equilibrium state was not reached in the melts.

Van Hook received his Ph.D. from Penn State University and subsequently joined the staff of the Research Division of the Raytheon Company of Waltham, Massachusetts where he continued his research career into the physics and chemistry of solid materials.

Paper 14 is by Godovikov and associates who made supplementary experiments of the PbS-Bi_2S_3 system using pyrosynthetic and hydrothermal methods designed to explain gaps in the earlier work of Van Hook as well as Graham et al. (1953). These investigations served to provide more precise data on the phase diagrams of the appropriate PbS-Bi_2S_3 systems. They demonstrated that care is necessary in comparing data obtained from pyrosynthetic and hydrothermal recrystallization under different conditions.

Godovikov et al. established evidence for a small amount of solid solution of PbS into Bi_2S_3 at temperatures above 671°C. More importantly, they showed that there is a combination of PbS-$2Bi_2S_3$ (cannizzarite-bonchevite) that breaks down into bismutine and galenobismutite at temperatures lower than 671° ± 5°C. Since the combination is formed at 737 ± 4°C, the region of its stability on the composition diagram is restricted to the narrow temperature interval of 671–737°C.

These investigators point out the possibility of using cannizzarite as a geological thermometer to provide a better understanding of its synthesis at a temperature above 671°C. Caution was urged regarding transfer of data obtained from laboratory methods to what might occur under natural conditions. A good summary of information in the Russian literature that was not easily obtainable previously is provided.

Craig in 1967 (Paper 15) gave both a concise summary and historical review of known problems regarding the phase relations of the Bi-Ag-Pb-S quaternary system and added significant new information for the system. Many minerals, described as existing in the Ag-Bi-Pb-S system, have been shown to be mixtures of phases. Where appropriate, Craig uses American terminology to discuss these (for example, matildite for $AgBiS_2$ instead of the term *schapbachite*). The system studied and reported in this paper is of special interest to mineralogists because it contains common sulfide minerals and the most abundant metallic native elements. In Craig's work all the phase assemblages are in equilibrium with vapor. Detailed isotherms and discussions for the system Ag-Bi-S are presented for 920°, 715°, 625°, 350°, and 325°C. This provides a simple way for one to follow the various reaction routes encountered in a cooling solution of the members of this system and duplicates to some degree the hydrothermal conditions (high to medium temperature) encountered in nature.

Craig reviews previous work in the various binary and ternary systems making up the Ag-Bi-Pb-S system and evaluates them in light of modern data. Similar data (isotherms) are summarized for the first time for the Bi-Pb-S system at temperatures of 920°, 400°, and 100°C using known phase relations in the appropriate binary and ternary systems and in selected planes within the quaternary system. Craig constructed tentative phase relations within the

Ag-Bi-Pb-S quaternary system. These data are represented and discussed briefly in three figures. The three figures are for high temperature (1050°C), intermediate temperature (400°C), and low temperature (100°C). His experimental investigations of natural cosalites suggest that this mineral is only stable below 425±25°C. The problem of cosalite stability is especially important. Although relatively common in many ore deposits, the phase has not been synthesized in the pure Bi-Pb-S system; consequently its exact position in the system is not clear. This fact points out clearly the value of this mineral as an indicator of the limiting thermal conditions at the time of mineralization. He notes that the general phase relations in the Ag-Bi-Pb-S system are now known but points up the fact that we still need to know the effects of pressure on the sub-solidus phase relations in the system.

While the effects of trace elements on phase equilibria in the system may be small, the effects of such key elements as Cu, Fe, and Zn, which are common in many sulfides, and of Se and Te, which can substitute for S, need to be investigated in a systematic fashion. This area of bismuth investigation badly needs attention. In the strictest sense, the conclusions reached by Craig are applicable only to natural systems in which a vapor phase was also present. The latest information on the phase relations of the various bismuth systems can be found in the diagrams given in Levin et al. (1964).

Craig began his studies on the Bi system as a student at Lehigh University, Bethlehem, Pennsylvania, where he received his Ph.D. in 1965. He continued his studies of phase relations at the Geophysical Laboratory, Carnegie Institution of Washington, Washington, D. C., and is now at the Virginia Polytechnic Institute at Blacksburg, Virginia.

REFERENCES

Graham, A. R., R. M. Thompson, and L. G. Berry. 1953. Studies of mineral Sulfo-salts. XVII: Cannozzarite. *Am. Mineral.* **38**:531–44.

Levin, E. M., C. R. Robins, and H. F. McMurdie. 1964. *Phase Diagrams for Ceramists.* American Ceramics Society, Columbus, Ohio. Supplements have been published regularly since 1964.

13

Reprinted from *Econ. Geology* **55**:759–788 (1960)

THE TERNARY SYSTEM Ag_2S-Bi_2S_3-PbS [1]

H. J. VAN HOOK

ABSTRACT

Phase equilibria in the systems Bi-S, Ag_2S-PbS, Ag_2S-Bi_2S_3, PbS-Bi_2S_3, $AgBiS_2$-PbS, and Ag_2S-Bi_2S_3-PbS have been studied by thermal analysis and quenching methods. The high solid solubility of bismuth and silver in the synthetic galena-type structures is correlated with the common occurrence of these elements in natural galena.

INTRODUCTION

STUDIES of argentiferous galena deposits (13) have shown that galena commonly contains minute inclusions of silver minerals that occur in galena as irregular inclusions or as lamellae. The lamellar intergrowths are commonly considered to represent unmixing during cooling as a result of decreasing solid solubility at low temperatures (27).

Individual silver minerals have not been found in some silver-producing ore deposits (21) and here the silver is probably held in solid solution in ore minerals such as galena and tetrahedrite.

Because of the minute size of the inclusions in most argentiferous galena, a positive identification of silver mineral, or minerals, is often not possible, and argentite has generally been assumed to be responsible for the silver content in these ores. The discovery that argentite is only slightly soluble in galena (22) left the question unanswered as to why argentite should be preferentially incorporated as minute disseminated bodies in a mineral in which it is so sparingly soluble.

Ramdohr (27) studied the mineralogy of argentiferous galena extensively and found that the most common inclusions are sulfosalts of silver, not argentite, as had been previously assumed. Furthermore, the sulfosalts commonly occur as lamellar inclusions parallel to certain crystallographic planes in galena, suggestive of unmixing, whereas argentite inclusions are generally irregular or rounded in shape. Intergrowths of galena and matildite ($AgBiS_2$) (Fig. 1) have provided a particularly good example of unmixing in galena solid solutions as specimens have been found in which the sulfosalt phase occurs in amounts sufficient to permit a positive identification. Ramdohr's studies (26) of these intergrowths indicated that solid solution of matildite in galena may be extensive, especially at high temperatures, and that the textures observed are due to a decrease in solubility related to a structural change in matildite at low temperatures.

The majority of the known sulfosalts can be grouped compositionally as

[1] Submitted in partial fulfillment for the Degree of Doctor of Philosophy, Pennsylvania State University.

FIG. 1. A natural intergrowth of β matildite (white lamellae) and galena (dark phase) solid solution. Ramdohr (1938), 70 ×.

sulfides of lead, copper, silver, and one or more of the group V_b elements: arsenic, antimony, or bismuth (25). Although structurally complex, they appear to be related to the simple sulfide structures such as galena or sphalerite by a distortion to varying degrees of these close-packed "basic" structures. Ross (30) and Hellner (15) have discussed the relationship between simple "basic" and variously complex "derivative" structures of sulfides and sulfosalts. The usefulness of this approach is best shown by those minerals that have inversions to high temperature polymorphs isostructural with one of the "basic" sulfide structures (4, 30). This is true of the silver sulfosalts matildite ($AgBiS_2$), and aramayoite ($Ag(BiSb)S_2$), and miargyrite ($AgSbS_2$), all of which have been reported to transform at about 200° C from complex low-temperature forms to high-temperature galena-type structures (11).

Trace element studies as reported by Fleischer (6) have shown that antimony, bismuth, and silver are the most common constituents in galena and, in some cases, are present in amounts up to several percent. Oftedahl (23) observed that the amount of these elements present in galena is greater in the "high temperature" deposits. He found (24) that solid solutions containing silver and bismuth in equal molecular proportions are indistinguishable from normal galena in polished section but that samples containing more bismuth than silver are distinctly anisotropic and exhibit octahedral parting in contrast to the usual (001) cleavage. These properties have been attributed to charge imbalance resulting from the atomic substitution of Bi^{+3} for Pb^{+2} (24), or to oriented exsolution intergrowths in the host mineral (34).

Taken together, the mineralogical and chemical studies suggest that argentite is only slightly soluble in galena, whereas bismuthinite and certain silver sulfosalts have much higher solubilities and that the lamellar intergrowths arise from decreasing solid solubility at low temperatures.

Despite the detailed mineralogical studies on galena intergrowths, no systematic experimental investigation of solubilities and phase relations has been made which might unify and coordinate these observations. Accordingly, phase equilibrium studies in the system Ag_2S-Bi_2S_3-PbS were undertaken to establish the solubility of silver sulfide and bismuth sulfide in galena and galena-type solid solutions.

Crystalline Phases in the Ternary System

Minerals reported in the system Ag_2S-Bi_2S_3-PbS are shown in Table 1 and Figure 2. The table has been compiled principally from Palache et al. (25) except where otherwise noted and includes only those minerals whose compositions fall within the ternary system, Ag_2S-Bi_2S_3-PbS. Other synthetic phases such as α $AgBiS_2$, $Ag_2S \cdot 3Bi_2S_3$, $(Ag_2S)_I$, are also given.

EXPERIMENTAL PROCEDURE

Mixtures used in this study were prepared from "zone purified" lead and bismuth (99.99% Pb and Bi), reagent grade silver powder, and Baker U.S.P. flowers of sulfur.

Ten-gram samples of the end-member compositions were first prepared by sealing sulfur and metal mixtures under vacuum in clear silica glass tubes and slowly heating to above the melting temperature. The sulfur in the silica tube was melted at about 120° C before sealing to reduce the volume of the sulfur plus metal reactants, thereby permitting a seal with a small space above the charge. Free space above the sulfide samples was kept as small as possible to avoid the necessity of adding excess sulfur to compensate for vapor loss. After complete melting, the products were examined by x-ray diffraction and in polished section for homogeneity before being combined in weighed amounts to give the desired binary and ternary compositions.

Liquidus, solidus, and some subsolidus temperatures in the system were determined by differential thermal analysis using apparatus described by Keith and Tuttle (18). The sulfide compositions previously synthesized were sealed in silica glass vials having a reentrant in the base for a differential couple. A heating rate of 4°/minute was maintained by a motor-driven variac. Cryolite (transition 563° C, 18) and sodium chlorite (melting point, 801° C) were used as calibration standards for the differential thermal heating experiments. The temperature of the phase transitions was taken as the beginning of the differential-thermal peak on heating. All of the transitions

TABLE 1

CRYSTALLINE PHASES REPORTED IN THE SYSTEM AgS-Bi_2S_3-PbS

Mineral	Composition	Crystal structure	Remarks
—	$(Ag_2S)_I$	Unknown	Stable above inversion temperature (592° C). For compositional range and corresponding inversion temperatures, see Kracek (20). Displacive inversion to $(Ag_2S)_{II}$(argentite).
Argentite	$(Ag_2S)_{II}$	Cubic, Im3m $a_0 = 4.88$ Å	Found only as pseudomorph: stable in temperature range 592–176° C. Displacive inversion to $(Ag_2S)_I$.
Acanthite	$(Ag_2S)_{III}$	Orthorhombic (?)	Stable naturally occurring form of Ag_2S. Pseudomorphs of argentite have the acanthite structure.
Galena	PbS	Cubic Fm3m $a_0 = 5.93$ Å	
α Matildite	$Ag_2S \cdot Bi_2S_3$]	Cubic Fm3m $a_0 = 5.64$ Å	Not found in nature. Stable form above 195° C; isostructural with PbS. For composition range, see Figure 12. Sluggish inversion to matildite.
β Matildite	$Ag_2S \cdot Bi_2S_3$	Orthorhombic or hexagonal	Studied by Graham (11) and Geller and Wernick (9) by single crystal x-ray methods.
—	$Ag_2S \cdot 3Bi_2S_3$	Monoclinic (*C*-centered)	Not reported in nature. Discovered in this study.
Schirmerite	$4AgBiS_2 \cdot PbS$	Unknown	Ramdohr (27) has reported a non-cubic pattern for the only known specimen.
Schapbacite	$2AgBiS_2 \cdot PbS$	Unknown	Original material found to be a mixture of galena and β matildite (27).
Beegerite	$6PbS \cdot Bi_2S_3$	Cubic (?)	Symmetry on the basis of morphology; no definite evidence of homogeneity. Ramdohr (27) reports that much, if not all beegerite is argentian.
Lillianite	$3PbS \cdot Bi_2S_3$	Orthorhombic (?)	Berry (2) reported lillianite to be a mixture of galenobismutite and galena; original material also a mixture of sulfides. Kobellite may be an antimonian variety.
Cosalite	$2PbS \cdot Bi_2S_3$	Orthorhombic Pbnm	A fairly common mineral in mesothermal and contact metamorphic deposits.
Wittite	$5PbS \cdot 3Bi_2S_3$	orthorhombic or monoclinic (?)	Needs further study; single known occurrence in amphibolite.
Cannizzarite	$6PbS \cdot 5Bi_2S_3$ (?)	Monoclinic	Graham (12) gave monoclinic indices for synthetic and natural material; possibly identical with galenobismutite.
Galeonobismutite	$PbS \cdot Bi_2S_3$	Orthorhombic Pnam	In volcanic fumeroles with bismuthinite and in other high temperature deposits.
Chivitite	$3PbS \cdot 4Bi_2S_3$	Orthorhombic (?)	Needs confirmation—some specimens are a mixture of bismuthinite and other sulfides.
Bonchevite	$PbS \cdot 2Bi_2S_3$	Orthorhombic (?)	Kostov (19); in contact metamorphic deposit.
Ustarasite	$PbS \cdot 3Bi_2S_3$	Unknown	Mentioned by Kostov (19).
Bismuthinite	Bi_2S_3	Orthorhombic Pbnm	Typically found in hydrothermal veins "formed at relatively high temperatures" and in pegmatites.

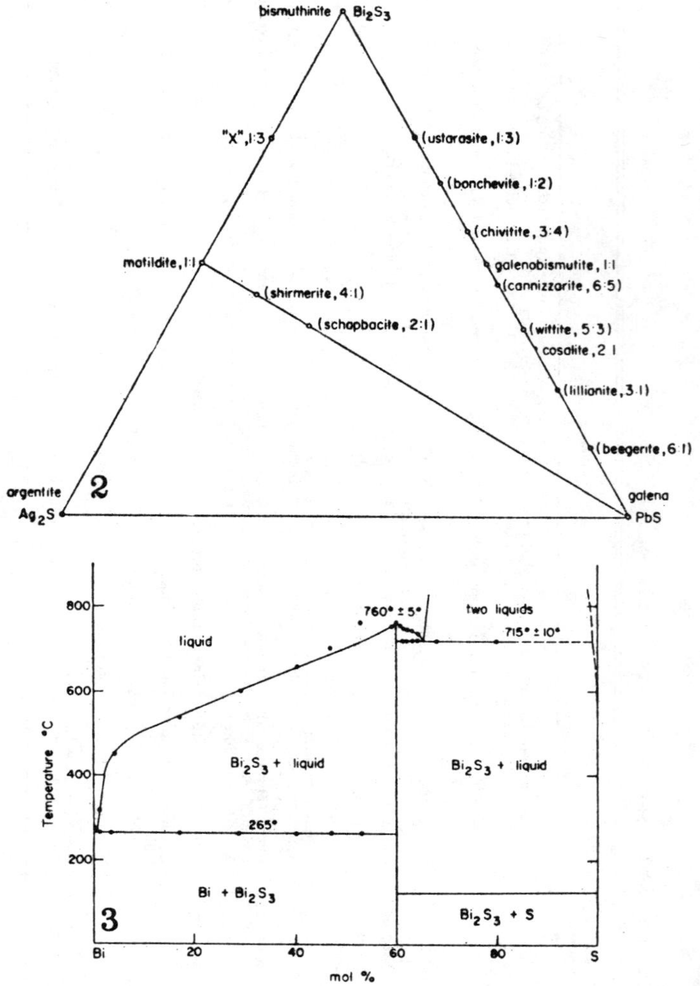

FIG. 2. Crystalline phases reported in the system Ag_2S-Bi_2S_3-PbS.
FIG. 3. Phase diagram Bi-S at the pressure of the system. Open circles are DTA temperatures, dots are from Schenck and Pardun (1933).

measured on cooling were found to be within 5° C of this value with the single exception of the Bi_2S_3 composition which supercooled about 50° C. The value recorded for the phase change was determined by the intersection of a line drawn through the trace of the background with a line through the first straight portion of the trace for the heat effect. All compositions were run at least twice and the results were found to be reproducible to within ± 2° C on heating.

Quenching runs were made in conjunction with the DTA investigation to identify the phases present and to establish solid solution limits. For this work samples were sealed in silica glass tubes and, after holding at a given temperature, were quenched by dropping the tube directly into water. The accuracy of the temperature measurement in these runs is probably ± 2° C.

Quench products were examined by x-ray diffraction using a Norelco Wide Range Diffractometer with filtered copper radiation and a separate silicon mount as a calibration standard. The accuracy of d-spacing measurements is believed to be ± 0.002 Å for reflections above 50° 2θ. Samples were also studied in polished sections using a standard metallographic microscope.

RESULTS OF THE INVESTIGATION

The System Ag-S

Kracek (20) studied the binary system Ag-S in detail using methods similar to those described above. He found a single compound, Ag_2S which melts congruently at 838 C ± 2° and extensive liquid immiscibility at melting temperatures for compositions on either side of the compound. The partial vapor pressure of Ag_2S in equilibrium with Ag° at the melting temperature is reported by Richardson and Jeffes (29) to be 10^{-6} at m., indicating that the decomposition by loss of sulfur to the vapor in sealed containers is negligible under these conditions.

TABLE 2

PHASE TRANSITION TEMPERATURES IN Ag_2S ON HEATING

	Transition $(Ag_2S)_{III}-(Ag_2S)_{II}$	Transition $(Ag_2S)_{II}-(Ag_2S)_{I}$	Melting point
Kracek	176.7° C ± 0.6°	593° C % 3°	838° C ± 2°
This study	175° C ± 5°	590° C ± 5°	835° C ± 5°

Experimental Results.—The phase transition temperatures in stoichiometric Ag_2S measured in this study are in essential agreement with those given by Kracek as is shown by a comparison of the results in Table 2. Several separately prepared samples of the stoichiometric Ag_2S composition showed a very small excess of silver which formed dendrites of hair silver when the charge was cooled to room temperature. Kracek reported similar results for silver-rich compositions. The excess silver present in these preparations, however, could be the result of errors in weighing or loss of sulfur vapor to the space above the charge and does not prove non-stoichiometric Ag_2S.

The System PbS

A study of this system by Bloem and Kröger (3) has shown that PbS is the only compound. Lead sulfide melts at 1077° C ± 5° at a total pressure ($P_{Pb} + P_{S_2}$) of 0.3 atm. whereas the maximum melting composition with 0.03 mol % excess Pb melts at 1,127° C ± 5° at 0.1 atm.

The melting point of PbS as measured in the present study (1,115° C ± 5°), is only in fair agreement with the reported value; this is probably due to the extreme variation in melting temperature for metal-rich compositions near PbS. The temperature recorded is presumably that of a composition intermediate between PbS (1,077° C) and $Pb_{1.0003}$ $S_{0.9997}$ (1127° C).

The System Bi-S

Phase equilibria in the system Bi-S are complicated by the relatively high vapor pressures generated at melting temperatures, particularly for the compound (Bi_2S_3) and for sulfur-rich compositions, and were not studied in previous work. The available data on Bi-Bi_2S_3 compositions (1) are shown in Figure 3.

Experimental Results.—The results of the differential thermal study of nine compositions are shown in Figure 3. The maximum melting (760° C ± 5°) composition corresponds to the compound Bi_2S_3, which melts at an estimated vapor pressure of 1–5 atm. Since the vapor above the charge is probably sulfur-rich, the composition of the condensed phases (solid and liquid Bi_2S_3) may be somewhat deficient in sulfur. Considering the known vapor pressure of sulfur, the mol fraction of sulfur in Bi_2S_3, and the molal volume of the container, i.e., the degree of filling of the vial, the change in composition of Bi_2S_3 at 760° C cannot exceed a loss of 0.5 mol % S and is probably much smaller than this value.

As sulfur is added to Bi_2S_3, the liquidus temperature is lowered to 715° C ± 10° and remains constant at this value for compositions having more than 65 mol % S. At high temperatures a vapor phase was observed above the condensed phases for all of these compositions. There appeared to be two possible phase relationships involving non-condensed binary equilibria which could give rise to the heat effect at 715° C: 1) Intersection of the three-phase surface, compound plus liquid plus vapor, or 2) intersection of the quadruple invariant point involving the compound, two liquids, and vapor.

Discussion of the Results.—The effect of the total pressure must be taken into account in any discussion of the phase relations because of the high vapor pressures generated in the melting range. The PT, TX, and PX projections of the assumed PTX (pressure-temperature-composition) space model of the system are shown in Figure 4. The system contains a congruently melting compound and a region of liquid immiscibility. The curves L_1CV and CL_2V in the PT projection represent univarient three-phase surfaces giving temperatures and pressures at which the compound Bi_2S_3 is stable together with a liquid and a vapor phase. The values given on the L_1CV surface were calculated from data by Schenck and Pardun (31). The curve C = L is the congruent melting curve of the compound. The univarient curves in the one-component systems, bismuth (17), and sulfur (16), are shown by dashed lines. The curves in the projection representing the surfaces of three coexisting binary condensed phases (SCL_2, $BiCl_1$ and CL_1L_2) are relatively insensitive to pressure changes and appear as straight lines essentially parallel to the pressure axis. This is in contrast with the pressure-sensitive CL_2V, L_1CV, and L_1L_2V surfaces where vapor is present. Isobaric sections ($(TX)_p$) through the PTX model are shown for several pressures in Figure

4. The TX section at P_1 is a schematic representation of the phase diagram at one atmosphere.

A discussion of the phase equilibria in terms of a PTX model does not apply directly to the experimental conditions in the present study as pressure is not an independently controlled parameter under conditions of constant volume. Pressure is generated internally and varies with temperature as well as with the molal volume of the phases present, and the relative molal volume of each phase is in turn dependent on the bulk composition (37). It is therefore necessary to consider the modifying effect of constant volume on the phase equilibria in PTX representation.

In isobaric ($(TX)_p$) sections of the binary system a composition such as X for pressure P_1 in Figure 4 is condensed at temperatures below the intersection with the three-phase surface (A), the containing vessel being collapsed by externally applied pressure. At the intersection temperature for any given pressure the assemblage changes isothermally from Bi_2S_3 plus liquid to Bi_2S_3 plus vapor (at A), and the container expands at constant pressure to accommodate the vapor phase.

In constant volume ($(TX)_v$) sections for the same binary mixture, however, vapor is present in the space above the sample over a wide range of temperatures and pressures, except for the limiting case where the thermal expansion of the condensed phase or phases reduces the available free space to zero. Assuming the liquid and vapor in these mixtures to be nearly pure sulfur, the CSV (compound + sulfur (S) + vapor) and CLV surfaces will probably parallel closely the L = V curve for pure sulfur in pressure and temperature (Fig. 4). Consequently compositions such as X (Fig. 4) at constant volume will consist of three phases whose total pressures are slightly less than those given by the L = V curve for pure sulfur at various temperatures, resulting in what is essentially a polybaric section through the binary system following this curve. The extent of departure of the CLV surface from the sulfur boiling point curve depends upon the amount of bismuth in the sulfur-rich vapor; the amount is unknown and may actually cause significant deviation, particularly at high temperatures.

It is evident from the preceding that the sulfur vapor pressure in equilibrium with bismuth sulfide-sulfur mixtures is not negligible at high temperatures and that the free space above the charge must be kept as small as possible to avoid significant losses of sulfur from the condensed phases. For this reason relatively large amounts of samples (1 gm) were used for the DTA study and an attempt was made to maximize the degree of filling of the vials.

For any given temperature and binary composition it is assumed that the three-phase surface, CLV, is limited to pressures below the boiling point curve for pure sulfur and to temperatures below the maximum melting composition (Bi_2S_3) at 760° C. Within this range of temperature and pressure, there appeared to be two possible phase relationships that might give rise to the heat effect observed at 715° C: (1) A continuous CLV solubility surface with a pressure maximum, or (2) the formation of two immiscible liquids involving a quadruple point Q (Fig. 4). The latter possibility is considered

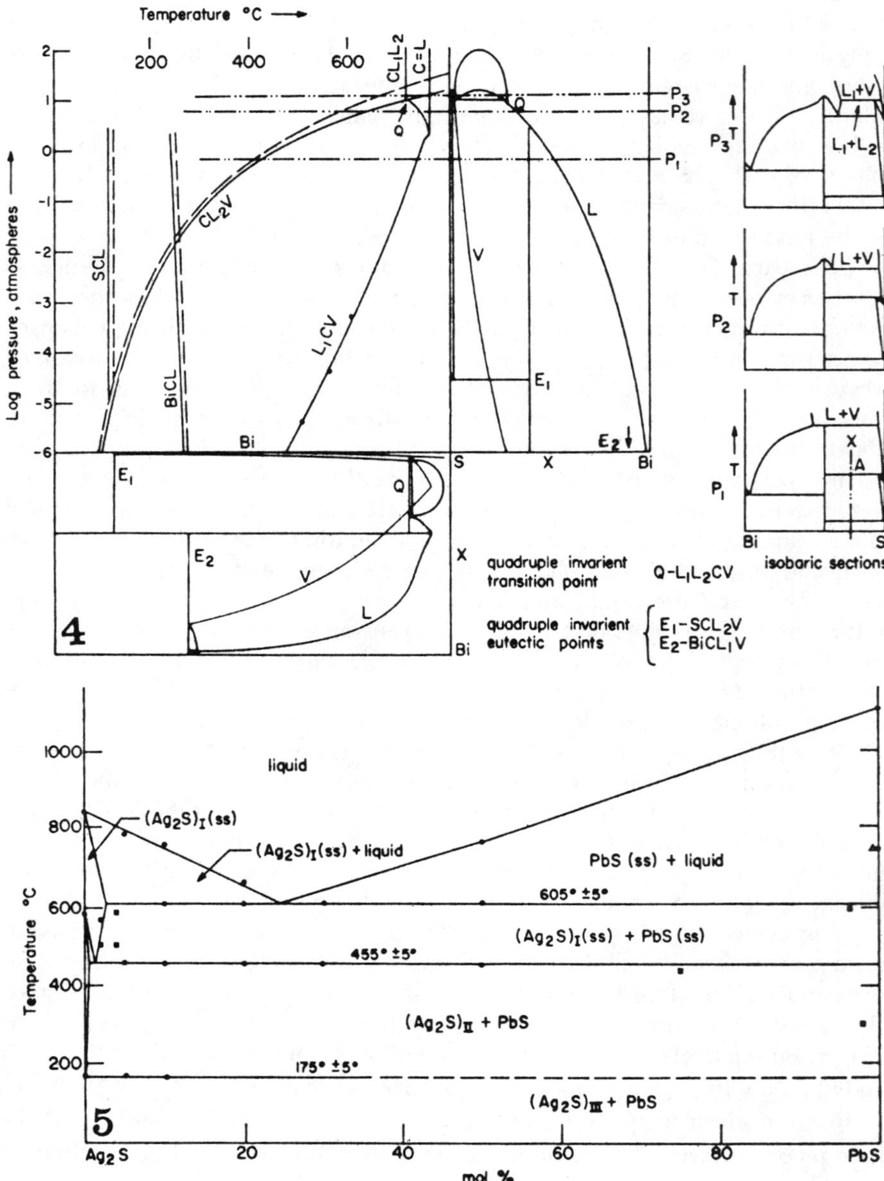

FIG. 4. Projections of the PTX model for the binary system Bi-S.

FIG. 5. Phase relations in the system Ag_2-PbS. In all phase diagrams in this paper, open circles (○) represent DTA temperatures, dots (·) one phase regions, triangles (△) liquid plus crystals, squares (□) two crystalline phases, and crosses (+ single phase boundary lines.

to be more likely for two reasons: 1) In the case of the continuous CLV surface, the intersection temperature at constant volume must change with increasing sulfur content and increasing pressure. The heat effect recorded, however, remained essentially invariant for all Bi_2S_3-S compositions as shown in Figure 3. The constancy of this transition temperature, on the other hand, is expected if liquid immiscibility is present, giving rise to a quadruple point *Q* (Fig. 4). 2) It is improbable that one liquid plus vapor are present above the reaction temperature because of the resulting high density required for the vapor phase. A calculation of the vapor density in equilibrium with one liquid phase with 65 mol % sulfur, (Fig. 3) gave a density value approximately 30 times as large as that given for pure sulfur vapor in the presence of liquid sulfur at the same temperature (715° C) indicating again that two liquids are more probable than one above 715° C for these sulfur-rich compositions.

The System Ag_2S-PbS

Liquidus relations in this system were studied by Friedrich (7). It is a simple binary system having a eutectic at 23 wt. % PbS. The eutectic temperature was reported as 630° C.

Nissen and Hoyt (22) investigated the solubility of Ag_2S in PbS by preparing a series of synthetic mixtures that were quenched from various temperatures; the fusion products were examined along with several argentiferous galena samples in polished sections. They identified very small rounded spots of argentite as a second phase under high magnification in both the natural and synthetic material. The maximum solubility was reported as less than 0.6 percent at 800° C in the synthetic samples, and approximately 0.1 percent in the galena specimens as determined by an assay for silver on the latter. More recent studies of argentite inclusions by Ramdohr (27) confirm the experimental findings of Nissen and Hoyt, although Ramdohr has suggested that smaller solubility limits are more probable, i.e., about 0.1 percent at high temperatures and 0.01 percent at low temperatures.

It should be noted here that very small amounts of bismuth (and possibly antimony, arsenic, etc.) as "impurities," in the starting materials used to make the binary compositions, would greatly increase the solubility of silver as indicated both by mineralogical observations and by the experimental results to be given presently.

Experimental Results.—The results of this investigation, presented in the form of a phase diagram (Fig. 5), are essentially in agreement with the findings of previous workers. The effect of PbS solid solution on the high-temperature heat effect in Ag_2S was found to be quite pronounced, lowering the inversion from 590° C for pure Ag_2S to 455° C for the Ag_2S solid solution. The differential thermal heat effect was also noticeably broadened by solid solution. The inversion in Ag_2S at 175° C, however, was not measurably changed by the presence of excess PbS.

The solubility limits of PbS in Ag_2S were studied by the quenching method using polished sections to identify the phases. The maximum solubility at the eutectic temperature (605° C) was found to be 3 mol % PbS. At 455° C, the inversion temperature of the maximum solid solution, the solubility is approximately 2 mol percent and below the inversion the solubility is considerably less than 2 percent as shown by exsolution blebs of PbS which appeared in Ag_2S quenched from above the inversion temperature.

The solubility of Ag_2S in PbS was also studied by the quenching method. A sample containing 0.5 mol % Ag_2S quenched from 700° C showed a trace of argentite in polished section, and comparison of X-ray data on this sample with PbS held at the same temperature showed a measurable difference in the lattice constants. Using a slow scanning speed on the X-ray diffractometer, the peak displacement indicated a change in the lattice constant of 0.0020 Å which is believed to represent a solubility of 0.4 mol % Ag_2S from $AgBiS_2$-PbS solubility data.

Discussion.—Solid solution of PbS in Ag_2S lowers the higher temperature inversion in Ag_2S by 135° C but has no measurable effect on the inversion at 175° C. This difference is probably a result of the relatively high solubility of PbS in $(Ag_2S)_I$ and the much lower solubility in $(Ag_2S)_{II}$ (argentite).

The solubility limits of Ag_2S in PbS are of particular interest in the present study and an X-ray method has been developed to determine the solubility avoiding the limitations of microscopic detection of a second Ag_2S phase. Since solid solution in PbS is very limited in the binary system, a direct determination of solubility using unit cell variation by preparng PbS-AgS mixtures was not feasible. A value of 0.4 mol % Ag_2S at 700° C was calculated by extrapolating the change in the lattice parameter of ternary Ag_2S-PbS-$AgBiS_2$ solid solutions and is probably close to the assumed maximum solubility at the binary eutectic temperature (615° C). These results are consistent with the limiting value of less than 0.6% Ag_2S proposed by Nissen and Hoyt (22).

The System Ag_2S-Bi_2S_3

Gaudin and McGlashan (8) studied a series of synthetic fusion products in the binary system Ag_2S-Bi_2S_3. The samples were heated to the melting range, cooled very slowly to room temperature, and then examined in polished sections. On the basis of the textural relationships found, the authors proposed a tentative phase diagram with a congruently melting "phase D" ($3Ag_2S \cdot Bi_2S_3$) corresponding to the rare mineral tapalpite, and the compound matildite ($Ag_2S \cdot Bi_2S_3$) which was believed to melt incongruently. Leutwein and Herrmann (21) prepared several synthetic Ag_2S-$AgBiS_2$ compositions but could find no indication of the "phase D" reported by Gaudin and McGlashan; they questioned the validity of its existence in the binary system, pointing out that the only known sample of tapalpite contains a considerable amount of tellurium.

The mineral matildite was first described in detail by Ramdohr (26). He found that the compound $Ag_2S \cdot Bi_2S_3$ has a polymorphic transition at 210° C, the low temperature form (β matildite) having orthorhombic symmetry and the high temperature form (α matildite) being cubic. The inversion temperature ($\beta \rightarrow \alpha$) was determined by heating the mineral (β $AgBiS_2$) on a high temperature X-ray unit. The inversion in matildite was re-examined by Graham (11) using both natural and synthetic specimens. Samples prepared by fusion and by hydrothermal techniques yielded α matildite which was converted to the low-temperature form by heating at 180° C for 9 days. Graham concluded that the change in symmetry is due to an inversion ($\alpha \rightarrow \beta$) and that the body-centered orthorhombic form is related to the high temperature face-centered cubic form by a slight distortion of the cubic lattice parameters. More recent work by Geller and Wernick (9) has indicated that the β form is hexagonal rather than orthorhombic.

Experimental Methods.—The phase boundaries of α $AgBiS_2$ and Ag_2S solid solutions were located by examination of polished sections. Variation of the cell dimensions of α matildite was used to confirm the results of the microscopic study and to extend the determination of Ag_2S solubility from the inversion in $AgBiS_2$ at 195° C to the melting point. The presence of excess Bi_2S_3 did not measurably affect the cell size of α $AgBiS_2$ and this phase boundry was located only approximately by microscopic examination.

Experiments dealing with the matildite inversion were carried out with samples in sealed tubes and with powdered material contained between two glass slides, i.e., open to the atmosphere. A comparison of X-ray patterns indicated that oxidation did not occur in the samples exposed to the atmosphere and that both of these annealing methods could be used to elucidate the phase relations at low temperatures. Grinding the samples after annealing was avoided since the reflections corresponding to Ag_2S are obliterated by this process; they reappear, however, upon further annealing.

Experimental Results.—The results of this study which are summarized in the phase diagram, Figure 6, are quite different from those of Gaudin and McGlashan (8). There was no indication of a "phase D" at melting temperatures and X-ray studies of runs held several months at 170° C have failed to show any evidence of this compound at low temperatures. A compound not reported by Gaudin and McGlashan having a composition $Ag_2S \cdot 3Bi_2S_3$ was found in the system; it appears to have no mineralogical equivalent. No evidence could be found for polymorphism or solid solution in this phase.

Matildite melts congruently at 801° C ± 5° and shows a large range of binary solid solution. Solid solution of Ag_2S in $AgBiS_2$ increases the cell constant from 5.640 Å to 5.706 Å (± 0.002 Å) at the eutectic temperature (615° C) where 10 mol % Ag_2S is dissolved in the crystalline phase, and decreases with decreasing temperature as shown in Figure 6 to 1.7 mol % Ag_2S at 200° C. Solid solution of 16 mol % Bi_2S_3 to stoichiometric $AgBiS_2$ does not change the size of the unit cell of the phase to a measurable degree (± 0.001 Å) although the diffraction peaks of quenched samples are more diffuse than the reflections for α $AgBiS_2$.

The solubility of Bi_2S_3 in $(Ag_2S)_1$ is 4 mol % at the eutectic (615° C). As in the system Ag_2S-PbS, solid solution lowers the high temperature transition in Ag_2S (from 590° to 522° C) but causes no measurable change in the argentite-acanthite inversion.

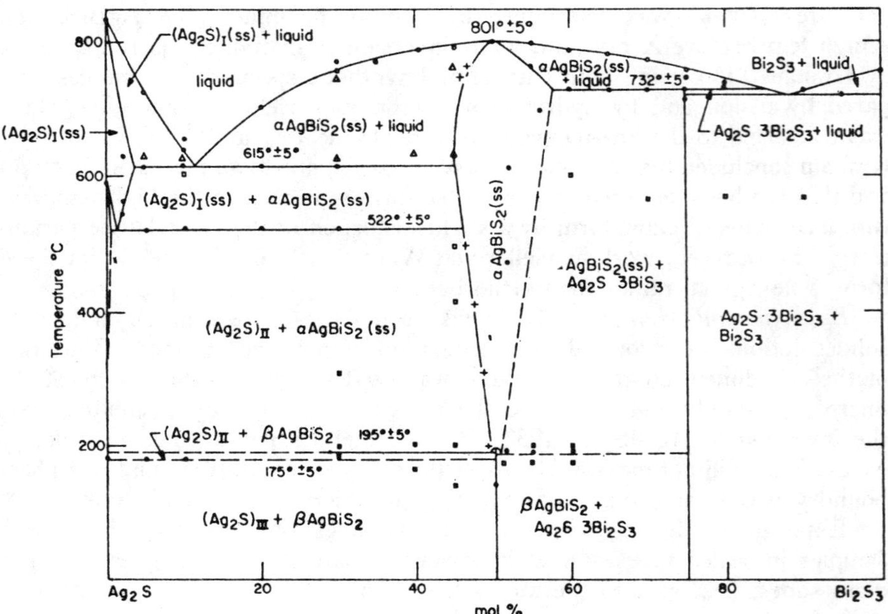

Fig. 6. The system Ag_2S-Bi_2S_3. Note extensive solubility in $\alpha AgBiS_2$.

The matildite inversion: The phase transition in matildite was studied using quenching techniques and thermal analysis. X-ray investigation of quenched samples indicated that the transition occurs reversibly at 195° C ± 5° in less than one week for $AgBiS_2$ and for binary mixtures of Ag_2S and α or β matildite. The conversion takes place more slowly (several weeks) for mixtures of α or β matildite and the $Ag_2S\cdot 3Bi_2S_3$ phase and here the inversion has been determined within the range 170°–200° C. Differential thermal analysis revealed a very small but reproducible heat effect in α matildite with excess Bi_2S_3 at 190° C for samples containing 52, 55, and 60 mol % Bi_2S_3 but not for $AgBiS_2$. Twin lamellae were also observed in these bismuth-rich matildite compositions at room temperature, but disappeared upon heating above 190° C.

Extra diffraction peaks were found in β matildite which do not correspond to d-spacings indexed by Graham (11) and may be identical to those which he attributed to uncombined Ag_2S and removed from the diffraction patterns of his synthetic product. These additional peaks were found both in the assemblages $\beta AgBiS_2 + (Ag_2S)_{III}$ (30, 40, and 45 mol % Bi_2S_3) and in mixtures identified as $\beta AgBiS_2 + Ag_2S\cdot 3Bi_2S_3$ (55 and 60 mol % Bi_2S_3) as well as in $\beta AgBiS_2$ (Fig. 6) and disappeared entirely when these compositions were held above the inversion temperature. The diffraction peaks in question are marked with an asterisk in Table 3 which gives a comparison of the d-spacings observed with results quoted by Graham. More recent X-ray studies by Geller and Wernick (9) indicate that $\beta AgBiS_2$ is hexagonal and not orthorhombic, their data compare favorably with the findings in this study. There was no evidence for the intermediate rhombohedral phase between α- and β-$AgBiS_2$ predicted by these authors.

The $Ag_2S \cdot 3Bi_2S_3$ phase: Single crystal oscillation and Weissenberg photographs were taken of cleavage fragments of $Ag_2S \cdot 3Bi_2S_3$ heated several days in a sealed tube with a KCl-NaCl salt mixture as a flux. The unit cell is monoclinic having $a_0 = 13.3$ Å, $b_0 = 4.05$ Å, $c_0 = 16.5$ Å, $\beta = 94°$; Weissenberg photographs of the zero, first, and second reciprocal lattice layers about *b* indicate that the cell is *C* face-centered, belonging to one of the three space groups, *C*2, *Cm*, or *C*2/*m*. The indexed powder diffraction data are given in Table 4. The density, measured on a one-half gram sample, was 6.8 ± 0.5, using the pycnometer method with methylene iodide ($\rho = 3.325$), as compared with a calculated density of 6.72 for a cell containing $4(AgBi_3S_5)$.

Fragments of $Ag_2S \cdot 3Bi_2S_3$ generally occur as thin plates that are elongated in the *b* direction with a good cleavage at about 15° to *b*. The direction of the longest repeat distance, taken as *c*, is approximately normal to the cleavage or parting surface responsible for the platy character. Cleavage is only fair in this direction (001), the surfaces of the tabular fragments being generally curved. In polished section $Ag_2S \cdot 3Bi_2S_3$ is strongly anisotropic and has a tendency to form equant grains as compared with the very much stronger anisotropism and acicular habit of Bi_2S_3 and the weak anisotropism of Bi-rich α $AgBiS_2$ solid solutions.

TABLE 3

COMPARISON OF POWDER DIFFRACTION DATA ON SYNTHETIC $AgBiS_2$

Graham (1951) "Ag_2S lines removed"			Present study	
d-spacing (kX)	Intensity (Cu Kα)	Calculated (hkl)	d-spacing Å	Intensity (Cu Kα)
			6.86*	2
			3.46*	2
3.29	8	011	3.30	8
3.16	1	101	3.17	3
2.82	10	110,002	2.83	10
			2.58*	1
			2.15*	0.5
2.03	6	020	2.03	6
1.967	5	200	1.969	6
			1.937*	0.5
			1.732*	2
1.708	3	121,013	1.714	3
1.670	0.5	211	1.672	1
1.648	0.5	022	1.650	2
			1.600*	0.5
1.586	0.5	202	1.583	0.5
			1.553*	0.5
1.411	1	004,220	1.414	1.5
1.316	1	031	1.319	1
1.282	1	130	1.281	1
1.247	1	310	1.248	1

* d-spacings not reported by Graham.

TABLE 4

SINGLE CRYSTAL DATA ON $Ag_2S \cdot 3Bi_2S_3$

Observed d-spacings (Å)	Intensity (CuKα)	Calculated d-spacings (Å)	Observed d-spacings (Å)	Intensity (CuKα)	Calculated d-spacings (Å)
5.45	3.5	5.49 (003)	2.585	1	$\bar{2}$.617 (206)
					2.560 (313)
5.32	1.5	5.40 (202)			
			2.460	2	2.477 (404)
4.08	3	4.06 (203)			
		4.11 (004)	2.338	1	2.352 (007)
		4.05 (010)			
			2.259	3.5	2.239 (405)
3.86	0.5	3.87 (110)			2.266 (116)
3.59	9	3.59 (20$\bar{4}$)	2.206*	1	2.208 (600)
					2.212 (406)
3.52	0.5	3.54 (11$\bar{2}$)			
			2.115	1	
3.46	4	3.47 (112)			
			2.014	3	
3.38	8	3.36 (204)			
			1.989	0.5	
3.31	3	3.31 (400)			
		3.30 (401)	1.393	0.5	
3.27	2	3.29 (005)	1.804	1	
3.22	0.5	3.20 (113)	1.728	1.5	
3.18	0.5	3.17 (402)	1.421	0.5	
		3.19 (401)	1.301	0.5	
3.12	0.5	3.16 (113)	1.166	0.5	
2.98	1	2.98 (310)			
2.955	3	2.967 (311)			
		2.953 (403)			
2.858	10	2.851 (205)			
		2.863 (31$\bar{2}$)			
2.761	1.5	2.754 (312)			
		2.744 (006)			
2.727	.5	2.732 (403)			

* Spacings were not calculated for reflections below this value.

Discussion.—The structure of $\alpha AgBiS_2$ can be derived from that of galena by substitution of Ag^{+1} + Bi^{+3} for $2Pb^{+2}$ where the silver and bismuth atoms are randomly distributed in the octahedral sites. Large amounts of either Bi_2S_3 or Ag_2S in excess of the 1 : 1 composition can be held in the structure at high temperatures, possibly as substitutional solid solution of the atomic constituents in the same octahedral positions, although the mechanism of solid solution is probably much more complex and may involve interstitial positions.

Assuming substitutional solid solution, the increase in cell dimensions of $\alpha AgBiS_2$ with added Ag_2S may be qualitatively accounted for by a comparison in size of the ionic radii (Ag = 1.12 Å and Bi = 1.1 Å, Goldschmidt, 1955) and covalent radii (Ag = 1.53 Å and Bi = 1.46 Å, Pauling and Huggins, 1934), assuming some intermediate ionic-covalent bonding in these structures. Using this reasoning, an excess of Bi_2S_3 should decrease the unit cell size: however, cubic $AgBiS_2$ with 16 mol % excess Bi_2S_3 has the same cell dimensions as does the pure compound. The assumption of substitutional solid solution with excess Ag_2S or Bi_2S_3 introduces the additional problem of accounting for an excess or deficiency of metal atoms in the octahedral positions. Alternatively, solid solution may be interstitial in the tetrahedral sites between four sulfur atoms (and four metal atoms), although the radii of bismuth and silver as cited above indicate a rather close fit in these positions. The sulfur-metal distance in $\alpha AgBiS_2$ is 2.44 Å in the tetrahedral positions as compared with 2.82 Å for octahedral coordination.

Addition of either Ag_2S or Bi_2S_3 to the compound requires a charge balance by some mechanism within the solid to maintain the original cubic structure, possibly by the creation of positive or negative "holes." The fact that such extensive electronic adjustments can be made is a further indication of the complex, non-ionic nature of the bonding.

The inversion in matildite: The temperature of the inversion (195° C) in $\alpha AgBiS_2$ is not measurably changed by the presence of excess Ag_2S or of Bi_2S_3 although the rate of inversion shown by quenching results is altered significantly. X-ray powder diffraction patterns of binary mixtures held for different periods of time both above and below the inversion interval have indicated that the inversion in Ag_2S-rich compositions and in $AgBiS_2$ requires several days to go to completion whereas with excess Bi_2S_3, the inversion takes several weeks.

Profuse twinning was observed in polished sections of α-matildite prepared in the presence of excess Bi_2S_3 and disappeared on heating above 190° C. At approximately this same temperature a heat effect was detected by differential thermal analysis. Both of these changes were observed on material which the X-ray powder diffraction patterns have shown to be cubic or nearly cubic and therefore do not represent changes that can be related to the sluggish $\alpha \leftrightarrows \beta$ inversion of matildite. These changes in Bi_2S_3 saturated matildite were not observed in stoichiometric $AgBiS_2$ or in Ag_2S saturated material. It is suggested that these changes are due to a metastable displacive transformation in Bi_2S_3 saturated αmatildite, which fortuitously takes place at a temperature near the stable matildite inversion.

The System PbS-Bi_2S_3

As the primary objective of this investigation was to study the solubility of silver in galena together with phase equilibrium relations among silver-bearing phases or minerals and galena, the system PbS-Bi_2S_3 was of secondary interest and only a few experiments were carried out with these compositions. The solubility of Bi_2S_3 in galena was studied in some detail, however, to locate the limits of solid solution along the PbS-Bi_2S_3 sideline.

Evidence for appreciable solubility of Bi_2S_3 in galena may be found in the mineralogical literature. Oftedal (24) described galena samples containing several per cent bismuth from ore deposits that are believed to have formed at high temperature. Leutwein and Herrmann (21) reported increasing bismuth content with depth in the Freiberg ore deposit without finding any discrete bismuth mineral to account for the increase. They concluded that bismuth is held in solid solution in the galena ore and that with increasing depth the higher temperature portion of the deposit is being reached in the mining operation. They prepared synthetic PbS compositions containing 2.5 and 5 wt. % Bi_2S_3 and reported a decrease in the galena cell constant with solid solution.

Experimental Methods.—The solubility of Bi_2S_3 in galena was determined by the quenching method using polished sections and x-ray diffraction techniques. The cell dimensions of galena were found to decrease approximately 0.004 Å for each 1 mol percent Bi_2S_3 held in solid solution for mixtures containing 0, 4, 7, and 10 mol % Bi_2S_3 quenched from 800° C. The composition of the galena solid solution at various temperatures was then deduced from a measure of the position of the diffraction peaks. For experiments at 500° and 400° C, a solid solution and a mixture of PbS and Bi_2S_3 were run together to assure that equilibrium solubility was being measured.

Experimental Results.—The solubility of Bi_2S_3 in galena is 9 mol % at 800° C and decreases at lower temperatures as shown in Figure 7. Equilibrium was not completely attained for the different starting materials held at 400° C and the solubility could only be given limits of 1–3 mol %. At 500° C the original solid solution and the mixture of PbS and Bi_2S_3 gave the same value for the galena lattice parameter.

Binary compositions higher in Bi_2S_3 in the system were studied by X-ray methods and by polished sections and are tentatively shown in Figure 7.

Discussion.—The mineralogical studies suggest a definite correlation between high formation temperatures and high bismuth content in galena. The experimental data support this hypothesis, as the results have shown that solid solution is appreciable at low temperatures and increases with rising temperature. The change in cell dimensions with solubility of Bi_2S_3 may be used as an indicator of minimum formation temperature only if the composition is strictly binary (PbS-Bi_2S_3) and if a second lead sulfosalt phase is present. Small amounts of silver have a pronounced effect on cell dimensions as will be shown presently, while the effect of other "impurities" such as copper, antimony, selenium, and tellurium (10) is unknown and may be significant. Solid solution of various elements may also stabilize galena beyond the solubility limits in this system. For example, a galena phase corresponding to the composition given for beegerite (14.3 mol % Bi_2S_3, Table 1) is not stable in binary compositions as the maximum solid solution at the eutectic is 9 mol % Bi_2S_3, but a ratio of PbS to Bi_2S_3 of 6: 1 is quite possible in ternary or quaternary compositions.

The octahedral parting of bismuthian galena in polished section described by Oftedal (23) and Wahlstrom (34) was not observed in the synthetic compositions. The most likely explanation for the parting in galena is that it results from growth of exsolution lamellae along octahedral directions. Apparently the lamellae did not develop sufficiently in synthetic mixtures to give rise to the octahedral parting or to permit microscopic identification. The fact that the lamellae were not visible, however, does not imply that equilibrium solubility was not measured, as solubility was established independently by

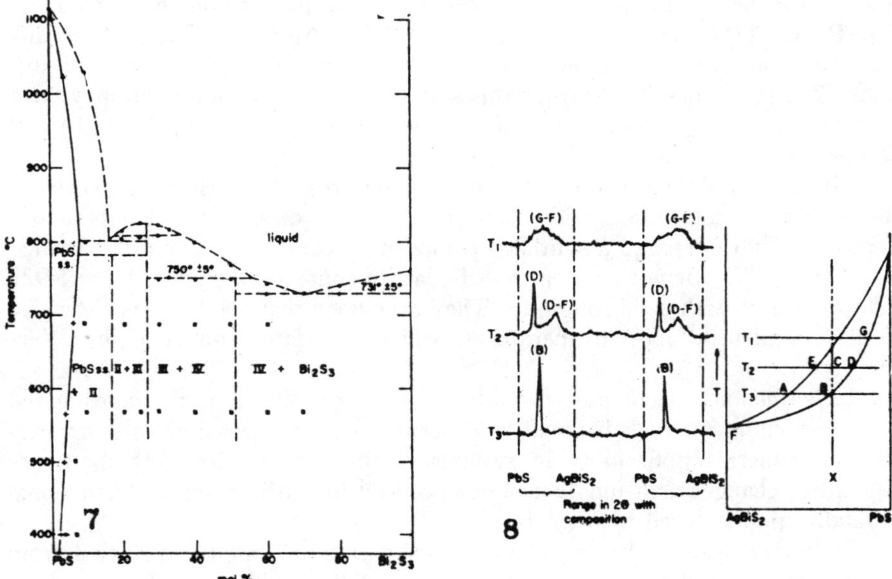

FIG. 7. The system $PbS-Bi_2S_3$.

FIG. 8. Schematic X-ray patterns for composition X in the system $AgBiS_2$-PbS quenched from different temperatures.

X-ray measurements on different starting materials held at the same temperature.

Additional experimental work is needed on the Bi_2S_3-rich portion of this system. The synthetic analogues of the known sulfosalt materials (Table 1) were not found in this study and apparently form by subsolidus reaction below the temperatures investigated. It is also possible that equilibrium was not attained in these compositions. In any event, the determination of the subsolidus stability relations would be useful in the interpretation of these phase assemblages in nature.

The System $AgBiS_2$-PbS

The mineralogical studies of matildite-galena ($AgBiS_2$-PbS) intergrowths by Ramdohr (26) have demonstrated that solid solution is extensive in this system. He described matildite-galena intergrowths in detail and proposed that unmixing of high temperature solid solutions is related to the matildite inversion. Although no systematic study of binary unmixing temperatures was made, Ramdohr assumed that the general form of the solvus could be deduced from the mineralogical information; he estimated that the solubility of $AgBiS_2$ in galena is approximately 14 mol % at a temperature below that of the inversion in matildite.

Chapman and Stevens (5) described a galena sample which contained: "numerous minute inclusions of a strongly anisotropic mineral which . . . are lath shaped and look as if they had been deposited from a solid solution in the host mineral." The composition of the sample was given as $Ag_2 \cdot Bi_2S_3 \cdot 11PbS$, corresponding to 15.4 mol % $AgBiS_2$. The powder diffraction photograph was shown, but no other X-ray data were given. Ramdohr (26) subsequently identified this second phase as matildite and may have used this specimen in his estimate of the solubility of $AgBiS_2$ in galena (14 mol %) at low temperature.

Leutwein and Herrmann (21) carried out some experimental work on the system in connection with their geological studies of the Freiberg ore deposit. They prepared a synthetic composition containing 2 mol % $AgBiS_2$ ($a_0 = 5.914$ kX) demonstrating that the lattice constant of galena ($a_0 = 5.922$ kX) decreases with solid solution. They concluded that solubility is extensive if not complete at high temperatures, which was later confirmed by Wernick (36).

Experimental Methods.—Liquidus-solidus equilibria were investigated by a quenching method that will be described in some detail since it appears to have general applicability in systems of this type. The method utilizes the lattice changes as a function of composition in conjunction with fractional crystallization induced by quenching.

In binary systems having complete solid solution, mixtures cooled from the region of solid plus liquid generally crystallize with some fractionation, the amount of fractionation depending on the rate of cooling. The liquid phase continuously changes composition with falling temperature during this process and approaches the composition of the lowest melting mixture. With rapid cooling, the residual liquid in a mixture such as "X" (Fig. 8) is intermediate between the composition of the residual liquid formed under equilibrium conditions (*A*, Fig. 8) and the final liquid resulting from complete fractionation (*F*, Fig. 8). Because of the pronounced change in the lattice constant with change in composition in this system, fractional crystallization in quenched charges is reflected in the X-ray diffraction patterns.

Applications to the present study: 1) Solidus temperatures were determined by holding solid solutions at successively higher temperatures until a change was detected in the position of the diffraction peaks of the solid phase on quenching. At higher temperatures the solidus could then be located from the composition of the primary crystals in quenched samples (viz. T_2, Fig. 8).

2) The "fractionation" quenching method was also used to decide whether the system $AgBiS_2$-PbS has a minimum on the liquidus. The existence of a binary minimum would be shown by the trend in composition of the liquid during fractional crystallization, i.e., towards the mixture having the lowest melting temperature.

3) For mixtures near $AgBiS_2$, the liquid phase crystallized essentially without fractionation giving relatively sharp diffraction peaks and it was possible to locate liquidus as well as solidus composition by measuring the position of the peaks corresponding to both of the quenched phases.

Experimental Results.—Results of the study of the binary system are shown in Figure 10. Miscibility at high temperatures is complete as was

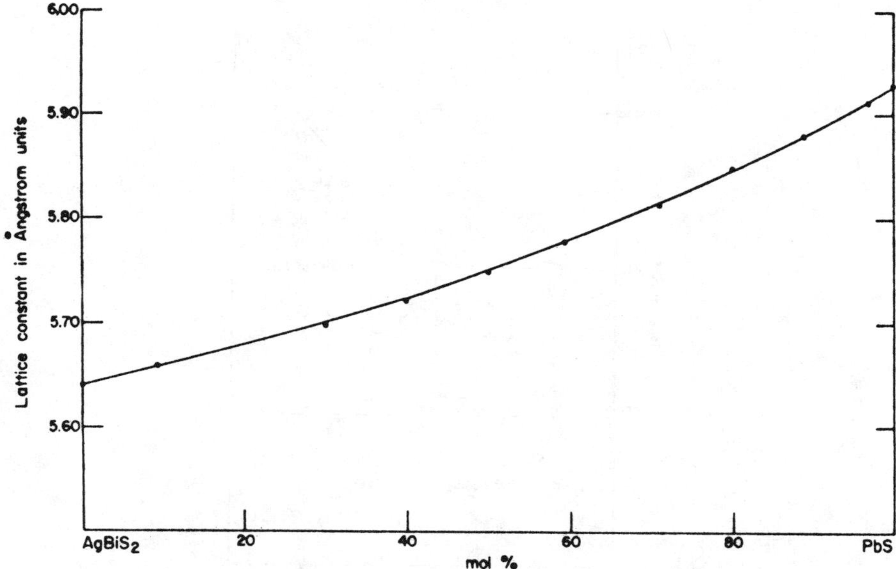

FIG. 9. Lattice constant of $AgBiS_2$-PbS solid solutions.

predicted by previous investigators. The variation in unit cell constant with composition for the binary solid solutions is shown in Figure 9. The system is represented as $AgBiS_2$-PbS to conform with the binary join the ternary system, although the assumed substitution, Ag^{+1}, $Bi^{+3} = 2Pb^{+2}$, indicates that the system should be considered as $AgBiS_2$-2PbS.[1] Liquidus and solidus curves were located by the quenching technique and X-ray analysis described in the preceding section. Differential thermal analyses were made as a check on the quenching results and were found to be in good agreement (Fig. 10). Diffraction patterns of mixtures near $AgBiS_2$ all indicated progressive crystallization with residual liquids approaching $AgBiS_2$ with falling temperature, indicating that there is no minimum in the system. For this particular investigation, it was necessary to cool the samples more slowly (in air rather than in water) because of the tendency of these liquids to quench without fractionation.

Runs were made at low temperatures for periods up to several months in an effort to determine more precisely the position of the solvus proposed by Ramdohr (26). In general, unmixing of the solid solutions is so sluggish that equilibrium could not be realized. Samples near $AgBiS_2$ in composition, however, did reveal some tendency to change to the stable low-temperature assemblage. A charge containing 3.9 mol % PbS held for three months at 170° C contained about 2 percent galena in polished section. A second sample with 10 mol % PbS was found by X-ray study to consist of a partially inverted high temperature solid solution along with smaller peaks corresponding

[1] Representation as $AgBiS_2$-PbS results in a positive deviation from Vegard's Law rather than negative as in Fig. 9.

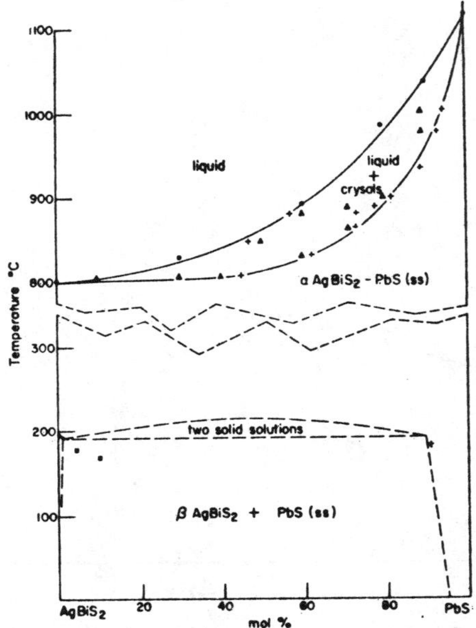

Fig. 10. The system $AgBiS_2$-PbS. Note break in the temperature axis.

to a second, galena-rich phase containing 10 mol % $AgBiS_2$. Compositions with a higher PbS content did not transform to the low-temperature assemblage.

Discussion.—Solid solution is complete at high temperatures in this system: solidus and liquidus temperatures rise continuously from the melting point of $AgBiS_2$ (801° C ± 5°) to that of PbS (1115° C ± 5°). The quenching technique described is considered to be preferable to thermal analyses as the results can be related to equilibrium conditions, avoiding possible errors due to superheating or supercooling of the sample. The principal error inherent in the method is that quenching may not be rapid enough to prevent reaction of the liquid with the primary crystals and that the measured solidus composition would then be higher in $AgBiS_2$ than the equilibrium value. Preliminary studies on the influence of quenching rate, however, showed that any partial reaction is reflected in a definite skewness of the diffraction peaks for the primary solid phase. A sharp peak for the solid separated by an interval from the diffuse peak representing the quench crystals was taken as evidence that no reaction had occurred between phases on quenching.

Unmixing of the solid solutions below solidus temperatures is extremely sluggish and was generally unsuccessful except for compositions near $AgBiS_2$. Ramdohr's studies (26) of galena-βmatildite intergrowths, however, have provided information on natural material from which the solubility in galena below unmixing temperatures can be calculated. He prepared X-ray powder photographs of a specimen from the Mayflower Mine, Ontario and found that the calculated cell constant for the galena phase in the intergrowth was

$a_0 = 5.89$ kX. He apparently assumed that this value should correspond to that of pure PbS ($a_0 = 5.93$ kX) and that the X-ray camera was inaccurate. He then "corrected" the calculated elements for the matildite phase by the factor 5.93/5.89. Graham (11) subsequently synthesized matildite and found that the cell dimensions ". . . agree poorly with those given by Ramdohr (26), after halving and interchanging his *a* and *b* periods." The discrepancy in the data is considerably reduced if Ramdohr's uncorrected values are compared with Graham's findings, viz.:
It is evident from these results that the correction which Ramdohr applied should not have been made and that the material examined was a galena solid solution whose lattice constant is near the measured value of 5.89 and not 5.93 kX, the "corrected" value. Taking 5.89 kX as the cell dimension and using the data compiled in the present study (Fig. 9), the galena phase in the intergrowth corresponds to a synthetic solid solution containing 8 mol % $AgBiS_2$. This value may be compared with the experimental measurement of 10 mol % at 170° C, and with the observation of Chapman and Stevens (5) that a specimen containing 15.4 mol % $AgBiS_2$ is an intergrowth of two phases.

The System Ag_2S-Bi_2S_3-PbS

Experimental Methods.—DTA methods were used to determine the melting relations in this system. Solid solution in matildite-galena structures was determined by examination in polished section of mixtures that were held at ternary solidus temperatures before quenching. In the composition triangle Ag_2S-$AgBiS_2$-PbS solidus temperatures are between 592° and 615° C depending on the composition whereas in the area Bi_2S_3-$AgBiS_2$-PbS, the ternary invarient temperatures are unknown and were estimated from a knowledge of the melting relations in the binary system PbS-Bi_2S_3.

The variation in lattice constant with composition, as determined by X-ray measurements of peak position on the quenched ternary solid solutions is shown in Figure 11.

Experimental Results.—The phase diagram (Fig. 12) summarizes the information obtained on the ternary system. In the composition triangle Ag_2S-$AgBiS_2$-PbS, the primary crystallization fields of argentite and galena-

TABLE 5

MEASUREMENTS ON MATILDITE

	Ramdohr (26), on natural material			Graham (11), synthetic product Measured values
	"Corrected" values	Measured values		
b/2	3.935	3.91	a	3.918
a/2	4.07	4.04	b	4.046
c	5.69	5.65	c	5.662

(It should be noted that Ramdohr's material is a solid solution in equilibrium with a galena phase, whereas Graham's data is on synthetic material without PbS present. This distinction is probably minor, however, as miscibility of PbS in β matildite is very small.)

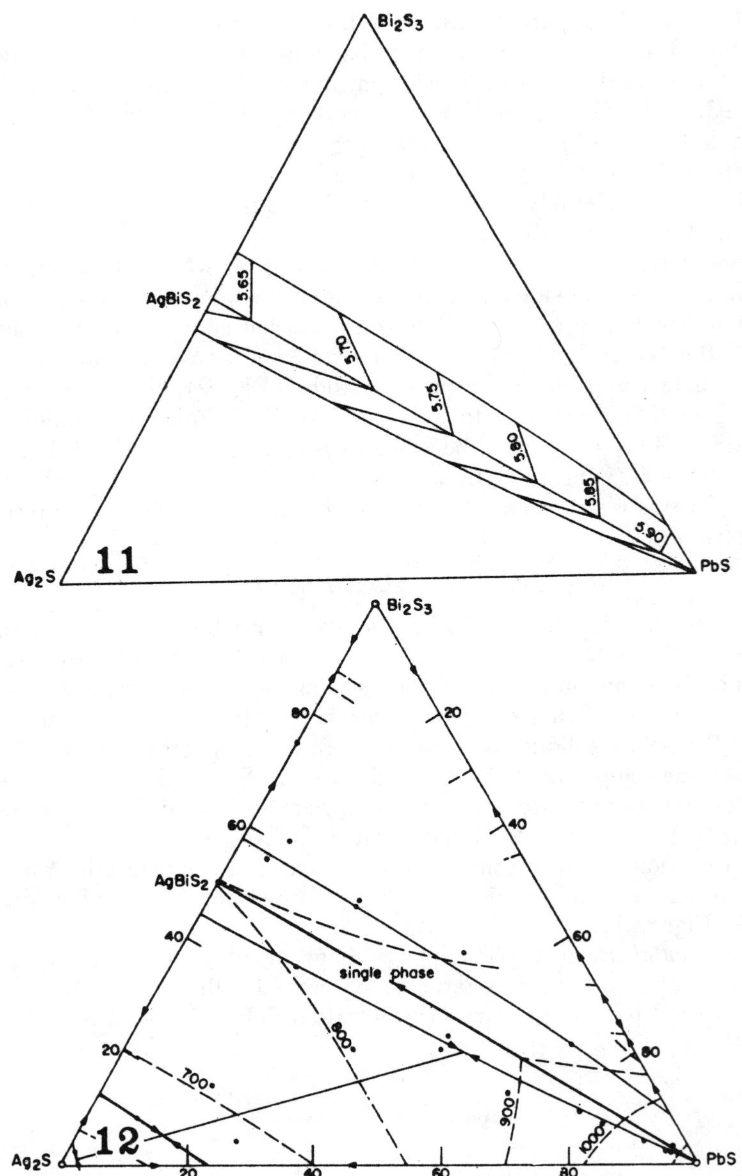

FIG. 11. Cubic lattice parameter of $\alpha AgBiS_2$-PbS ternary solid solutions.
FIG. 12. The system Ag_2S-Bi_2S_3-PbS.

matildite solid solutions are separated by the eutectic boundary curve which has a minimum at 592° C ± 5° at the composition $81Ag_2S$-$5Bi_2S_3$-$14PbS$. A three-phase tie line connects this minimum point on the boundary curve with minima on the phase boundary of the argentite and galena-matildite solid solutions as shown in Figure 12.

The solubility of Ag_2S at the solidus in the galena-matildite solid solutions decreases continuously from 10 mol % (45 mol % Bi_2S_3, 55 mol % Ag_2S) on the Ag_2S-$AgBiS_2$ side line to 0.4 mol % in the system PbS-Ag_2S whereas the solubility of Bi_2S_3 at the solidus is 16 mol % in AgBiS and decreases to 9 mol % in PbS. The variation in cell dimensions for these ternary solid solutions is shown in Figure 11. Matildite-galena solid solutions containing excess Bi_2S_3 were anisotropic in polished section and showed diffuse, poorly defined reflections in X-ray power patterns. The twinning lamellae found in the $AgBiS_2$ phase with excess Bi_2S_3 were not observed in these ternary compositions. Solid solutions containing excess Ag_2S over $AgBiS_2$-PbS compositions, on the other hand, were all optically isotropic and exhibited relatively sharp diffraction peaks.

Discussion.—The extensive ternary solubility in the galena-matildite structure is probably the most significant feature of the experimental results in this system. Apparently substitutional solid solution of $Ag^{+} + Bi^{+3}$ for $2Pb^{+2}$ need not be quantitatively satisfied at high temperatures, although the addition of excess Ag_2S or Bi_2S_3 must involve some structural accommodation to the resulting charge differences and cation deficiency or excess in order to maintain the cubic form. At lower temperatures the solubility of Ag_2S and Bi_2S_3 in PbS and $\alpha AgBiS_2$ decreases rapidly (Figs. 6, 7) and a similar relationship probably also holds true in the ternary system resulting in solid solutions which are restricted more closely to the $AgBiS_2$-PbS join. The range of these cubic structures is further diminished below the intersection with the $AgBiS_2$-PbS solvus, giving rise to the βmatildite-galena exsolution intergrowths (Fig. 1) described by Ramdohr (26). The schematic isothermal sections in Figure 13 gives the general form of phase relations which would be expected at low temperatures.

The experimental results have indicated that schapbacite and schirmerite are not distinct mineral species (Table 1) but are members of a continuous series of solid solutions and are unstable below the solvus. The galena phase is continuous with these high-temperature solid solutions and the assignment of any compositional limit separating them is arbitrary: perhaps PbS solid solutions with less than 10 mol % $AgBiS_2$ should be termed galena, retaining the name schapbacite for minerals with higher $AgBiS_2$ content.

The experimental results have shown that the solubility of argentite in galena is very small in the binary system (0.4%) but may be large, even at low temperatures, in the presence of bismuth. Moreover, silver and bismuth may occur in unequal molecular proportions in the cubic solid solutions without exsolution of a second phase. It is interesting to note in this connection that the argentiferous galena samples from the Freiberg deposit analyzed spectroscopically by Leutwein and Herrmann (21) were all richer in Ag_2S than in Bi_2S_3 (Fig. 14). The galena was reported to be homogeneous in polished section and essentially free of elements other than in bismuth and silver.

The solubility of bismuth in galena is similarly dependent on the presence of silver but is perhaps less striking since solubility is appreciable (9%) in

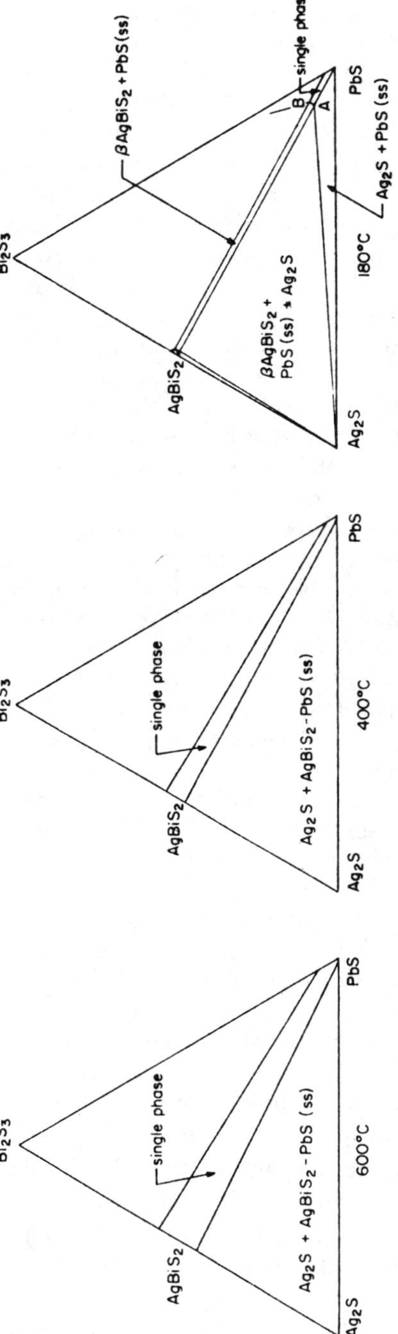

FIG. 13. Subsolidus equilibria for three isothermal sections in the system Ag_2S-Bi_2S_3-PbS.

the binary system PbS-Bi_2S_3. The spectroscopic analyses given by Oftedal (24) of galena samples from Norwegian localities are of interest here since the minerals studied were also reported to contain only bismuth and silver as major impurities. In some of these samples a second anistotropic lead-bismuth sulfosalt was observed in polished section, appearing as lamellae along octahedral directions in the galena. The bulk composition was, therefore, believed to correspond to homogeneous galena formed at higher temperatures Sample 1 (Fig. 14) appeared in polished section as galena plus a second phase, samples 2, 3, and 4 contained only a trace of the second phase, and compositions 5, 6, and 7 appeared as homogeneous galena without visible exsolution lamellae of the second phase. Assuming 0.1% of a second Ag_2S or Bi-sulfosalt phase can be detected in polished section, these natural galenas can be used to locate the approximate solubility limits at "low-temperature" shown by the cross-hatched area in Figure 4 within the limits of high-temperature solid solutions. These observations compare favorably with the assumed subsolidus relations shown in Figure 13. The analyses of these natural galena samples show no particular tendency for silver and bismuth to combine in equal amounts, i.e., as solid solutions of $AgBiS_2$ in PbS. This variation in the ratio Ag/Bi for trace amounts in galena has lead some investigators to assume that solution of silver in galena has no relation to the presence of bismuth. Goldschmidt (10, p. 407) observed: "Strangely enough, all galena rich in bismuth (about 10.000 ppm Bi) also contains thallium (100–500 ppm) and is always rich in silver, so that it has been suggested that a silver sulphobismuthite such as matildite, $AgBiS_2$, may be substituted for 2PbS. The author considers it to be more likely that silver and bismuth are taken up quite independently into the galena lattice." Chemical analyses of βmatildite-galena intergrowths (26, 5) had led to the opposite view—that solid solution is essentially that of $AgBiS_2$ in PbS and that large excesses of Ag_2S or Bi_2S_3 in galena are not observed. The reason for this difference in opinion can be seen by a study of the 180° C isothermal plane in Figure 13. In any mixture of galena and $\beta AgBiS_2$ with or without a third phase, the composition of the galena must be given by some point along the line AB. The isothermal section indicates the general case of a line AB that generates a surface with varying temperature. The surface, however, is probably very narrow and may actually be a point in isothermal section and a line in the polythermal model. At any rate, the position of the surface or line will change, but will probably remain on or near the $AgBiS_2$-PbS join, yielding a galena phase in equilibrium with $\beta AgBiS_2$ having essentially a 1 : 1 ratio of Ag/Bi. On the other hand, the galena samples with trace amounts of silver and bismuth studied by Goldschmidt and others were free of additional ternary phases and there is no restriction on composition within the solubility limits of the single PbS phase. In this case the availability of silver or bismuth exerts the controlling effect on composition independent of the requirements of the phase relationships. Nevertheless, even in trace amounts, one may assume that the presence of silver in galena favors the capture of bismuth (or antimony, etc.) and vice versa, as this would result in a more favorable charge relationship in the structure.

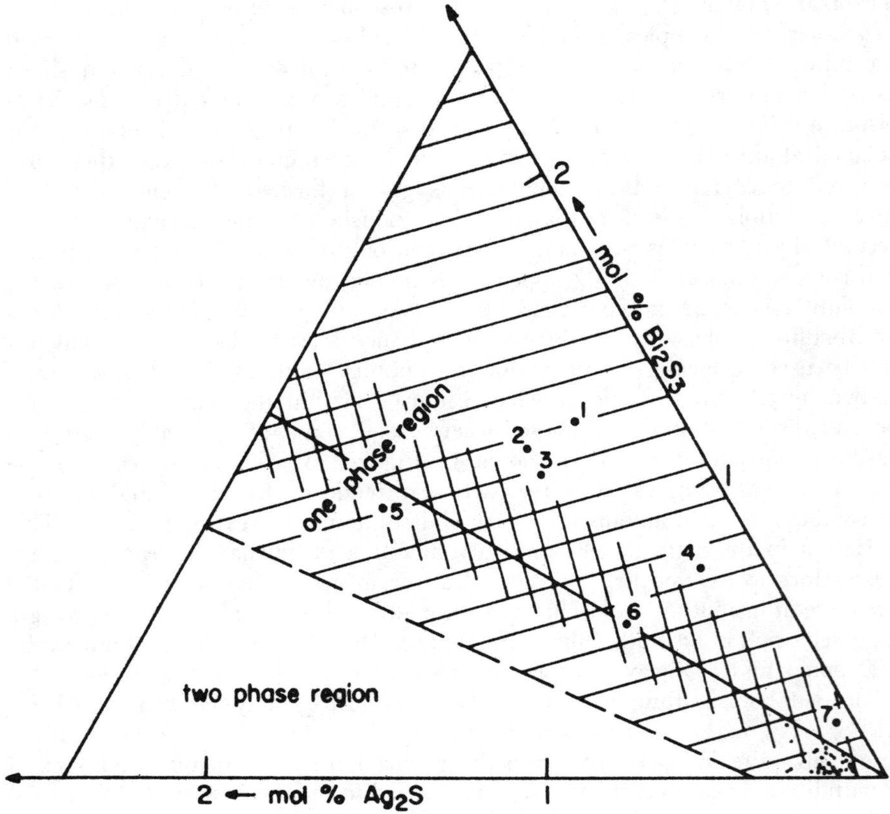

FIG. 14. Ternary composition of some natural galenas (small dots are data from Leutwein and Herrmann, 21; numbered dots from Oftedal, 24. See text for discussion.

Argentiferous varieties of the rare mineral beegerite ($6PbS \cdot Bi_2S_3$) have been reported to contain 10–15 wt. % Ag (27) and have been described as cubic on the basis of morphology. The experimental results indicate that a single cubic phase is stable for beegerite compositions containing more than 3.8 mol % Ag_2S (3.0 wt. % Ag) and that a higher silver content (approaching the $AgBiS_2$PbS join) would increase the stability of the phase at high temperatures. Below the temperature of the $AgBiS_2$-PbS solvus, "beegerite" containing Ag_2S will unmix into two (galena + Pb-Bi sulfosalt) or three (βmatildite + Pb-Bi sulfosalt + galena) phases, depending on the amount of Ag_2S present. Ramdohr (27) has noted that the only property distinguishing beegerite from galena is the anisotropism. Since anisotropism was also observed in all synthetic galena compositions having excess Bi_2S_3, it seems probable that beegerite is nothing more than a high-temperature galena solid solution. The cubic morphology of beegerite also suggests a single galena phase as originally formed at high temperatures.

SUMMARY

The silver content in many ore deposits has been traced to small amounts of silver minerals in the form of minute inclusions in galena ore. The intergrowths, which are generally disseminated evenly throughout the specimen, have been interpreted as exsolution lamellae resulting from unmixing in the solid state. The most common silver mineral, argentite, however, is only sparingly soluble in galena, and the validity of the interpretation of the inclusions as unmixing intergrowths depended on the discovery of other silver minerals having a greater solubility in the galena structure.

Ramdohr (26) investigated intergrowths of galena and the silver sulfosalt matildite ($AgBiS_2$) and found that matildite is dimorphic, having a high-temperature form isostructural with galena. He proposed that the intergrowths resulted from unmixing of homogeneous solid solutions with falling temperature arising from the sudden decrease in solubility below the inversion temperature in matildite.

Studies of trace elements in galena have shown further that silver and bismuth are very common constituents and that galena rich in bismuth from "high-temperature" ore deposits is almost invariably high in silver (10).

Experimental studies in the ternary system Ag_2S-Bi_2S_3-PbS were undertaken to determine the solubility of Ag_2S and Bi_2S_3 in galena and in ternary solid solutions and to study the effect of the matildite inversion on ternary phase equilibria.

Previous studies of the binary system Ag-S and Pb-S, together with the present experimental results on the system Bi-S, have shown that the three sulfides Ag_2S, Bi_2S_3, and PbS melt congruently in sealed containers at the pressure of the system. Compositional changes due to loss of sulfur to the condensed phases can be minimized by keeping the space above the charge as small as possible. This is especially important in the system Bi-S where congruent melting of Bi_2S_3 occurs at pressures of a few atmospheres. It was concluded that mixtures of these sulfides could be studied as condensed equilibria in sealed containers neglecting the small compositional changes due to sulfur vaporization.

The limiting binary systems Ag_2S-PbS, Ag_2S-Bi_2S_3, PbS-Bi_2S_3, the binary join $AgBiS_2$-PbS, and the ternary system Ag_2S-Bi_2S_3-PbS were investigated using differential thermal analysis combined with quenching methods. The products were examined in polished section and by X-ray diffraction to identify the phases and to determine their limits of solid solution. All samples were sealed in evacuated silica glass tubes to prevent decomposition and oxidation at high temperatures.

The results of the experimental work are summarized in Figure 12. The solubility of PbS in $(Ag_2S)_I$ is 3 mol % at the eutectic temperature (605° C) and that of Bi_2S_3 in $(Ag_2S)_I$ is 4 mol % at the Ag_2S-$AgBiS_2$ eutectic (615° C). Solubility in the high-temperature form ($(Ag_2S)_I$) lowers the inversion in Ag_2S at 590° C to 455° C in the system Ag_2S-PbS and to 522° C in the system Ag_2S-Bi_2S_3. The solubility of PbS in $(Ag_2S)_{II}$ (argentite) is probably less than 1 mol % in both binary systems and in the ternary system also.

Two binary compounds have been synthesized in the system Ag_2S-Bi_2S_3

(Fig. 6). Matildite ($AgBiS_2$) is dimorphic, having a high-temperature galena-type structure (α) that transforms reversibly to a lower symmetry (β) form at 195° C ± 5°. X-ray powder diffraction data on the β modification are compared with previous results on synthetic $\beta AgBiS_2$ (11, 9). The cubic form melts congruently at 801° C ± 5° and is stable over a range of binary compositions. A second binary compound, $Ag_2S \cdot 3Bi_2S_3$, is monoclinic and has no known naturally occurring counterpart.

Miscibility of Ag_2S in PbS is limited to 0.4 mol % Ag_2S in binary compositions whereas significantly higher amounts of Bi_2S_3 (9 mol % at 800° C) may be held in the galena phase in the binary system PbS-Bi_2S_3. Addition of both Ag_2S and Bi_2S_3 to the galena structure, has a profound effect on solubility and a continuous series of solid solutions are formed along the join between PbS and $\alpha AgBiS_2$. The compositional range of the galena phase, moreover, is not restricted to this join at high temperatures, but extends to ternary compositions containing up to 10 mol % Ag_2S or 16 mol % Bi_2S_3 in excess of $AgBiS_2$-PbS mixtures. With decreasing temperature, the stability region of the galena phase decreases, being limited to compositions closer to the $AgBiS_2$-PbS join (Fig. 13). Solid solution is further diminished below intersection with the solvus, related to the structural change in $AgBiS_2$ at 195° C, and gives rise to the unmixing textures observed in matildite-galena intergrowths (Fig. 1). Although the unmixing reflects a large reduction in solid solubility, considerable $AgBiS_2$ remains dissolved in galena at low temperatures. Experimentally, 10 mol % $AgBiS_2$ was found in PbS at 170° C in the presence of excess $AgBiS_2$ and the galena phase in an intergrowth described by Ramdohr (26) contained 8 mol % $AgBiS_2$ (3.76 wt. % Ag, or approximately 1000 troy oz/ton of Ag) as determined using data from the present investigation. It is concluded that the solubility of silver in galena may be extensive, even at low temperatures, when bismuth is present. A similar relationship in the system Ag_2S-Sb_2S_3-PbS is suggested by the available data on miargyrite ($AgSbS_2$) affording a possible explanation for the common occurrence of Ag, Bi, and Sb in natural galenas.

RAYTHEON COMPANY,
RESEARCH DIVISION,
WALTHAM, MASS.,
Aug. 10, 1959

ACKNOWLEDGMENT

The author is indebted to Dean O. F. Tuttle for his guidance and encouragement throughout the investigation, to Dr. L. D. Glasser for her assistance in the single crystal X-ray study, and to Dr. R. Roy for many helpful discussions. The National Carbon Company provided a fellowship under which this work was carried out.

REFERENCES

1. Aten, A. H. W., 1905, Über phasen gleichgewichte im system: wismut und schwefel: Z. Anorg. Chem., v. 47, p. 387–398.
2. Berry, L. G., 1940, Studies of mineral sulpho-salts: IV—Galeno-bismutite and "Lillianite": Am. Mineralogist, v. 25, p. 726–734.
3. Bloem, J., and Kröger, F. A., 1956, The p-T-x-phase diagram of the lead-sulfur system: Zeitschr. Physikal Chemie, v. 7, p. 1–14.
4. Buerger, M. J., 1947, Derivitive crystal structures: Jour. Chem. Phys., v. 15, p. 1–16.

5. Chapman, E. P., and Stevens, R. D., 1933, Silver and bismuth-bearing galena from Leadville: Econ. Geol., v. 28, p. 678–685.
6. Fleischer, M., 1955, Minor elements in some sulfide minerals: Econ. Geol., 50th Anniv. Vol., P. 2, p. 970–1024.
7. Friedrich, K., 1907, The melting point curves of the systems galena-pyrrhotite and galena-silver sulfide: Metallurgie, v. 4, p. 479–486.
8. Gaudin, A. M., and McGlashan, D. W., 1938, Sulfide silver minerals—a contribution to their pyrosynthesis and to their identification by selective iridescent filming: Econ. Geol., v. 33, p. 143–193.
9. Geller, S., and Wernick, J. H., 1959, Ternary semiconducting compounds with sodium chloride-like structure: $AgSbSe_2$, $AgSbTe_2$, $AgBiS_2$, $AgBiSe_2$: Acta Cryst., v. 12, p. 46–54.
10. Goldschmidt, V. M., 1954, Geochemistry: Oxford Univ. Press, 730 pp.
11. Graham, A. R., 1951, Matildite, aramayoite, miargyrite: Am. Mineralogist, v. 36, p. 436–449.
12. Graham, A. R., Thompson, R. M., and Berry, L. G., 1953, Studies of mineral sulfo-salts: XVII—Cannizzarite: Am. Mineralogist, v. 38, p. 536–544.
13. Guild, F. N., 1917, A microscopic study of silver ores and their associated minerals: Econ. Geol., v. 12, p. 297–353.
14. Hansen, M., 1936, Der aufbau der zweistoffliegerungen: Julius Springer, Berlin, 1,100 pp.
15. Hellner, E., 1958, A structural scheme for sulfide minerals: Jour. Geology, v. 66, p. 503–525.
16. Jackson, F. L., Thaete, E. H., et al., 1954, Sulfur: Encyclopedia of Chem. Tech., v. 13, p. 358–373.
17. Kelly, K. K., 1935, Contributions to the data on theoretical metallurgy, III: U. S. Bur. Mines Bull., 383, p. 1–132.
18. Keith, M. L., and Tuttle, O. F., 1952, Significance of variation in the high-low inversion of quartz: Am. Jour. Sci., Bowen Volume, p. 203–280.
19. Kostov, I., 1958, Bonchevite, $PbBi_4S_7$, a new mineral: Mining Mag., v. 31, Na 241, p. 821–828.
20. Kracek, F. C., 1946, Phase relations in the system sulfur-silver and the transitions in silver sulfide: Am. Geophys. Union Trans., v. 27, p. 364–374.
21. Leutwein, F., and Herrmann, A. G., 1954, Kristallchemische und geochemische untersuchungen uber vorkommen und verteilung des wismuts im bleiganz der kiesig-blendigen formation des freiberger ganggreviers: Geologie, v. 3, p. 1039–1056.
22. Nissen, A. E., and Hoyt, S. L., 1915, On the occurrences of silver in argentiferous galena ores: Econ. Geol., v. 10, p. 172-179.
23. Oftedal, I., 1940, Untersuchungen über die nebenbestandteile von erzmineralein norwegischer zinkblendeführender vorkommen: Skrifter Norske Videnskaps-Akad., Oslo, I. Mat.-naturrklasse, No. 8.
24. Oftedal, I., 1943, Om betingelsene for oktaldrisk delbarhet hos vismutrik blyglans: Norsk. Geol. Tids., v. 22, p. 61–68.
25. Palache, C., Berman, H., and Frondel, C., 1944, The system of mineralogy, vol. 1: John Wiley and Sons, Inc., New York, 834 pp.
26. Ramdohr, P., 1938, Über schapbacit, matildit, und den silber- und wismut-gehalt mancher bleiglanze: Sitzungberichte der Preuss, Akad. Wiss., Phys.-Math. Klasse, p. 71–91.
27. Ramdohr, P., 1955, Die erzmineralein und ihre verwachsungen: Akademie-Verlag, Berlin, 875 pp.
28. Ricci, J. E., 1951, The phase rule and heterogeneous equilibrium: D. Van Nostrand Co., New York, 505 pp.
29. Richardson, F. D. and Jeffes, J. H. E., 1952, The thermodynamics of substances of interest in iron and steel making, III—Sulfides: Jour. Iron and Steel Inst. (London), v. 171, p. 165–175.
30. Ross, Virginia, 1957, Geochemistry, crystal structure, and mineralogy of the sulfides: Econ. Geol., p. 755–774.
31. Schenck, R., and Pardun, H., 1933, Untersuchungen über die chemischen systeme der Lenard phosphore, I: Zeitschr. anorg. allgem. Chem., v. 211, p. 209–221.
32. Schwartz, G. M., 1942, Progress in the study of exsolution in ore minerals: Econ. Geol., v. 37, p. 345–364.
33. Tuller, W. N., 1954, The sulfur data book: McGraw-Hill Co., New York, 143 pp.
34. Wahlstrom, E. E., 1937, Octahedral parting in galena from Boulder County, Colorado: Am. Mineralogist, v. 22, p. 906–911.
35. Warren, H. V., 1932, Relation between silver content and tetrahedrite in the ores of the North Cananea Mining Co., Cananea Sonora, Mexico: Econ. Geol., v. 27, p. 737–743.
36. Wernick, J. H., 1959, Constitution of the $AgSbS_2$-PbS, $AgBiS_2$-PbS, and $AgBiS_2$-$AgBiSe_2$ Systems: Am. Mineralogist, in press.
37. Zernike, J., 1955, Chemical phase theory: N. V. Uitgevers-Maatschappij E. E. Kluwer, Antwerp, 493 pp.

14

EXPERIMENTAL STUDY OF THE PbS-Bi_2S_3 SYSTEM

A. A. Godovikov, V. A. Klyakhin, Zh. N. Fedorova, and R. M. Leibson

This article was translated expressly for this Benchmark volume by Leonard J. Stanten, University of Kansas, Lawrence, from the original Russian article in Inst. Geologii i Geofiziki Trudy **5**:10–33 (1967).

Investigation of the PbS-Bi_2S_3 system is important for a proper understanding of the complex and difficult to analyze mineral groups--the lead-bismuth sulphosalts. Twenty different mineral types can be distinguished among them at the present time, and the number is constantly growing. Recently, giessenite has been added (Greaser 1963). The possible existence of two modifications of lillianite (Syritso and Senderova 1964) has also been proposed.

In addition, an analysis of data on the chemical makeup of lead-bismuth sulphosalts (Godovikov 1965) indicates a great many of the minerals related to them have not been distinguished on sufficiently valid grounds. At the same time, certain investigators (e.g., Kostov 1957) consider the discovery of many new minerals among the group as being quite probable.

Taking into account the great difficulties in identifying discrete minerals of the group (by optical and X-ray methods) because of the intimate intergrowth of minerals with each other (Kolonin 1965), it must be recognized that without experimental investigation of corresponding systems, success in this area is unlikely.

The lead-sulfide-bismuth sulfide system was first studied experimentally by Graham, Thompson, and Berry (1953). In addition to PbS and Bi_2S_3, five intermediate phases were successfully derived, one of which was clearly identified as galenobismutite. X-ray powder diagrams of the remainder were examined and compared to those of natural minerals. They were conditionally described as phases 1, 2, 3 and 4 of the author's work mentioned. In the work of these same authors, accessible to us, phases 1 and 4 are not described at all. Phase 2 is identified as "cannizzarite" by Wolfe (1938), which, as was shown earlier (Godovikov 1965), is identical to bismutoplagionite. It is not at all, properly speaking, cannizzarite of the composition PbS-$2Bi_2S_3$ (Zambonini, Corrobi, and Foire 1925).

Another compound--phase 3--has, according to these authors, a Debye powder pattern close to that of a fibrous component, weibullite, although certain differences do exist between them, as the authors themselves note.

Due to intimate intergrowth with one another, not one of the phases has been subjected to a quantitative chemical analysis. Pb, Bi, and S are determined in them only by qualitative experiments.

To solve the problem of the means by which silver enters into Galena, the PbS-Bi_2S_3 system was studied by Van Hook using a pyrosynthesis method (Van Hook 1960); his basic attention was concentrated on the part near PbS. He established that under high PbS temperatures, up to 9 molar % Bi_2S_3 dissolves and that with a lowering of temperature this quantity is reduced, reaching approximately 4 molar percent at 500°C. The composition diagram introduced by Van Hook should be considered as tentative. The number of points in it is insufficient, and what is most important, as the author himself notes, in many instances he did not achieve an equilibrium state. In this part of the composition diagram, Van Hook did not go lower than 550°C. He determined three intermediate phases: galenobismutite, lillianite (25 molar percent Bi_2S_3), and a phase, close in makeup to 5PbS-Bi_2S_3 (16 molar percent Bi_2S_3), i.e., a bismuth analog of jordanite (5PbS-As_2S_3) and schultzite (5PbS-Sb_2S_3). Concerning the possibility of the existence of beegerite, Van Hook writes: "...the galena phase, corresponding to the composition given for beegerite (14 molar % Bi_2S_3) is unstable in binary compositions because the maximal solubility in the solid state for the eutectic is equal to 9 molar % Bi_2S_3, but a compound of PbS:Bi_2S_3, equal to 6:1, is quite possible in ternary and quarternary systems" (p. 775). He assigns a special role to silver here, emphasizing that natural beegerite is known, and it contains a significant quantity of silver.

In the summary conclusion on the PbS-Bi_2S_3 system, Van Hook emphasizes: "Supplementary experimental investigations are necessary in the Bi_2S_3 part of this system. Synthetic analogs of known sulphosalts are not to be found in present research, apparently, they are formed through subsolidus reactions below the temperatures which have been studied" (p. 776).

Research of the fourth reciprocal system PbS-PbSe-Bi_2S_3-Bi_2Se_3 (Malevskii, Rikhter, and Veres 1963), employing a pyrosynthesis method, resulted in the isolation therein of ten intermediate phases: (1) an unbroken series of cubically solid solutions PbS-PbSe; (2) a phase analogous to natural gungarrite ($Pb_4Bi_2(S_xSe_{1-x})_7$ - needle-shaped; (3) a cosalite-lillianite solid solution ($Pb_{3-y}Bi_2$(S Se_{1-x})6-y - needle-shaped phase; (4) a laminated phase of the same composition; (5) a needle-shaped phase ($PbBi_2(S_xSe_{1-x}$); (6) a laminated phase of the same composition, analogous to weibullite; (7) a needle-shaped phase $PbBi_4(S_xSe_{1-x})_7$, including cannizzarite; (8) a laminated phase of the same composition, including the selenium analogue of cannizzarite; (9) Bismuthin; and (10) a laminated, trigonal Bi_2Se_3. Thus, only in the PbS-Bi_2S_3 system has there been established (besides PbS and Bi_2S_3) four intermediate phases, analogous to cannizzarite,

Table 1. Intermediate phases of PbS-Bi_2S_3 system produced by various researchers.

Graham, Thompson, Berry	Schenck et al.	Van Hook	Malevskii et al.
-	6PbS·Bi_2S_3	-	-
-	-	5PbS·Bi_2S_3	-
-	-	-	4PbS·Bi_2S_3
-	3PbS·Bi_2S_3	3PbS·Bi_2S_3	3PbS·Bi_2S_3-
5PbS·4Bi_2S_3*	-	-	-2,5PbS·Bi_2S_3 (Solid Solution)
PbS·Bi_2S_3**	PbS·Bi_2S_3	PbS·Bi_2S_3	PbS·Bi_2S_3
			PbS·2Bi_2S_3

* Composition adduced conditionally taking into consideration the resemblance of an X-ray powder diagram of phase 2 with an X-ray powder of bismutoplagionite.

** Phase 3, close to weibullite, is not included in the table insofar as weibullite is a selenium analogue of galenobismutite and should be examined in the series PbSe-Bi_2Se_3.

galeno-bismutite, a solid solution of the composition cosalite-lillianite and gungarrite.

The PbS-Bi_2S_3 system was studied employing a method of reduction restoration in a current of H_2 by Schenck et al. (1939). Their work established the existence of three intermediate combinations at 500°C: beegerite, lillianite and galenobismutite, and also a solid solution of Bi_2S_3 in PbS (up to 9.3 molar %).

A study of the results of experimental investigations (Table 1) shows that significant discrepancies exist between the data of various authors.

Thus, a phase analogous to beegerite was established only by Schenck and his colleagues, and one characteristic of cannizzarite only by Malevskii, Rikhter, and Veres. These same men established both gungarrite and a solid solution in a lillianite base. Lillianite was confirmed by Van Hook, Schenck and other investigators, but was not produced by Graham in hydrothermal experiments. The single phase that was synthesized by all without exception is galenobismutite. Besides this, phase 2 was produced by Graham. It is a possible analog of bismutoplagionite.

Taking all this into consideration, supplementary experiments of the PbS-Bi_2S_3 system were necessary, both by pyrosynthetic and hydrothermal methods.

RESULTS OF THE STUDY OF THE PbS-Bi_2S_3 SYSTEM USING A PYROSYNTHESIS METHOD

For an investigation of synthetic PbS and Bi_2S_3, a series of alloys at intervals of every 2.5 molar percent was prepared. References for PbS and Bi_2S_3 were: sulphur of the class A-2, bismuth of the class A-2 and as especially pure lead of the class B-3. Specimens were prepared in quartz ampoules filled before sealing with carbon dioxide gas placed in silit ovens which were heated and cooled according to a specified program (Table 2).

After this, a part of the specimens was taken for investigation, and the remaining material was annealed in the course of a month. The derived specimens were studied, a part of them was put into a second annealing, having developed, on the whole, up to 2.5 or 3

Table 2. Compositions of Samples and Methods of their Production.

Sample No.	Content, Mol % Bi_2S_3	System of Production
90	100	Heating for 3 hours to 800°,* 2 hour exposure, cooling at 50°/hour; at 400° the oven is left to cool freely.
541	97.5	
501	95	
542	92.5	
502	90	
543	87.5	
503	85	
544	82.5	Heating to 900° at 200°/hour, cooling to 700° at 5°/hour, then to 500° at 10°/hour, then at 20°/hour.
504	80	
545	77.5	
505	75	
546	72.5	
506	70	
547	67.5	
507	65	
548	62.5	
508	60	
549	57.5	
509	55	
550	52.5	
510	50	
551	47.5	
511	45	
552	42.5	
512	40	
553	37.5	
513	35	Heating to 1100° at 200°/hour, cooling to 750° at 5°/hour, then to 500° at 10°/hour, and then at 20°/hour.
554	32.5	
514	30	
555	27.5	
515	25	
556	22.5	
516	20	
557	17.5	
517	15	
558	12.5	
518	10	
559	7.5	
519	5	
560	2.5	
PbS	0	Heating at 300°/hour to 1150°, exposure at that temperature 1 hour, cooling at 300°/hour.
84	66.67	Heating to 770°, sharp cooling in water, annealing at 710° for 2 weeks.

*All temperatures are degrees Centigrade.

Table 2. (Continued)

Sample No.	Content Mol % Bi_2S_3	System of Production
84-z 600-z 601-z		Samples 84, 600, 601, produced according to above method, annealed at 650°* for a week and chilled at that temperature.
547-z 506-z		Samples 547 and 506 annealed at 690° for 2 weeks and chilled.
591 592	29 28	Heating to 1100° at 200°/hour, cooling at 200°/hour, annealing at 730° for 2 months.
504-0 506-0 550-0		Samples 504, 506, and 550, annealed at 710° for a month.
551-0		Sample 551, annealed at 730° for 1 month.
$551\text{-}0_2$ $511\text{-}0_2$ $552\text{-}0_2$ $512\text{-}0_2$ $553\text{-}0_2$ $513\text{-}0_2$ $554\text{-}0_2$ $514\text{-}0_2$ $555\text{-}0_2$ $515\text{-}0_2$ $556\text{-}0_2$		Samples 551, 511, 552, 512, 553, 513, 554, 514, 555, 515, 556, annealed at 730° for 2 months.
516-0 557-0 558-0		Samples 516, 557, 558, annealed at 790° for 2 months.
81	25	Chilling in water at 920°, annealing at 710° for 2 weeks, annealing at 730° for two months, annealing at 795° for 10 days.
82	33.3	Chilling at 860°, annealing at 710° for 2 weeks, annealing at 720° for 10 days, divided, ground, pressed into tablets. Annealing at 710° for 12 twelve-hour periods.
574	25	Chilling at a temperature above the melting point. Annealing at 795° for 2 weeks.
84-2 600-2		Samples 84 and 600 annealed at 670° and chilled at this temperature.

*All temperatures degrees Centigrade.

Table 2. (Continued)

Sample No.	Content Mol % Bi_2S_3	System of Production
88	50	Heating for 4 hours to 800°*, Cooling: 800-780° at 2°/hour; 780-750° at 0.5°/hour; exposure at 750° for 4 hours. Cooling to 750-715° at 3°/hour, subsequent cooling to room temperature at 20°/hour. Annealing at 730° for 20 days. The sample is worn down, pressed into tablets, and left to anneal. Annealing for a month.
92	50	Heating to 800° for 4 hours, exposure at this temperature for 2 hours, cooling with shut-down of oven. Annealing at 730° for a month. Ingot divided, ground, and compressed into tablets, left to anneal by a progressive temperature increase from 700-730°. Annealing for a month.
561-1	2.5	Chilling at 1050°.
562-1	5	
563-1	7.5	
564-1	10.00	
561-2	2.5	Chilling at 1000°.
562-2	5	
563-2	7.5	
564-2	10	
561-3	2.5	Chilling at 950°.
562-3	5.0	
563-3	7.5	
564-3	10.0	
561-4	2.5	Exposure for a week at 900° and chilling at 900°.
562-4	5.0	
563-4	7.5	
564-4	10.0	
561-5	2.5	Exposure at 830° for a week and cooling at 830°.
562-5	5.0	
563-5	7.5	
564-5	10.0	
597	99	Heated above melting point and sharply cooled in water. Annealing at 710°C for 2 weeks, chilling at 710°.
599	98	
600	66.67	Heated above melting point, chilled. Annealing at 690° for 10 days.
601	67	
612	65.5	
613	68	
614	69	

*All temperatures degrees Centigrade.

months. Polished sections of the derived specimens were studied under a MIN-8 microscope; their hardness was measured on a microhardness meter of an MeF microscope; X-rays were taken with RDK type cameras with a diameter of 57.3 mm on a URS-60 instrument and several on a TURM-60; and thermograms were derived on a Kurnakov photopyrometer. Chemical analyses were made under the direction of V. I. Bogdanova. The density of specimens 517, 557, and 509 was measured by V. Yakushev using a micromethod (Klyakhin, Yakushev 1966), and the density of the remaining specimens was measured pycnometrically by O. A. Karpushina.

As a result, a composition diagram was set up (Illustration 1,a), which differs somewhat from the diagram set up by IMGRE colleagues (cf. Illustration 1,b), especially in the part between phases $PbS \cdot Bi_2S_3$ and Bi_2S_3. In the diagram there are: (1) a wide area of solid solutions under high temperatures on a galena base (at 834±6°C Bi_2S_3 solubility in PbS reaches 10% and lowers to 2.5-3 molar % at low temperatures); (2) a laminated phase of the composition $9PbS \cdot 2Bi_2S_3$, forming itself by peritectic action at 834±5°C; (3) an area of lillianite based solid solutions, forming themselves peritectically at 816±4°C--at 750°C its limits are from 26 to 31% Bi_2S_3; at low temperatures there exists only a narrow area of a solid solution in the neighborhood of 26% Bi_2S_3; (4) a $PbS \cdot Bi_2S_3$ combination, formed peritectically at 750±30°C; (5) a high temperature $PbS \cdot Bi_2S_3$ combination, formed peritectically at 737±4°C and breaking down at 671±3°C into bismuthine and galenobismutite; (6) eutectically at a temperature of 713±6°C of a composition 21% PbS; and (7) a very thin area of solid solutions on a bismutite base, containing the temperature of the eutectic, 0.8 molar % PbS. Below is given a short description of each of the phases established.

<u>The area of solid solutions on a galena base</u>. In the majority of samples of solid solutions of Bi_2S_3 in PbS crystallites were produced of a size between 1 and 1.5 mm and with good facets (Figure 2). The crystals of pure PbS have only cubic faceting. Of those containing 2.5 and 5% Bi_2S_3, the tops of the cube are cut by the surfaces of an octahedron. In a sample with 7.5% Bi_2S_3 crystalline particles also have an octahedron form. In a sample containing 10% Bi_2S_3, there was no success in producing faceted crystals. With a change of composition, hardness changes sharply (Table 3).

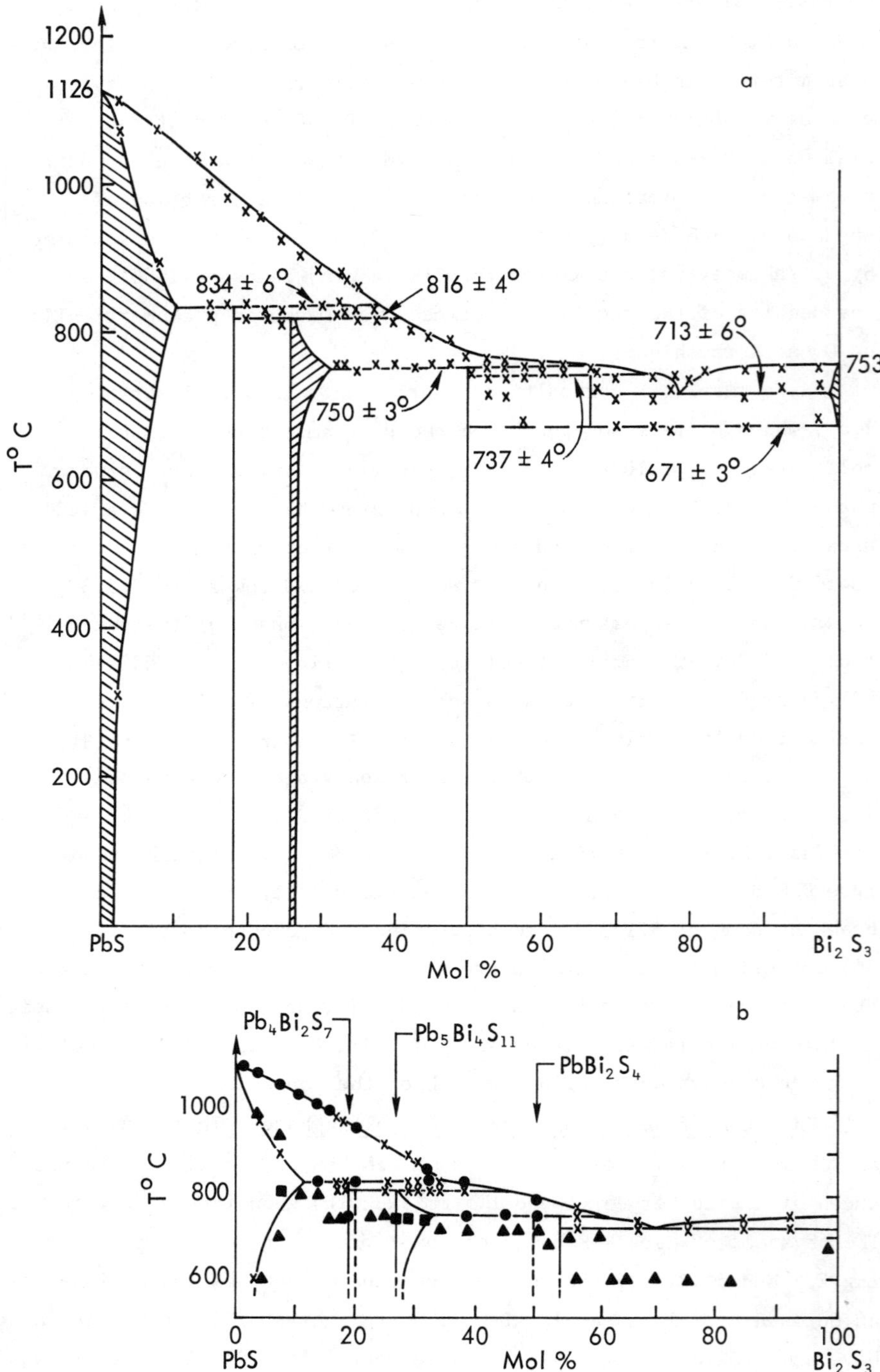

Figure 1. Our diagram of composition of PbS-Bi_2S_3 system (A). IMGRE's diagram of composition of PbS-Bi_2S_3 system (B).

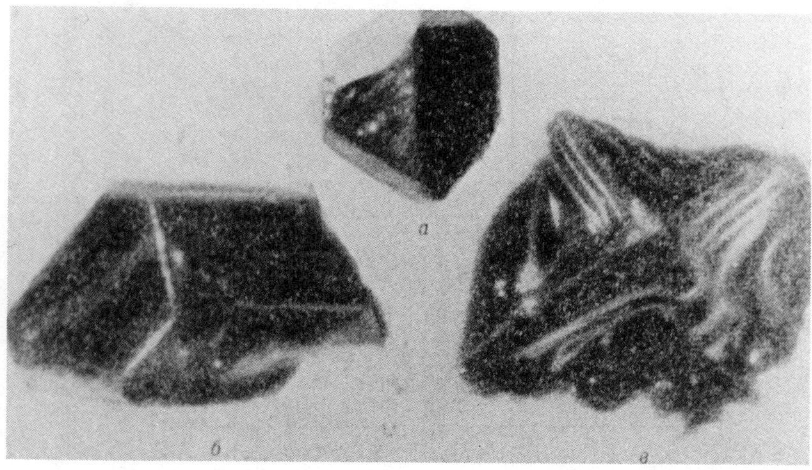

Figure 2. Crystals from solid solution region on galena base.
A - with 2.5 mol % Bi_2S_3 magnified 20X.
B - with 5 mol % Bi_2S_3 magnified 20X.
C - with 7.5 mol % Bi_2S_3 magnified 40X.

Table 3. Vickers hardness (Hv) y solid solutions of Bi_2S_3 in PbS of various makeup.

Content Mol % Bi_2S_3	KG/MM^2
0	74
2.5	119
5	140
7.5	156
10	174

Figure 3 also indicates that with an addition of an admixture to galena, hardness rises sharply, and later with a perceptible Bi_2S_3 volume, it changes linearly according to the formula (103±7.1x)kg/mn, where x is the content of Bi_2S_3 in solid solution in molar %.

In polished sections, all of the samples described are very similar to one another. They have qualities typical of galena, differing from it only in the absence of "painted" triangles. The action of standard etching agents is illustrated in Table 4.

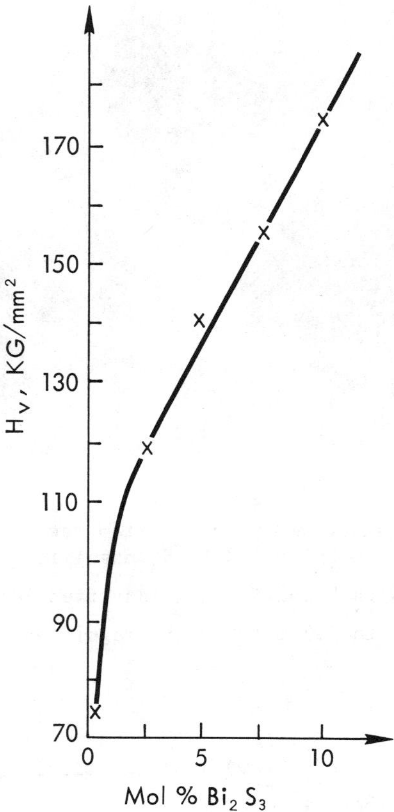

Figure 3. Diagram of change of hardness with change of makeup of solid solution Bi_2S_3 in galena.

The lattice parameter α, like hardness, changes proportionally with composition, following essentially the equation:

$$a = 5.9330\text{--}0.0029x \pm 0.0017A,$$

where x is the content of Bi_2S_3 in molar % (Table 5, Figure 4).

According to the lattice parameters, it is possible to calculate the volume of Bi_2S_3 in solid solution with an accuracy up to ±0.6 molar %. The lattice parameter was calculated for a sample containing at first 5 molar % in Bi_2S_3. After slow cooling to room temperature, separation into thin layers of gungarrite and a solid solution, poorer in bismuthine occurs. For the solid solution formed,

Table 4. Results of diagnostic etching.

Phase \ Etchant	Na_2S (Saturated Solution)	KOH (20%)	$FeCl_3$ (20%)	HCl (1:1)	HCl (concentrated)	HNO_3 (1:1)	HNO_3 (concentrated)
Bismutine	Bluish or yellow spot, depending on orientation	no etching	Pale, grayish brown spot	no etching	Drop dries quickly, no etching	Etches	Etches
Bonchevite	Pale yellow spot, barely visible	"	Pale brown spot	Barely noticeable	Brown spot (for various sections of various intensity	"	"
Galenobismutite	no etching	"	Intense brown spot	"	Brownish spot, lightly eroded, leaving a weak, gray-brown spot	"	"
Lillianite	"	"	Iridescent spot	Corroded brown spot	the same	"	"
Gungarrite	"	"	Brown spot	Gray spot	Etches very strongly	"	"
Solid Solution Basis	"	"	Brownish spot of middle intensity	Momentarily leaves a black spot	Black spot	"	"

Table 5. Change of lattice parameters of solid solutions Bi_2S_3 in PbS and dependency on composition.

Bi_2S_3 Mol %	a_{mean} Å
0	5.9347
2.5	5.9241
5.0	5.9170
7.5	5.9095
10	5.9055

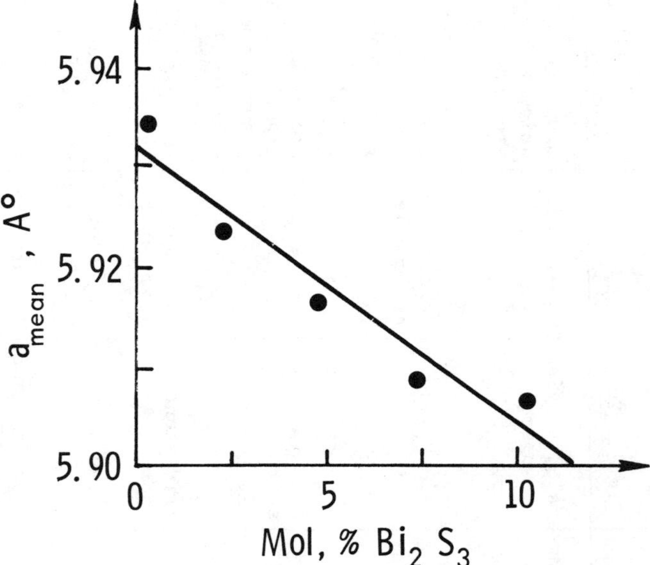

Figure 4. Graph of change of lattice parameters of solid solution Bi_2S_3 in PbS with change of composition.

once again the lattice parameter is equal to 5.924 Å. Calculating the dependence of the lattice parameter from the composition of the solid solution, it is possible to determine that at low temperature the solid solution contains 3.0±0.6% bismuthine.

The composition of $9PbS \cdot 2Bi_2S_3$. In the annealing of samples containing from 22.5 to 12.5% Bi_2S_3 (in ampoules) well faceted laminated crystalline particles up to 3 mm in size (drawing 5) are formed.

A microchemical analysis of these lamina (Table 6) gives the formula $9PbS \cdot 2Bi_2S_3$. A homogenized alloy of 18% Bi_2S_3 annealed over

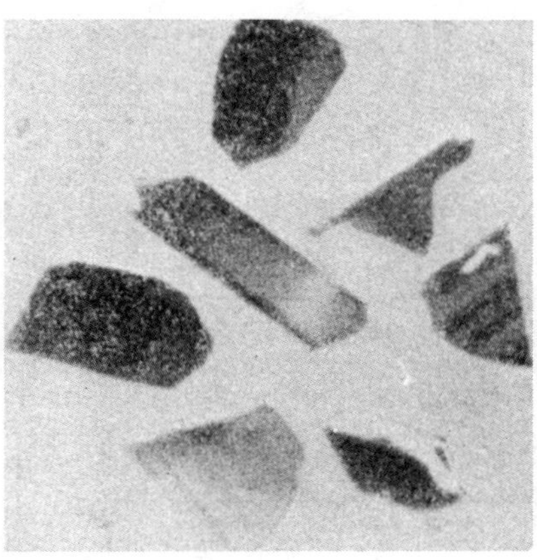

Figure 5. Crystals of combination $9PbS \cdot 2Bi_2S_3$ (artificial gungarrite). Magnified 20X.

two months at 780° has a shiny whitish color with a light blue tint; cleavage parallel to the lamina is also present. In reflected light a double reflection is distinctly visible. Under crossed Nicols, the material is strongly anisotropic and has a color from bright yellow to dark brown.

The density of the crystalline particles as measured is 7.2 g/cm_3. Its hardness is 223 kg/mm^2. On a background of galena this combination is brighter, with a yellowish tint. The results of the etching are given in Table 4.

X-rays of various samples (Table 7) are identical to each other, and to an X-ray of synthetic gungarrite as well, and also to laminated "lillianite," as described by Siritso and Senderova (1964). Its formula, however, differs from that usually attributed to gungarrite--$4PbS \cdot Bi_2S_3$.

<u>The region of solid solutions on a lillianite base</u>. In the region from 26 to 31% Bi_2S_3 at temperatures of about 750°C, solid solutions are produced that break down with slow cooling to galenobismuthite and lillianite, containing 26% Bi_2S_3. The boundaries of the solid solution regions were calculated on the basis of results derived from measurements of the areas of each phase identified in thin sections with the help of an integrated microscope stage. The samples have a barely perceptible cleavage. They have a more light

Table 6. Results of microchemical analysis of synthesized samples.

Samples	Theoretical formula	Hanging mg	Composition, weight, %			Σ	Atomic quantity			Design equation
			Pb	Bi	S		Pb	Bi	S	
601	$PbS \cdot 2Bi_2S_3$	100	15.45	66.96	17.81	100.22	746	3204	5554	$PbS \cdot 2.15Bi_2S_3$
84	$PbS \cdot 2Bi_2S_3$	101	14.51	67.59	18.06	100.16	700	3234	5632	$PbS \cdot 2.31Bi_2S_3$
92	$PbS \cdot Bi_2S_3$	99	26.27	56.13	17.30	99.70	1268	2686	5395	$PbS \cdot 1.05Bi_2S_3$
555	Lillianite	101	45.86	37.30	16.49	99.65	2213	1785	5142	$5PbS \cdot 2Bi_2S_3$
517-0 (crystal-line)	Gungarrite	52	58.52	25.76	14.89	99.17	2824	1232	4643	$9PbS \cdot 2Bi_2S_3$
558 (parti-cles)										

Table 7. X-rays of gungarrite type combinations.

	$9PbS \cdot 2Bi_2S_3$ Co-target		Syritso and Senderova's laminated "lillianite"		$4PbS \cdot Bi_2S_3$ IMGRE's data	
Line No.	I	d/n	I	d/n	I	d/n
1	5	3.71	2	3.67	1.5	3.67
2	4	3.51	1	3.49	1	3.52
3	10	3.38	10	3.36	10	3.39
4	5	3.33	-	-	-	-
5	3	3.08	1	3.06	1	3.10
6	4	2.98	2	2.98	-	-
7	9	2.93	6	2.91	5	2.94
8	9	2.84	4	2.82	5	2.84
9	2	2.61	-	-	-	-
10	1	2.40	-	-	-	-
11	2	2.31	-	-	-	-
12	2	2.22	-	-	-	-
13	7	2.17	3	2.16	2.5	2.18
14	5	2.14	2	2.14	-	-
15	10	2.09	9	2.09	4	2.10
16	7	2.06	-	-	-	-
17	8	2.03	6	2.03	2	2.04
18	4	1.958	3	1.941	1.5	1.962
19	4	1.893	2	1.884	3	1.892
20	4	1.842	1	1.838	1	1.848
21	5	1.827	-	-	-	-
22	6	1.780	-	-	-	-
23	9	1.762	10	1.764	6	1.764
24	3	1.740	-	-	-	-
25	4	1.717	2	1.712	1	1.713
26	4	1.701	-	-	-	-
27	4	1.687	2	1.688	-	-
28	4	1.671	2	1.665	1	1.669
29	1	1.625	-	-	-	-
30	2	1.554	-	-	-	-
31	1	1.519	-	-	-	-
32	4	1.487	-	-	-	-
33	5	1.468	3	1.467	-	-
34	5	1.449	-	-	1	1.450
35	4	1.419	1	1.409	1	1.422
36	7	1.398	6	1.380	-	-
37	2	1.388	-	-	-	-
38	2	1.380	-	-	-	-
39	5	1.359	-	-	-	-
40	8	1.349	8	1.339	3	1.347
41	6	1.341	-	-	-	-
42	2	1.320	-	-	-	-
43	4	1.313	4	1.304	3	1.311

Table 7. (Continued)

$9PbS \cdot 2Bi_2S_3$ Co-target			Syritso and Senderova's laminated "lillianite"		$4PbS \cdot Bi_2S_3$ IMGRE's data	
Line No.	I	d/n	I	d/n	I	d/n
44	3	1.304	-	-	-	-
45	4	1.299	2	1.295	-	-
46	4	1.240	2	1.231	1	1.240
47	3	1.212	2	1.207	2	1.211
48	7	1.207	2	1.97	2	1.191
49	7	1.191	3	1.187	-	-
50	3	1.170	-	-	-	-
51	4	1.155	1	1.149	-	-
52	4	1.135	-	-	-	-
53	3	1.130	-	-	1	1.130
54	4	1.121	3	1.123	-	-
55	1	1.113	-	-	-	-
56	1	1.101	-	-	-	-
57	2	1.089	-	-	-	-
58	3	1.065	-	-	-	-
59	2	1.058	-	-	-	-
60	6	1.045	1	1.041	-	-
61	7	1.029	2	1.027	2	1.031
62	2	1.021	-	-	-	-
63	7	1.017	-	-	-	-
64	7	1.008	2	1.003	1	1.007
65	2	1.005	-	-	-	-
66	2	0.994	-	-	-	-
67	3	0.991	-	-	-	-
68	10	0.987	8	0.986	3.5	0.989
69	1	0.978	-	-	-	-
70	1	0.976	-	-	-	-
71	5	0.970	2	0.970	1	0.972
72	2	0.959	-	-	-	-
73	2	0.949	-	-	-	-
74	2	0.944	-	-	-	-
75	4	0.932	-	-	-	-
76	4	0.929	-	-	-	-
77	3	0.927	-	-	-	-
78	4	0.924	-	-	-	-
79	8	0.922	2	0.923	2	0.923
80	6	0.919	-	-	-	-
81	3	0.912	-	-	-	-
82	4	0.905	-	-	-	-
83	2	0.903	-	-	-	-
			2	0.891	1	0.892
			2	0.881	1.5	0.882
			-	-	1.5	0.818

blue tint to them in comparison with galenobismutite, but are more yellow than galena. Under the microscope, the phase next to galenobismutite has a grayish color with noticeable double-reflection. Anisotropy is strong with colors from grayish-yellow to grayish-brown. Hardness for various samples range from 177 to 195 kb/mn^2 and on the average comprises 184 kg/mn^2. Density is 7.06±0.01 g/cm^3. For the results of etching tests, see Table 4.

Microchemical analysis of sample 555 (cf. Table 6) suggest the combination $5PbS \cdot 2Bi_2S_3$. X-rays from samples of this region (Table 8) do not have any regular change depending upon composition. They are identical to the X-rays of the combination $5PbS \cdot 2Bi_2S_3$ and of natural lillianite described by Ontoev (1959).

Galenobismutite $PbS \cdot Bi_2S_3$. Homogeneous galenobismutite is obtained with great difficulty. A two-month annealing with slow cooling does not give homogeneous material.

Table 8. X-rays of synthetic and natural solid solution on lillianite base.

Sample 555 Co-target			IMGRE's data		Ontoev's lillianite	
Line No.	I	d/n	I	d/n	I	d/n
1	2	4.07	1	4.06	-	-
2	2	3.62	1	3.67	-	-
3	8	3.47	10	3.51	4	3.50
4	7	3.36	3	3.40	8	3.36
5	3	3.308	-	-	-	-
6	3	3.02	-	-	-	-
7	5	2.97	1.5	3.00	5	2.97
8	8	2.88	5	2.90	10	2.89
9	5	2.75	3	2.77	7	2.76
10	3	2.67	1	2.70	-	-
11	4	2.33	1	2.36	2	2.35
12	9	2.13	6.5	2.15	9	2.13
13	10	2.04	10	2.06	10	2.03
14	8	1.956	5.5	1.971	3	1.946
15	6	1.871	1	1.892	-	-
16	6	1.814	1	1.833	-	-
17	5	1.760	1.5	1.775	-	-
18	5	1.742	1.5	1.754	4	1.757
19	1	1.707	-	-	-	-
20	1	1.685	-	-	-	-

Table 8. (Continued)

Sample 555 Co-target			IMGRE's data		Ontoev's lillianite	
Line No.	I	d/n	I	d/n	I	d/n
21	2	1.652	-	-	-	-
22	3	1.621	1	1.638	-	-
23	3	1.519	1	1.528	-	-
24	2	1.483	-	-	-	-
25	4	1.458	1.5	1.456	-	-
26	6	1.438	1	1.427	3	1.449
27	2	1.414	-	-	-	-
28	2	1.393	1	1.393	-	-
29	4	1.379	-	-	-	-
30	2	1.349	-	-	-	-
31	8	1.326	2.5	1.336	6	1.326
32	5	1.294	1	1.304	3	1.288
33	3	1.266	-	-	-	-
34	5	1.219	1.5	1.228	-	-
35	4	1.198	1	1.207	-	-
36	4	1.182	1	1.192	-	-
37	5	1.169	1	1.177	-	-
38	4	1.127	-	-	-	-
39	4	1.121	-	-	-	-
40	4	1.107	-	-	-	-
41	4	1.095	-	-	-	-
42	4	1.063	1	1.068	-	-
43	4	1.050	1	1.057	-	-
44	1	1.034	-	-	-	-
45	6	1.029	1	1.034	-	-
46	5	1.017	-	-	-	-
47	3	1.004	-	-	-	-
48	4	0.993	-	-	-	-
49	5	0.981	-	-	-	-
50	5	0.979	-	-	-	-
51	5	0.975	-	-	-	-
52	3	0.968	-	-	-	-
53	2	0.052	-	-	-	-
54	2	0.943	-	-	-	-
55	5	0.933	1	0.937	-	-
56	2	0.919	1.5	0.921	-	-
57	6	0.917	1	0.915	-	-
58	3	0.912	-	-	-	-
59	3	0.910	-	-	-	-
-	-	-	1	0.879	-	-
-	-	-	1	0.838	-	-

After a month-long annealing at 730°C, sample material was ground and then compressed into tablets and annealed again for a month. The sample was of a lead-gray color, resembling bismutine, except that bismutine has a yellowish tint. The cleavage of galenobismutite as expressed is significantly less strong than that of bismutine. During the annealing process, long, needle-shaped crystals of the same composition grew on the surface of the tablets.

Under microscopic examination galenobismutite is white. Double-reflection is barely noticeable. Anisotropy is weak with colors from grayish-yellow to bright gray-brown. The density of the material equals 7.136±0.007 g/cm^3. The results of diagnostic etching are given in Table 4. A microchemical analysis of sample 92 was recalculated as $PbS \cdot 1.05Bi_2S_3$ (cf. Table 6). Its X-ray pattern (Table 9) is identical to that of galenobismutite included in Mikheev's handbook (1957).

Cannizzarite (bonchevite) $PbS \cdot 2Bi_2S_3$. This phase is homogeneous above 670°C. It melts at 750°C and at 671°C it breaks down into bismutine and galenobismutite.

Separate, well-faceted crystals of this combination were not produced. An igot had a rayed structure, resembling bismutine a great deal, except that bismutine has a yellowish tint. It has perfect cleavage along the c axis. Under the microscope, a polished section has a yellowish tint and a very strong double-reflection from yellowish-white to pale gray. It is strongly anistropic, of a color of yellow to dark gray with a brown tint, but weaker than in bismutite, where a reddish tint is also characteristic. Hardness (the mean of two samples) is 153 kg/mm^3. Density for two samples (by pycnometer) was 6.611±0.015 (sample 601) and 6.488±0.027 (sample 600). Measured by a micromethod it gave a density of 6.8 g/cm^3 (sample 509). For the results of diagnostic etching, see Table 4.

Microchemical analysis of several of the samples from this region are given in Table 6. They show that separate samples are somewhat different in makeup from one another.

An X-ray of samples from the high-temperature phase (Table 10) does not resemble X-rays of bismutine or galenobismutine, but rather brings to mind an X-ray of the very same combination produced by colleagues at IMGRE. Data from natural combinations with similar

Table 9. X-rays of synthetic and natural $PbS \cdot Bi_2S_3$.

Sample 92 Co-target			Mikheev's		IMGRE's data	
Line. No.	I	d/n	I	d/n	I	d/n
1	6	3.62	4	3.67	2	3.62
2	10	3.44	10	3.47	8.5	3.45
	–	–	1	3.39	–	–
	–	–	3	3.25	–	–
3	5	3.00	7	3.03	3	3.00
	–	–	–	–	2	2.87
4	4	2.74	4	2.79	2.5	2.75
5	2	2.63	3	2.66	1	2.65
6	6	2.45	7	2.47	3	2.45
7	2	2.38	6	2.38	1.5	2.35
8	4	2.23	3	2.26	1.5	2.24
9	4	2.19	3	2.19	1.5	2.18
10	9	2.05	7	2.06	10	2.04
	–	–	3	2.02	–	–
11	10	1.959	7	1.975	9	1.95
12	–	–	3	1.905	–	–
12a	3	1.873	6	1.882	3	–
13	6	1.753	7	1.662	4	1.77
14	2	1.721	3	1.729	4.5	1.74
15	2	1.700	3	1.699	–	–
16	1	1.641	–	–	1	1.63
17	2	1.502	–	–	1	1.557
18	2	1.522	3	1.510	1	1.523
18a	–	–	–	–	1	1.498
19	5	1.449	6	1.451	4	1.446
20	3	1.411	6	1.415	1	1.407
21	3	1.378	6	1.378	1.5	1.379
22	2	1.355	–	–	1.5	1.350
23	3	1.329	–	–	1.5	1.329
24	2	1.301	–	–	1	1.291
25	3	1.240	–	–	–	–
26	2	1.202	–	–	1	1.201
27	1	1.173	–	–	1.5	1.172
28	3	1.097	–	–	1	1.097
29	3	1.057	–	–	2	1.055
30	1	1.037	–	–	2	1.034
	–	–	–	–	1	1.014
31	1	0.991	–	–	1	0.992
32	4	0.978	–	–	–	–
33	1	0.952	–	–	–	–
34	2	0.941	–	–	–	–

Table 10. X-rays of synthetic combinations of $PbS \cdot 2Bi_2S_3$

Sample 601 Co-target			IMGRE's data	
Line No.	I	d/n	I	d/n
1	2	5.84	1	5.90
2	2	4.88	1	4.94
3	1	3.88	–	–
4	10	3.74	10	3.78
5	3	3.45	1	3.44
6	6	3.28	1.5	3.31
7	2	3.18	–	–
8	2	3.08	–	–
9	8	2.93	2.5	2.95
10	3	2.84	–	–
11	3	2.79	–	–
12	3	2.75	1.5	2.75
13	3	2.69	–	–
14	3	2.49	–	–
16	7	2.40	1	2.42
17	2	2.34	–	–
18	8	2.30	3	2.30
19	6	2.22	1	2.20
20	5	2.15	1	2.16
21	3	2.11	–	–
22	3	2.09	–	–
23	3	2.07	–	–
24	3	2.04	–	–
25	4	2.00	–	–
26	4	1.99	–	–
27	4	1.96	2	1.96
28	4	1.93	–	–
29	5	1.89	1	1.89
30	1	1.82	–	–
31	6	1.77	2	1.77
32	4	1.74	1	1.74
33	4	1.707	1	1.713
34	3	1.661	–	–
35	2	1.560	–	–
37	1	1.544	–	–
38	1	1.524	–	–
39	4	1.514	1	1.511
40	2	1.492	–	–
41	5	1.482	1	1.484

Table 10. (Continued)

Sample 601 Co-target			IMGRE's data	
Line No.	I	d/n	I	d/n
42	3	1.457	1	1.451
43	3	1.428	-	-
44	3	1.415	1	1.409
45	2	1.403	-	-
46	2	1.378	1	1.373
47	2	1.346	-	-
48	2	1.333	-	-
49	2	1.305	-	-
50	2	1.280	1	1.276
51	2	1.258	-	-
52	3	1.243	1	1.241
53	3	1.228	-	-
54	1	1.217	-	-
55	2	1.210	-	-
56	2	1.203	1	1.202
57	2	1.194	-	-
58	2	1.185	-	-
59	3	1.178	-	-
60	3	1.159	-	-
61	4	1.151	1	1.154
62	1	1.124	-	-
63	1	1.113	-	-
64	1	1.105	-	-
65	2	1.097	1	1.096
66	3	1.066	-	-
67	3	1.053	-	-
68	1	1.037	-	-
69	3	1.033	1	1.032
70	2	1.015	-	-
71	4	1.003	-	-
72	2	0.994	-	-
73	4	0.973	1	0.973
74	2	0.966	-	-
75	2	0.964	-	-
76	2	0.951	-	-
77	3	0.937	-	-
78	2	0.930	-	-
79	3	0.922	-	-
80	3	0.913	-	-
81	1	0.911	1	0.899

X-rays do not exist at present. The X-ray mistakenly introduced for so-called bonchevite by I. Kostov is closer to an X-ray of bismutine than to an X-ray of the combination described.

Bismutine Bi_2S_3. It is produced quite simply by normal melting of Bi and S in a stoichiometric relationship in sealed ampoules. An ingot of a silvery-white color, with a strong metallic glitter, had perfect cleavage along the c axis. It is striated along the cleavage perpendicular to this axis on account of the formation of duplicates. Under a microscope a polished section is white with a slight yellowish tint; on a background of bismuth it is gray. A double-reflection is strong even in the air--with colors from white-yellowish to gray. It is strongly anisotropic, with an anisotropic color from golden-yellow to gray-brown. In distinction from sulfo-salts, bismutine's anisotropic color is of a reddish tint. Furthermore, for bismutine the presence of polysynthetic duplicates is characteristic. Hardness is 132 kg/mm^2. Results of diagnostic etching are presented in Table 4. Up to 2.5 molar % PbS dissolves in bismutine at a temperature of 710°C. An X-ray of bismutine is identical to an X-ray of natural Bi_2S_3 in Mikheev's handbook (1957).

All of the phases described are well distinguished by X-rays and thermograms. In the presence of two or more sulfo-salts, it is easy in one thin section to tell them apart by the means of etching with HCl (1:1). Phases more rich in PbS etch more intensively. Phases are well distinguished by etching with $FeCl_3$ (20%). It is necessary that the concentration of the etchant be precisely 20% by weight (6.9% Fe). With the use of a solution of a different concentration the color of the etched parts is non-characteristic. Furthermore, for diagnostic purposes it is convenient to use a saturated solution of Na_2S. It intensively etches Bi_2S_3. Depending on the orientation of granules in the etched spot it has a light-bluish or brown-yellowish color. The remaining phases, with the exception of $PbS \cdot 2Bi_2S_3$, are not etched by this reagent, but $PbS \cdot 2Bi_2S_3$ gives a pale yellow, barely perceptible spot.

The Results of Hydrothermal Recrystallization

Hydrothermal recrystallization was produced in a 1M solution of NH_4Cl at a temperature of 420°C and with a pressure of 500 atm. (space factor 57.38%). In all experiments, with the exception of experiment 63, $PbCl_2$ was added to compensate for the shift of the relationship $\frac{Pb^{+2}}{Bi^{+3}}$ in the initial period of crystallization. The quantity of Pb^{+2} was calculated tentative data regarding the solubility of galena under these circumstances.

For recrystallization, non-homogeneous ingots from bonchevite to beerite composition which had been produced by pyrosynthesis were used as a source material. For the composition of source materials, see Table 11. Source material was placed at the bottom of an insert. The methodology of the experiment has been described in greater detail earlier (Godovikov, Il'iasheva, Kliakhin et al. 1965).

The experiments were carried out in autoclaves with a floating titanium insert with an internal volume of 8.5 cm^3. The autoclave was positioned in a cylindrical oven designed to receive it. After exposure for several days, the autoclave was chilled in cold water.

As a result of transfer, aggregates (clusters) of crystals grew on the stopper and on the walls of the insert. In all, four phases were produced: bismutine, galenobismutite, lillianite and galena. Furthermore, in the precipitate of experiment 60, X-rays showed Graham's phase 2 (Graham et al. 1953), which was located in mixture with galena. Formation of hydrogen sulfide in the inserts did not occur. The appearance of this or that phase during recrystallization depends on the composition of the ingot, and in certain cases on the duration of the experiment (see Table 11). Below is given a description of the phases produced and of their conditions of formation.

Bismutine was produced in a recrystallization of ingots of a composition from Bi_2S_3 to $PbS \cdot 3Bi_2S_3$. It forms comparatively macrocrystalline aggregates of columnar to needle-shaped crystals growing on the stopper (Figure 6) and walls with individual crystals measuring up to 8 mm along the length axis and with a width of several tenths of a millimeter. The crystals are usually positioned in several fan-shaped groups; the position most advantageous for growth is with an orientation of the c axis parallel to the surface of the stopper. The crystals formed in such conditions have dimensions. When the c axis is oriented near the corner toward the surface of the stopper, shortened crystals grow. In section, the crystals are tetragonal to flattened-hexagonal. They are either cleaved toward the peak, or have skeletal type cavities at the peak (tubular, see Figure 7); less

Figure 6. Fan-shaped outgrowths of bismutine crystals on stopper of insert. Magnified 3X.

Figure 7. Close to square section of a crystal of bismuthine with zoned center (polished section with partially crossed nicols). Magnified 260X.

frequently, they are faceted. Here, primarily one facet is developed, and the peak of the crystal has a sharp-ended form. Along their extension, the crystals are strongly striated.

Table 11. Summary table of experiments in hydrothermal re-crystallization of $PbS \cdot Bi_2S_3$ system samples.

No.	Experiment No.	Source Material	Temp. of Autoclave Bottom °C	Initial Weight of Ingot G	Loss of Ingot Weight	Composition of Produced Phase
1	68	Ingot with gross weight $PbS \cdot Bi_2S_3$	420	1.22	0.54	Bismutine
2	69	Same	420	1.25	0.52	"
3	80	"	420	0.71	0.49	"
4	81	"	420	0.69	0.69	"
5	82	Gross weight $2PbS \cdot 3Bi_2S_3$	420	0.69	0.46	"
6	83	Same	420	0.70	0.42	"
7	64	Gross weight $PbS \cdot Bi_2S_3$	420	1.25	0.80	Galenobismutite
8	65	Same	420	1.19	0.62	"
9	96	"	420	0.80	0.16	"
10	97	"	420	0.83	0.17	"
11	116	Mechanical mixture of galenite and bismutite $PbS+Bi_2S_3$	420	1.00	-	(Galenite (Gungarrite
12	117	Gross weight $5PbS \cdot 4Bi_2S_3$	420	0.70	-	Galenobismutite
13	108	Same	500	0.61	0.45	(Bismutine (Bismuth

Table 11. (Continued)

No.	Experiment No.	Source Material	Temp. of Autoclave Bottom $0^{o}C$	Initial Weight of Ingot G	Loss of Ingot Weight	Composition of Produced Phase
14	52	Gross weight $2PbS \cdot Bi_2S_3$	420	2.18	0.18	Lillianite
15	53	Same	420	2.12	0.18	"
16	60	"	420	2.38	0.75	(" (Galenite**
17	78	"	420	1.40	0.22	Lillianite
18	79	"	420	2.28	0.22	"
19	93	"	420	0.62	0.44	"
20	118	"	420	2.00	2.00	(Bismutine (Bismuth
21	62	Mechanical mixture of galenite and bismutine $2PbS+Bi_2S_3$	420	2.5	*	(Lillianite (Galenite
22	114	Gross weight $5PbS \cdot Bi_2S_3$	420	0.68	0.11	Same
23	58	Gross weight $3PbS \cdot Bi_2S_3$	420	1.88	0.39	"
24	59	Same	420	2.21	0.62	"
25	54	Gross weight $4PbS \cdot Bi_2S_3$	420	2.26	*	Lillianite
26	56	Gross weight $6PbS \cdot Bi_2S_3$	420	3.23	2.82	(Lillianite (Galenite
27	57	Same	420	3.34	2.41	Same
28	86	"	420	0.70	0.25	"
29	87	"	420	0.53	0.27	"

Note 1. Experiments were conducted in inserts of 8.5 ml in volume at pressure of 500 atm, from 2 to 30 days in duration.

Note 2. In experiment 118a 50 cm^3 insert was used.

* Not weighed

** Galenite = galena

Optical and diagnostic etching data are analogous to the same data for pyrosynthetic bismutine. A section of several crystals produced in experiment 82 in an etching of $FeCl_3$ (20%) reveals a heterogeneous composition. Along one border, such crystals are etched somewhat more strongly than the whole crystal. This, perhaps, is connected with the fact that a similar streak was added by some sulfasalt (galenobismutite [?]). The existence of a sulfasalt is confirmed by chemical analysis which showed a high lead content (Table 12); X-rays did not reveal the presence of another phase.

Galenobismutite was successfully produced only with the recrystallization of an ingot of galenobismutite composition (Experiments 64, 65).

Crystals have a columnar to needle-shaped habit, and not infrequently form fan-shaped outgrowths (Figure 8). The size of separate crystals is up to 1.5-2 mm in length with a width of up to 0.1-0.2 mm. In section they present the form of a distorted, flattened rhombus, or of a sharp-cornered parallelogram. As with bismutine, the peaks of the crystals are hollow. The are strongly straited parallel to their length, and on account of perfect cleavage along it they easily cleave into separate needles.

Figure 8. Outgrowths of needle-shaped to columnar crystals of galenobismutite (polished section). Magnified 20X.

Table 12. Results of microchemical analysis of combinations, synthesized hydrothermically.

Experiment No.	Theoretical formula	Composition weight %			Σ	Atomic Quantity			Calculated Formula	Density g/cm^3
		Pb	Bi	S		Pb	Bi	S		
84	Bismutine Bi_2S_3	3.00	78.17	18.41	99.58	167	3740	5741		6.77
83	"	1.45	79.70	18.51	99.66	70	3813	5772		6.76
64	Galenobismutite $PbS \cdot Bi_2S_3$	29.28	54.28	16.70	100.26	1413	2597	5208	$PbS \cdot 1.09Bi_2S_3$*	not determined
65										
93	Lillianite	47.38	36.90	15.63	99.91	2287	1765	4874	$5.18PbS \cdot 2Bi_2S_3$*	not determined
									$(5PbS \cdot 1.85Bi_2S_3)$	
56	Same	49.45	35.35	15.23	100.20	2386	1691	4784	$5.64PbS \cdot 2Bi_2S_3$*	7.14
									$(5PbS \cdot 1.77Bi_2S_3)$	
56	Galenite (Galena)	83.97	2.19	14.04	100.20	4052	105	4378	$Pb_{0.975}Bi_{0.025}S_{1.042}$	not determined

Optical and diagnostic etching data are analogous to existing data on pyrosynthetic galenobismutite. The results of etching the phase in thin sections with HCl (1:1) and $FeCl_3$ (20%) reveal the homogeneity of the material produced. For data on a chemical analysis of galenobismutite, see Table 12.

X-rays of hydrothermal galenobismutite (Table 13) are practically identical to those of pyrosynthetic galenobismutite.

Lillianite is formed by the recrystallization of ingots whose makeup corresponds to cosalite, lillianite, and beegerite. Furthermore, lillianite is produced by the recrystallization of a mechanical mixture of bismutine-galena, corresponding in its sum composition to

Table 13. X-ray of galenobismutite (experiment 65). Camera diameter 57.3 mm, CuK_α, Sample diameter 0.5 mm.

No.	I	d/n	No.	I	d/n
1	1	5.35	24	3	1.690
2	3	4.41	25	1	1.635
3	1	4.04			
4	2	3.79	26	1	1.573
5	6	3.54	27	2-3	1.544
			28	3	1.508
6	10	3.38	29	3	1.499
7	1	3.30	30	5	1.444
8	5	2.96			
9	2	2.85	31	4	1.406
10	5	2.70	32	5	1.373
			33	2	1.348
11	3	2.61	34	1	1.341
12	6	2.43	35	4	1.325
13	4	2.35			
14	2	2.22	36	3	1.309
15	4	2.17	37	2	1.300
			38	3	1.282
16	5	2.04	39	3	1.247
17	5	1.996	40	6	1.231
18	9	1.939			
19	6	1.843	41	4	1.094
20	3	1.809	42	6	1.055
			43	3	1.037
21	3	1.776	44	3	1.022
22	5	1.738	45	4	1.014
23	5	1.705			

Note: Correction of sample width not made.

cosalite. Lillianite is formed in paragenesis with galena; crystals of lillianite grow upward on the galena and on the walls of the insert.

Lillianite crystals are columnar, resembling galenobismutite crystals in form. They reach 3 mm in length with a width of 0.1-0.2 mm and, as a rule, have a streak from the peak to the interior of the crystal (Figure 9). In cross-section the crystals have variegated form: from parallelograms to complex figures that owe their appearance to well-developed streaking and, apparently, to doubling parallel to length. Perfect cleavage along the length was not discovered. There was conchoidal fracture during crushing. The action of hot HCl causes a cleavage near the corners 60-80% toward the length-axis to appear on the facets.

Optical and diagnostic etching data of the described sulfasalt corresponds to analogous data for pyrosynthetic lillianite. Etching by HCl (1:1) and $FeCl_3$ (20%) show a homogeneity of the examined material. Data from a chemical analysis of lillianite are given in Table 12. X-rays of the sulfosalt produced in all experiments are identical to each other and to an X-ray of pyrosynthetic lillianite (Table 14).

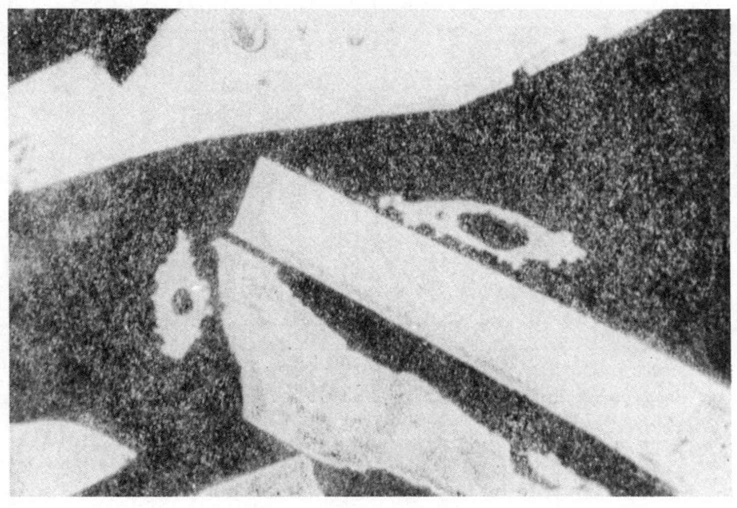

Figure 9. Cross section of two columnar-shaped crystals of lillianite. Internal cavities are visible (polished section). Magnified 260X.

Table 14. X-ray (Debye) of lillianite. Camera diameter 57.3 mm CoK_α, Sample diameter 0.5 mm.

Experiment 60					
Line No.	I	d/n	Line No.	I	d/n
1			28	2	1.657
2			29	2	1.627
3	2	4.02	30	2	1.523
4	2	3.61	31	3	1.487
5	10	3.46	32	5	1.463
			32a	1	1.449
6	5	3.36	33	1	1.417
7	2	3.28	34	3	1.398
8	1	3.08	35	2	1.383
9	2	3.02	36	3	1.355
10	4	2.96	37	3	1.329
11	7	2.87	38	3	1.302
12	6	2.74	39	3	1.292
13	4	2.67	40	6	1.267
14	1	2.55	41	3	1.220
15	1	2.42	42	3	1.201
16	5	2.33	43	6	1.189
17	8	2.13	44	3	1.174
18	10	2.04	45	3	1.131
19	5	1.962	46	3	1.124
20	5	1.947	47	6	1.055
21	3	1.879	48	3	1.033
22	3	1.814	49	2	1.021
23	1	1.781	50	2	1.010
24	6	1.763	51	6	0.995
25	6	1.741	52	2	0.982
26	1	1.677	53	2	0.971
27	1	1.667	54	5	0.955
			55	1	0.935
			56	5	0.920
Experiment 56					
Line No.	I	d/n	Line No.	I	d/n
1	1	4.34	37	6	1.330
2	1	4.18	38	4	1.298
3	3	3.96	40	2	1.271
3a	2	3.70	40a	1	1.246
4	2	3.61	41	5	1.221
5	10	3.42	42	4	1.198

Note: Correction of sample width not made.

Table 14. (Continued)

Line No.	I	d/n	Line No.	I	d/n
6	5 wide	3.29	43	4	1.188
10	3	2.93	44	6	1.172
11	5	2.84	48	6	1.031
12	4	2.74			
13	3	2.64	49	4	1.018
14	2	2.51	50	2	1.005
15	1	2.43	51	2	0.996
16	4	2.33	52	6 wide	0.982
16a	1	2.19	55	4	0.936
17	8	2.13			
18	9-10	2.03	56	6	0.920
19	6	1.951	57	4	0.913
21	4	1.868			
22	3	1.812			
25	6 wide	1.747			
26	1	1.682			
29	1	1.632			
29a	1	1.596			
30	3	1.528			
31	2	1.488			
32	3	1.459			
32a	4	1.447			
33	1	1.411			
34	3	1.396			
35	3	1.378			
36	3	1.354			

Note: Correction of sample width not made.

Galena is formed in the same experiments as the lillianite-type sulfosalt. But its precipitation takes place some time after the start of the crystallization of the sulfasalt, approximately from the time when there is a dissolving and transfer of 0.2-0.3 g of the original ingot of lillianite composition, galena crystals at first grow around the stopper, but afterward the area occupied by galena increases and galena gradually covers the entire surface of the stopper, gripping and cementing the lillianite crystals without any traces of corrosion.

Lillianite crystals grow upward on the galena and on the walls of the insert. Galena crystals formed a solid crust which inpeded a determination of crystallographic form, but in some instances it was possible to establish that its crystals have a cubic form. Galena is also formed at the moment of the cooling of the autoclave. Here, a part of the galena grows epitaxically on the bismutine as well as on the sulfosalts--galenobismutite and lillianite.

Graham's phase 2 was established by X-rays (Table 15) in a microcrystalline precipitate produced in experiment 60 (the recrystallization of an ingot of cosalite composition) alongside galena. The formation of this phase occurred at a lower temperature, apparently, in the interval 300-400°C. This same phase, in appearance an outgrowth of fine needles, was produced by us in experiments 88 and 89, which repeated the conditions of the phase's formation by Graham, Thomson, and Berry. Together with phase 2, in this experiment bismutine was produced in the shape of fine-grained, flat growths. No transfer of this phase or of bismutine onto the stopper was noted.

CONCLUSION

The investigations of the PbS-Bi_2S_3 system that were carried out served, above all, to provide more precise data of the phase diagrams. We have essentially supported a part of the phase diagram by Malevski et al. (1963), included between PbS and PbS·Bi_2S_3, although the composition of individual intermediate phases turned out to be closer to stoichiometric in the Pb:Bi relationship. This concerns galenobismutite especially. This, apparently, can be explained by the experimental use of purer source materials than in Malevski's experiments.

At the same time, a part of the diagram between PbS·Bi_2S_3 and Bi_2S_3 differs noticeably from the ones introduced by Malevski. Here a small solid solution of PbS into Bi_2S_3 is established at temperatures above 671°C and what is more important, there appears a combination of PbS-2Bi_2S_3 (cannizzarite-bonchevite), which breaks down into bis-

Table 15. X-ray (Debye) phase, analogous to Graham's phase 2 (1953). Camera diameter 57.3 mm., CuK_{α}, Sample diameter 0.35 mm.

Phase 2		Small-grained deposit formed in chilling (Exp.60)		Galenite (Mikheer, 1957)		Phase 2		Small-grained deposit formed in chilling (Exp. 60)		Galenite (Mikheer, 1957)	
I	d/n	I	d/n	I	d/n	I	d/n	I	d/n	I	d/n
1	7.38					5	2.03	10	2.08	10	2.09
2	5.13							5	1.925		
10	3.82	6	3.79			4	1.910	5	1.880		
3	3.49	5	3.52			3	1.791	10	1.777	9	1.780
3	3.38	8	3.36	9	3.44	1	1.734				
1	3.29							8	1.703	8	1.707
	--					2	1.684	2	1.668		
	--					1/2	1.618	6	1.629		
6	3.01	10	2.92	10	2.96	1	1.573	4	1.571		
5	2.87	2	2.88			1/2	1.526				
2	2.76	5	2.73					3	1.505		
6	2.68					1	1.488	3	1.489	5	1.480
2	2.54	5	2.47					6	1.475		
1/2	2.39	2	2.35			1/2	1.441	3	1.450		
5	2.22	5	2.23					3	1.434		
		5	2.19					4	1.417		
		3	2.13			1	1.384	8	1.394		

Note: Correction of sample width not made.

mutine and galenobismutite at lower than 671±5°C. This combination is formed peritectically at 737±4°C. Thus, the region of its stability on the composition diagram is limited by the very narrow temperature interval between 671°C and 737°C. With the plotting of thermal analysis data on the diagram, difficulties were indeed encountered in the segment $PbS{\cdot}Bi_2S_3$-$PbS{\cdot}2Bi_2S_3$. So, the majority of the samples were characterized by a low temperature peak at 713±6°C, corresponding to the eutectic $PbS{\cdot}2Bi_2S_3$. At the same time only one sample in the section $PbS{\cdot}Bi_2S_3$--$PbS{\cdot}2Pb_2S_3$ had a temperature "hold" at 671±5°C, corresponding to the breakdown of cannizzarite into galenobismutite and bismutine. This circumstance is most likely connected with the great difficulties in reaching equilibrium for these samples and also with the weakness of the thermal effect of the breakdown of cannizzarite.

Juxtaposition of the data of pyrosynthetic and hydrothermal recrystallization, carried out in the conditions described above, show that there is not a full analogy between the one and the other. Besides galena and bismutine, lillianite, galenobismutite, and gungarrite were produced by the hydrothermal process. At the same time, cannizzarite was not established. On the contrary, in the hydrothermal recrystallization there appeared a new phase, analogous to Graham's second phase, not produced by pyrosynthesis.

All of this warns of the necessity to observe great caution in the transferral of data about pyrosynthetic investigation of sulfide systems to natural samples, and it places under suspicion the categorical conclusions of Malevski and other investigarors about the existence or the absence in nature of this or that sulfasalt. In this direction supplementary investigations are necessary--most prominently ones for determining many unsolved problems connected with the existence of cosalite and beegerite.

Particular interest is presented by a synthesis of cannizzarite in hydrothermal conditions at a temperature above 671°C with an aim of clarifying the possibilities of using it as a geological thermometer. The formation of cannizzarite at high temperatures in nature was, as is well-known, indicated by Zabonini and others, who first described this mineral from fumaroles with temperatures above 500°C.

In conclusion the authors express gratitude to Bogdanov, Kuminova et al. for their assistance and consultation.

Literature

Godovikov, A. A. Minerals of the Bismutine-Galena Series. Novosibirsk: USSR Academy of Sciences, 1965.

Godovikov, A. A., Iliasheva, N. A., Kliakhin, V. A., et al. New data on physico-chemical study of Bi-Se system. Materials on Genetic and Experimental Mineralogy, vol. 3. Novosibirsk: USSR Academy of Sciences, 1965.

Kliakhin, V. A., Yakishev, V. G. Determination of the density of a mineral from microhanging with the Rudenko and Vasilevsky method. Materials on Genetic and Experimental Mineralogy, vol 5. Novosibirsk: USSR Academy of Sciences, 1965.

Kolonin, G. R. The reduction structures of lead-bismuth sulfasalts. Materials on Genetic and Experimental Mineralogy, vol 3. Novosibirsk: USSR Academy of Sciences, 1965.

Kostov, I. K. The problem of isomorphism in mineral sulfasalts. Annals of the All-Union Mineral Society, vol. 86, no. 3, 1959.

Malevski, A. Yu., Rikhter, T. L., Veres, T. I. Lead-bismuth sulfasalts and isomorphic substitutions of sulphur by selenium in them. Proceedings of IMGRE, Issue 18, 1963.

Mikheev, V. I. An X-ray mineral locater. Gosgeolizdat, 1957.

Ontoev, D. O. Lillianite of the Bukukin Deposit and the conditions of its formation. Reports of the USSR Academy of Sciences, vol. 126, no. 4, 1959.

Syritso, L. F., Senderova, V. The problem of the existence of lillianite. Annals of the All-Union Mineral Society, 2nd Series, vol. 93, no. 4.

Greaser, V. S. Giessenit-ein neues Pb-Bi sulefasalts [sic] aus dem Dolomit des Bin. natales. Schweiz. Mineral. und Petregr. Mitt. 43, no. 2, 1963.

Graham, A. R., Thompson, R. M., Berry, L. G. Study of mineral sulphosalts: XVII--cannizzarite. Am. Mineral. 38, no. 5-6, 1953.

Van Hook, H. J. The ternary system Ag_xS-Bi_xS_3-PbS. Econ. Geol. 55, no. 4, 1960.

Kostov, I. K. Benchevite, a new mineral. Min. Mag. 31, no. 241, 1958.

Shenk, R., Hoffman, I., Knepper, W., Vegler, H. Gleichgewichtsstudien Sulphide. Zts. anorg. chemic. Bd. 240, Hf. 2, 1939.

Wolfe, C. W. Cannizzarite and bismuthinite from Vulcano. Am. Mineral. 23, no. 11, 1938.

Zambonini, F., de Fiore, O., Carrobi, G. Su un solfobismutite di plombe di Vulcano (Isole Eolie). Rendiconte dell Accademia delle Scienze fisiche et matematiche. Ler. 3[a] 31, Fascicolo 1-3, 1925.

15

Reprinted from *Mineralium Deposita* **1**:278–306 (1967)

Phase Relations and Mineral Assemblages in the Ag-Bi-Pb-S System

James R. Craig
Geophysical Laboratory, Carnegie Institution of Washington, Washington, D.C.

Phase relations within the Ag–Bi–S, Bi–Pb–S, and Ag–Pb–S systems have been determined in evacuated silica tube experiments. Integration of experimental data from these systems has permitted examination and extrapolation of phase relations within the Ag–Bi–Pb–S quaternary system. In the Ag–Bi–S system liquid immiscibility fields exist in the metal-rich portion above 597 ± 3°C and in the sulfur-rich portion above 563 ± 3°C. Ternary phases present correspond to matildite ($AgBiS_2$) and pavonite ($AgBi_3S_5$). Throughout the temperature range 802 ± 2°C to 343 ± 2°C the assemblage argentite (Ag_2S) + bismuth-rich liquid is stable; below 343°C this assemblage is replaced by the assemblage silver + matildite. Five ternary phases are stable on the PbS–Bi_2S_3 join above 400°C phase II (18 mol-% Bi_2S_3), phase III (27 mol-% Bi_2S_3), "cosalite" (33.3 mol-% Bi_2S_3), phase IV (51 mol-% Bi_2S_3), and phase V (65 mol-% Bi_2S_3). Phase IV corresponds to the mineral galenobismutite and is stable below 750 ± 3°C. Phases II, III, and V do not occur as minerals, but typical lamellar and myrmekitic textures commonly observed among the Pb–Bi sulfosalts and galena evidence their previous existence in ores. Phase II and III are stable from 829 ± 6°C and 816 ± 6°C, respectively, to below 200°C; Phase V, stable only between 730 ± 5°C and 680 ± 5°C in the pure Bi–Pb–S system is stabilized to 625 ± 5°C by the presence of 2% Ag_2S. Experiments conducted with natural cosalites suggest that this phase is stable only below 425 ± 25°C in the presence of vapor. – In the Ag–Pb–S system the silver-galena assemblage is stable below 784 ± 2°C, whereas the argentite + galena mineral pair is stable below 605 ± 5°C. – Solid solution between matildite and galena is complete above 215 ± 15°C; below this temperature characteristic Widmanstätten structure-like textures are formed through exsolution. Schematic phase relations within the quaternary system are presented at 1050°C, at 400°C, and at low temperature.

Die Phasenbeziehungen in den Systemen Ag–Bi–S, Bi–Pb–S und Ag–Pb–S wurden durch Versuche in evakuierten Quarzglasröhrchen bestimmt. Die Auswertung aller experimentellen Daten gestattete eine Extrapolation der Phasenbeziehungen im quaternären System Ag–Bi–Pb–S. – Im System Ag–Bi–S besteht ein Zwei-Schmelzenfeld im metallreichen Teil über 597 ± 3°C und im schwefelreichen Teil über 563 ± 3°C. Die ternären Phasen entsprechen den Mineralien Schapbachit ($AgBiS_2$) und Pavonit ($AgBi_3S_5$). Zwischen 802 ± 2°C und 343 ± 2°C ist die Paragenese Silberglanz (Ag_2S) + Bi-reiche Schmelze stabil; unterhalb 343°C wird sie jedoch ersetzt durch die Paragenese Silber + Schapbachit. – Fünf ternäre Phasen sind stabil im Schnitt PbS–Bi_2S_3 oberhalb von 400°C: Phase II (18 Mol-% Bi_2S_3), Phase III (27 Mol-% Bi_2S_3), „Cosalite" (33.3 Mol-% Bi_2S_3), Phase IV (51 Mol-% Bi_2S_3) und Phase V (65 Mol-% Bi_2S_3). Phase IV entspricht dem Mineral Galenobismutit und ist stabil unterhalb 750 ± 3°C. Die Phasen II, III und V kommen zwar nicht in der Natur vor, jedoch weisen typische myrmekitische und lamellare Gefüge, die man häufig in Pb–Bi-Sulfosalzen und deren Verwachsungen mit Bleiglanz beobachtet, auf die ehemalige Existenz solcher Phasen in diesen Erzen hin. Die Phasen II und III sind stabil von 829 ± 6°C bzw. 816 ± 6°C bis unter 200°C. Die Phase V, die im reinen System Bi–Pb–S zwischen 730 ± 5°C und 680 ± 5°C auftritt, wird in Gegenwart von 2% Ag_2S stabilisiert bis herab zu 625 ± 5°C. Versuche mit natürlichen Cosaliten lassen darauf schließen, daß diese Phase nur unterhalb 425 ± 25°C in Gegenwart einer Gasphase stabil ist. – Im System Ag–Pb–S ist die Paragenese Silber-Bleiglanz unterhalb von 784 ± 2°C stabil, die Paragenese Silberglanz-Bleiglanz dagegen unterhalb 605 ± 5°C. – Die Mischkristallreihe von Schapbachit und Bleiglanz ist vollständig oberhalb 215 ± 15°C; unterhalb dieser Temperatur entstehen charakteristische Entmischungsgefuge ähnlich den Widmannstättenschen Figuren. Für das quaternäre System werden schematische Phasenbeziehungen für 1050°C, 400°C und eine noch tiefere Temperatur gegeben.

Introduction

Sulfosalt minerals,[1] generally overlooked in the examination of ore deposits, are common accessories in many areas of mineralization. Although the principal sulfide and sulfide-type mineral constituents of many deposits have undergone systematic investigations in the field and in the laboratory, the sulfosalts have generally been regarded as mineralogical curiosities and have received little serious consideration. However, since the sulfosalt minerals do constitute a portion of many hypogene ores, they may provide information regarding conditions of ore formation.

Such information may be provided by "geologic indicators" – that is, minerals or mineral assemblages whose absolute P–T stability relations are known. Knowledge of stability fields and hence the value of any given mineral or mineral assemblage as a geologic indicator can be determined, however, only through laboratory investigation of the phase relations of the appropriate chemical systems.

Among the more common of the sulfosalt minerals in ores are those which have compositions lying within the Ag–Bi–Pb–S system. Investigation of phase relations in this system can therefore provide knowledge which is directly applicable to the understanding of the P–T conditions under which ore formation occurs.

The Ag–Bi–Pb–S system is especially interesting because in addition to sulfosalts it contains common simple sulfide minerals and the most abundant metallic native elements. The minerals whose compositions fall within the Ag–Bi–Pb–S system are silver (Ag), bismuth (Bi), lead (Pb), sulfur (S), argentite and acanthite (Ag_2S), bismuthinite (Bi_2S_3), galena (PbS), matildite[2] (or schapbachite) ($AgBiS_2$), pavonite ($AgBi_3S_5$), cosalite ($Pb_2Bi_2S_5$), and galenobismutite ($PbBi_2S_4$). The position of each of these phases within the Ag–Bi–Pb–S compositional tetrahedron is shown in Fig. 1; also shown is a metallic Pb–Bi phase ε which is not known as a mineral.

The geologically significant portion of this system is the distorted five sided polyhedron with bounding apexes of Ag–Bi–Bi_2S_3–Ag_2S–PbS; this portion of the quaternary system is outlined in heavy lines in Fig. 1.

Fig. 1. Minerals and phases in the Ag–Bi–Pb–S system. The geologically significant portion of this system is outlined with heavy lines. (Mol-% diagram.)

Numerous additional minerals have been described as existing in the Ag–Bi–Pb–S system

[1] Sulfosalt minerals have been defined by *Berry* (1965) as "multiple sulfides A_m, B_n, X_p in which A is one or more metals, B is a semi-metal [such as Bi, As, Sb] and X is S or Se; sulfosalts always contain As, Sb, Bi, rarely Ge, Sn, or V."

[2] The $AgBiS_2$ phase is generally referred to in American literature as "matildite" but in European literature as "schapbachite" (*Ramdohr*, 1960). Consistent with American usage the name "matildite" is employed in this paper.

especially on the $PbS-Bi_2S_3$ join, but have subsequently been shown to be mixtures of phases. Discussions of such phases are given below under the respective constituent systems.

Experimental Approach

Experimental investigation of the Ag–Bi–Pb–S system was carried out in four steps: 1. detailed investigation of phase relations in the Ag–Bi–S system, 2. detailed investigation of the $PbS-Bi_2S_3$ join with subsequent projection into the Bi–Pb–S system, 3. general investigation of the Ag–Pb–S system, 4. general investigation of the Ag–Bi–Pb–S system.

Reactants

Elemental bismuth, lead, silver, and sulfur used in the synthetic study were obtained from the American Smelting and Refining Co. and were of 99.999+% purity according to the supplier's spectrographic analysis. Bismuth in the form of pellets was reduced in a stream of hydrogen at 230–250°C for 6–10 hours prior to use. Silver obtained in the form of 2-ounce bars was filed and then reduced in a stream of hydrogen at 700 to 800°C for 6–10 hours. Lead, after removal of the outer oxidized coating, was filed and used directly. No catalysts were used.

Techniques

Most experimental data were obtained through the use of sealed, evacuated silica glass tubes as described by *Kullerud* and *Yoder* (1959). Experiments were heated in nichrome-wound resistance furnaces in which the temperature was regulated to ±2°C; temperatures were measured by means of chromel-alumel thermocouples and a Leeds and Northrup potentiometer. High-pressure experiments were carried out in welded gold tubes heated in cold-seal pressure vessels. No reaction of any elements or phases with the gold tubes was observed. Differential thermal analysis (D.T.A.) procedures were identical with those described by *Kullerud* (1963).

Samples were rapidly cooled in cold water at the termination of each experiment. Quench products were examined and identified by means of X-ray powder diffraction and reflected light microscopy. Precise lattice parameter determinations were made using silicon (a_0 = 5.4306 Å) as internal standard; a minimum of eight oscillations at a goniometer speed of $^1/_2$° 2θ per minute and a chart rate of 1° 2θ per inch were used for each determination.

Ag–Bi–S System

Previous Work

Ag–S System. – Although the general features of the Ag–S system were determined by *Roessler* (1895), the system was not known in detail until *Kracek*'s (1946) extensive study by differential thermal analysis. The single sulfide of silver, Ag_2S, exists in three forms below its congruent melting point of 838 ± 2°C. The low-temperature form, stable below 176°C, is monoclinic (*Ramsdell*, 1943); the intermediate form, stable between 176°C and the high-temperature transition (622 ± 3°C in the presence of excess S, 586 ± 3°C in the presence of Ag (*Kracek*, 1946) is body-centered cubic (*Emmons* et al., 1926; *Rahlfs*, 1936); the high-temperature form, stable between the second transition and the melting point of Ag_2S, is reported to be face-centered cubic (*Djurle*, 1958).

Kracek (1946) reports the limits of solid solution of the highest temperature form of Ag_2S as less than 2.7 ± 0.2 mol-% toward S at 740°C and less than 1.3 ± 0.3 mol-% toward Ag at 804°C. Maximum solid solution of the low-temperature form is less than 0.02 mol-% toward Ag or S (*Wagner*, 1953).

Two large regions of liquid immiscibility are present within the system; the one between Ag and Ag_2S compositions has a maximum width of 25.3 mol-% and occurs above 906°C, whereas that between Ag_2S and S composition has a maximum width of 64.0 mol-% and exists above 740°C (*Kracek*, 1946).

Bi–S System. – Phase relations in the $Bi-Bi_2S_3$ portion of this system were determined by *Aten* (1905) and in the Bi_2S_3-S portion of the system by *Van Hook* (1960). The only compound, Bi_2S_3, is stoichiometric at all temperatures below its melting point of 760 ± 5°C (*Van Hook*, 1960).

Ag–Bi System. – The Ag–Bi system has been examined by many workers and is summarized in *Hansen* and *Anderko* (1958). The system contains a simple eutectic at 262 ± 1°C at 95.3 mol-% Bi (*Petrenko*, 1906); no intermediate compounds are reported.

Ag–Bi–S System. – The first synthesis of a ternary Ag–Bi–S phase was by *Roessler* (1895), who crystallized $AgBiS_2$ from a metallic melt. The first systematic work carried out within the ternary system was by *Gaudin* and *McGlashan* (1938), who investigated the $Ag_2S-Bi_2S_3$ join. They report two compounds – "Phase D" ($3\,Ag_2S \cdot Bi_2S_3$), melting congruently at about 650°C, and matildite ($Ag_2S \cdot Bi_2S_3$), melting incongruently to Bi_2S_3 + liquid at about 560°C.

The mineral matildite was first described in detail by *Ramdohr* (1938), who reported an inversion from the low-temperature (β) orthorhombic form to the high-temperature (α) cubic form at 210°C. *Graham* (1951) confirmed the structure of the β form on the basis of single crystals synthesized by hydrothermal techniques; X-ray diffraction peaks which could not be indexed on the basis of an orthorhombic symmetry were regarded as due to Ag_2S present as an impurity. More recently, *Geller* and *Wernick* (1959) reported that β-$AgBiS_2$ is hexagonal rather than orthorhombic.

Schenck et al. (1939) conducted a series of experiments at 510°C; the only ternary phase recognized was $AgBiS_2$. In a series of solid state diffusion experiments with Ag_2S and Bi_2S_3 wafers, *Ross* (1954) reported formation of $AgBiS_2$, $5\ Ag_2S \cdot 4\ Bi_2S_3$, and $3\ Ag_2S \cdot 4\ Bi_2S_3$.

Nuffield (1954) reported the existence of a new sulfosalt, pavonite ($AgBi_3S_5$), and its synthetic analogue. An extensive study of the Ag_2S–Bi_2S_3 join was conducted by *Van Hook* (1960), who presented phase relations on this join as shown in Fig. 2. *Van Hook* synthesized $AgBiS_2$ and $AgBi_3S_5$; apparently unaware that $AgBi_3S_5$ had previously been described as a mineral, he referred to this as "Phase X."

Table 1. *Experimental data at 920°C*

Composition, mol-%			Time	Products
Ag	Bi	S	[hours]	
80.1	14.0	5.9	4	Liq
79.1	13.0	7.9	4	2 Liq
64.7	15.1	20.2	1.5	2 Liq
60.8	18.3	20.9	1.5	2 Liq
59.1	32.9	8.0	2	Liq
54.1	12.2	33.7	1.5	2 Liq
40.0	46.1	13.9	4	2 Liq
40.0	26.2	33.8	5	2 Liq
29.8	34.9	35.3	2	Liq
29.0	50.8	20.2	2	2 Liq
27.0	60.0	13.0	3	Liq
21.0	51.6	27.4	4	Liq

Experimental Data and Discussion

Phase relations within the Ag–Bi–S system were determined in the temperature range 920

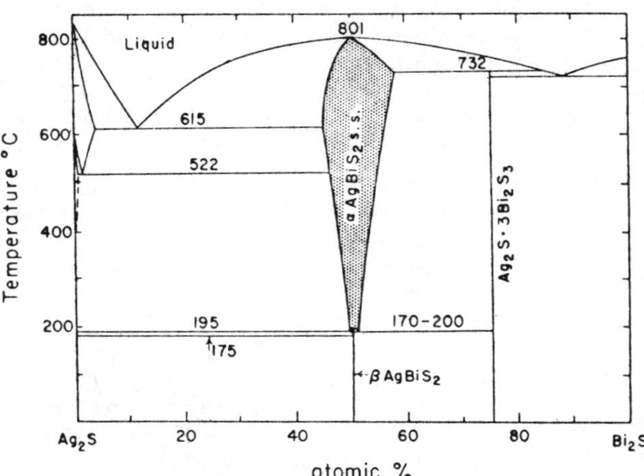

Fig. 2. The Ag_2S–Bi_2S_3 join (from *Van Hook*, 1960).

to 200°C and are presented in a series of isothermal sections at 920, 715, 625, 350, and 325°C. The results of silica tube experiments, upon which the isothermal sections are based, appear in Tables 1, 2, 4–10. D.T.A. experiments are listed in Table 3. Relations within this system are discussed below in a succession of events from high to progressively lower temperatures. This manner of presentation parallels to some degree the conditions which prevail in a cooling ore mass and permits clear understanding of the developments occurring within the system from simple high-temperature to more complex low-temperature relations.

920°C Isotherm. – Phase relations within the Ag–Bi–S system as determined at 920°C are shown in Fig. 3. These relations are based upon experimental data listed in Table 1. The data presented in this table and others that follow are limited to those experiments bearing most directly upon the observed phase relations. Additional experiments, at the temperatures of the isotherms presented and at other temperatures which corroborate the relations illustrated, are omitted here for the sake of brevity; the entire bulk of experimental data derived during this study is presented in *Craig* (1965).

At this temperature the "condensed"[3] system contains 1. a small S-rich liquid field which contains less than 1% metal; 2. a large S-rich two-liquid field which spans the entire system and which is bounded by the approximate composition of 1 and 60 mol-% Ag on the Ag–S join (*Kracek*, 1946) and 1 and 34 mol-% Bi on the Bi–S join (*Van Hook*, 1960); 3. a large homogeneous liquid field which extends across the central portion of the system and which closes along the Bi–S join to extend along the Ag–Bi join; 4. a large metal-rich two-liquid field which extends approximately 54 mol-% Bi into the ternary system from the Ag–S join

[3] All phase diagrams presented below are "condensed," that is, all phase assemblages are in equilibrium with a vapor. This usage is consistent with that of recent sulfide investigations (e.g. *Kullerud* and *Yoder*, 1959)

on which the bounding compositions are 68 and 92 mol-% Ag; 5. a small region in which Ag and liquid are stable.

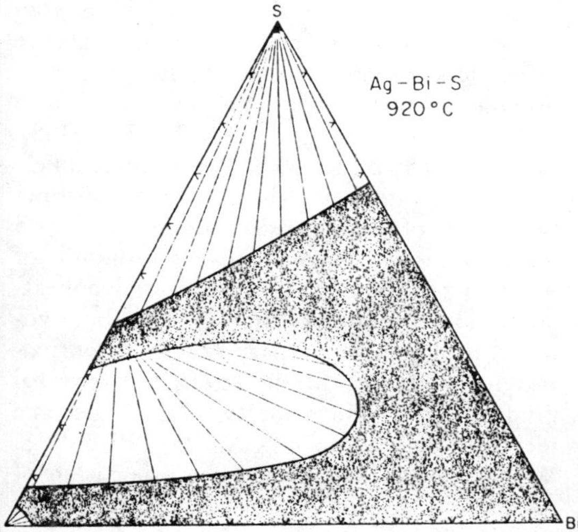

Fig. 3. The 920°C isotherm of the Ag–Bi–S system. Homogeneous liquid is stippled. (Mol-% diagram.)

Homogeneous Liquid Regions. – All homogeneous liquids in this system, with the possible exception of the nearly pure sulfur-liquid, crystallize even on the most rapid cooling to form intricate mixtures of phases stable at lower temperatures. Silver-rich liquids crystallize on cooling to form granular masses of silver in poorly developed cubic-like patterns in a matrix of bismuth. Bismuth-rich liquids typically develop dendrites and "star-like" skeletal crystals (Fig. 4) of $AgBiS_2$ and Ag within the bismuth matrix on cooling. Liquids near Ag_2S or $AgBiS_2$ compositions crystallize on cooling to form complex dendrites (Fig. 5) and/or delicate myrmekitic intergrowths of Ag_2S and/or $AgBiS_2$ with interstitial Ag and/or Bi. Liquids deficient in Ag and approaching $AgBi_3S_5$ or Bi_2S_3 composition form intricate lamellar networks of random or suboriented laths (Fig. 6) of $AgBi_3S_5$ and/or Bi_2S_3 within a Bi matrix.

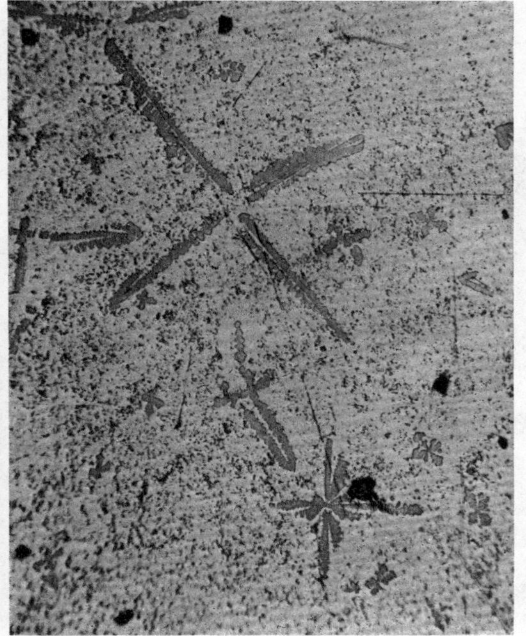

Fig. 4. "Stars" of $AgBiS_2$ (grey) developed in Bi matrix (light grey) on quenching of homogeneous liquid from 400°C. (× 400 oil.)

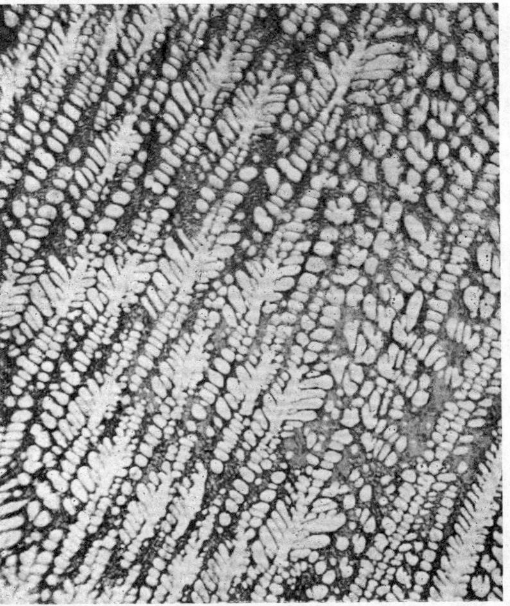

Fig. 5. Texture formed on quenching homogeneous Ag_2S–$AgBiS_2$ rich liquid from 715°C. $AgBiS_2$ dendrites (white) in Ag_2S matrix (grey). (× 715 oil.)

Two-Liquid Regions. – Recognition of the sulfur-rich two-liquid field has been by visual observation of experimental charges immediately upon withdrawal from the furnace (time elapsed in withdrawal of charge from furnace to point of observation was less than $^1/_4$ second). As a charge is withdrawn from the furnace, there is clearly evident a metallic liquid overlain by a

Fig. 6. Laths of $AgBi_3S_5$ (grey) developed in Bi matrix (white) on quenching homogeneous liquid from 715°C. (× 335 oil.)

deep reddish sulfur-rich liquid. As the charge cools, the metal-rich liquid crystallizes and the sulfur-rich liquid assumes a yellowish color.

The extent of the metal-rich two-liquid field into the ternary system from the Ag–S join has been determined by examination of quenched samples in polished sections. When appreciable quantities of both liquids are present a distinct gravitational separation occurs with the Ag–Bi rich liquid settling below the Ag_2S–$AgBiS_2$ rich liquid (see Fig. 7). Although the actual specific gravities of the two liquids at 920°C are not known, the relative differences may be estimated from the specific gravities of the phases which crystallize from each liquid on cooling: Ag_2S, 7.2–7.3; $AgBiS_2$, 6.9; Ag, 10.5; Bi, 9.7–9.8. The greatest difference in densities and thus the most distinct separation of liquids exists along the Ag–S join. Increased Bi content results in liquids of increasingly similar composition and appearance. The maximum extension of the two-liquid field into the ternary system is approximately 54 mol-% Bi; however, the terminal point, point at which the two liquids become identical, lies at about 30% Ag, 35% Bi, 35% S. Within the two-liquid field the orientation of tie-lines as shown in Fig. 3 is based upon estimated liquid compositions and upon extrapolation upward in temperature of tie lines whose directions were accurately determined at temperatures.

Developments between 920 and 715°C. – Below 920°C the phase relations remain essentially as shown in Fig. 3 to 906 ± 1°C. Here the metal-rich two-liquid field withdraws from the Ag–S binary (*Kracek*, 1946) but persists within the ternary system. Below 906°C recession of the Ag-rich member of this two-liquid field into the ternary system necessitates creation of the univariant assemblage Ag + Ag_2S-rich liquid + Ag-rich liquid + vapor.

Ag_2S crystallizes at 838 ± 2°C (*Kracek*, 1946). At 804 ± 2°C (*Kracek*, 1946) the liquid field between Ag and Ag_2S withdraws into the ternary system, permitting Ag and Ag_2S to exist stably together and resulting in the establishment of the univariant assemblage Ag + Ag_2S + Ag_2S-rich liquid + vapor. However, in the interval 804–800°C the high-temperature assemblage Ag + Ag_2S-rich liquid + vapor is replaced by the assemblage Ag_2S + Ag-rich liquid + vapor through an invariant reaction. The ternary phase $AgBiS_2$ appears at its congruent melting point of 801 ± 4°C and Bi_2S_3 crystallizes at 760 ± 3°C.

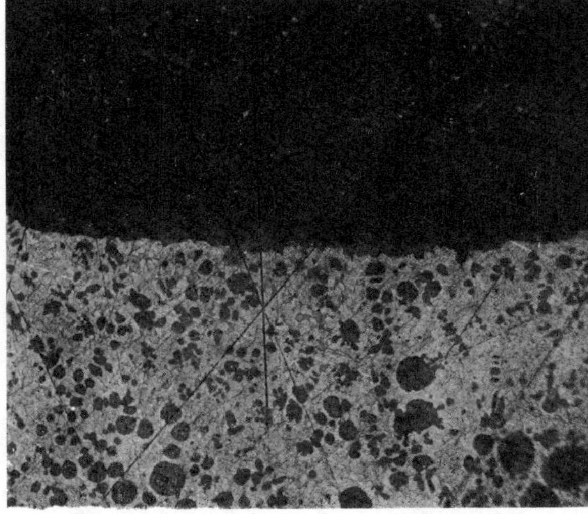

Fig. 7. Sharp boundary between Ag–Bi liquid (light) and Ag_2S–$AgBiS_2$-liquid (dark) in specimen quenched from 920°C. (× 490 oil.)

As the temperature decreases to 740 ± 2°C (*Kracek*, 1946) the sulfur-rich two-liquid field withdraws from the Ag–S join, resulting in the formation of the univariant assemblage Ag_2S + Ag_2S-rich liquid + S liquid + vapor. The $AgBi_3S_5$ phase appears at 732 ± 4°C, the temperature at which it melts incongruently to $AgBiS_2$ + liquid in the presence of vapor. Slightly below the temperature of appearance of $AgBi_3S_5$ and within the limits of 723 ± 3°C, the three separate reactions occur:

1. Liquid → $AgBiS_2$ + S liquid
2. Liquid → $AgBi_3S_5$ + Bi_2S_3
3. Liquid → Bi_2S_3 + S liquid

1. This reaction designates the establishment of tie lines between $AgBiS_2$ and S liquid. These tie lines divide the S-rich two-liquid field into two separate two-liquid fields.

2. This reaction designates "eutectic" relations between $AgBi_3S_5$ + Bi_2S_3.

3. The third reaction refers to the disappearance of the S-rich two-liquid field from the Bi–S join; procession of this liquid into the ternary system results in the creation of the univariant assemblage Bi_2S_3 + Bi_2S_3-rich liquid + S liquid + vapor. Determination of the temperature of this reaction was by visual observation of experimental charges at reaction temperatures. Below 723°C the charge visibly contains needle-shaped Bi_2S_3 crystals + S liquid; above this temperature two liquids are visible. Cooling of an experimental charge, of a composition with the limits of the two-liquid field, from above 723°C results in abrupt formation of Bi_2S_3 crystals.

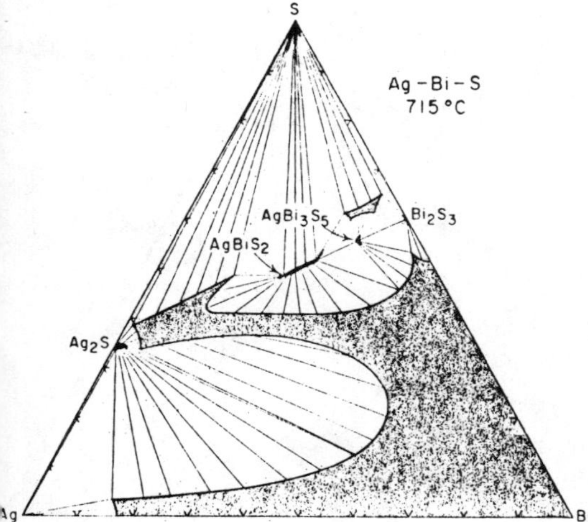

Fig. 8. The 715°C isotherm of the Ag–Bi–S system. Homogeneous liquid regions are stippled. (Mol-% diagram.)

715°C Isotherm. – The sequence of events described above results in phase relations at 715°C as shown in Fig. 8. Experimental data at 715°C appears in Table 2. At this temperature the size of the metal-rich two-liquid field is considerably smaller than at 920°C. The decrease in size is primarily due to the rapid increase in area of the Ag_2S + Ag-Bi liquid field in the interval between 802 ± 2°C and 715°C.

Table 2. *Experimental data at 715°C*

Composition, mol-% Ag	Bi	S	Time [hours]	Products
76.9	8.1	15.0	43	Ag + Ag_2S + AgBi liq
75.1	10.1	14.8	20.5	Ag_2S + AgBi liq
75.0	23.0	2.0	111	AgBi liq
61.2	5.0	33.8	44	Ag_2S + 2 liq
56.8	5.0	38.2	117	Ag_2S-rich liq
54.4	20.2	25.4	29	Ag_2S + AgBi liq
47.3	27.0	25.7	124	2 liq
43.0	12.0	45.0	117	Liq
39.9	16.9	43.2	44	$AgBiS_2$ + liq
33.0	58.0	9.0	43	Liq
32.6	12.8	54.6	118	2 liq
32.0	54.2	13.8	111	Ag_2S + liq
30.9	52.9	16.2	44	2 liq
26.5	16.7	56.8	118	$AgBiS_2$ + 2 liq
19.9	21.4	58.7	74	$AgBiS_2$ + liq
16.4	24.0	59.6	74	$AgBiS_2$ + 2 liq
15.0	44.9	40.1	22.5	Liq
14.8	39.8	45.4	22.5	$AgBiS_2$ liq
8.4	28.0	63.6	74	2 liq
2.9	35.9	61.2	111	Bi_2S_3 + liq
2.1	48.2	49.7	111	Liq

Developments between 715 and 350°C. – The phase assemblages present at 715°C remain unaltered except for minor changes in the liquidus surfaces as the temperature decreases to 676 ± 3°C. At this temperature the two-liquid field in the $AgBiS_2$–Bi_2S_3–S portion of the system disappears as the metal-rich member is consumed in a ternary monotectic reaction. At 676 ± 3°C the composition of this liquid lies on or very close to the $AgBi_3S_5$–S join. DTA experiments on mixtures of $AgBi_3S_5$ + S and

Ag_2S $AgBiS_2$ join, the liquidus appears as a "saddle" with its maximum on the join. Thus Ag_2S-rich liquids remain on both sides of the join below 615 °C. The narrow two-liquid field remaining in the metal-rich portion of the system persists to 597 ± 3 °C, at which temper-

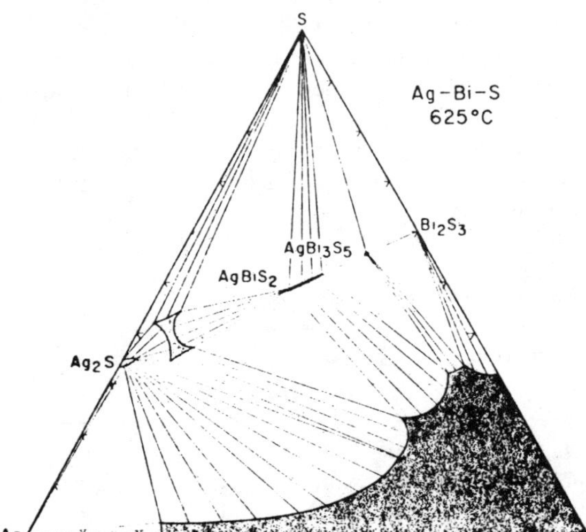

Fig. 9. The 625°C isotherm of the Ag–Bi–S system. Homogeneous liquid regions are stippled, as is Ag_2S. (Mol-% diagram.)

ature the Ag_2S-rich liquid member disappears in a ternary monotectic to $Ag_2S + AgBiS_2$ + AgBi liquid as indicated by the experimental data presented in Table 6. At the monotectic temperature the Ag_2S-rich liquid contains 53.5 ± 1 mol-% Ag, 9.5 ± 1 mol-% Bi, 37.0 ± 1 mol-% S.

The two-liquid field in the sulfur-rich portion of the system persists to 563 ± 2°C, at which

Table 6. *Experimental data on monotectic reaction at 597± 3°C*

Composition, mol-% Ag	Bi	S	Temp. [°C]	Time [hrs.]	Products
27.3	50.0	22.7	625	24	2 liq
55.0	8.0	37.0	600	88	Ag_2S + 2 liq
53.0	12.0	35.0	600	88	Ag_2S + 2 liq
52.0	11.0	36.9	600	88	$AgBiS_2$ + 2 liq
33.4	43.1	23.5	595	45	$Ag_2S + AgBiS_2$ + AgBi liq
33.9	42.9	23.2	588	38	$Ag_2S + AgBiS_2$ + AgBi liq

Table 7. *Experimental data on monotectic reaction at 563± 2°C*

Composition, mol-% Ag	Bi	S	Temp. [°C]	Time [hrs.]	Products
50.3	5.7	44.0	565	40	$AgBiS_2$ + 2 liq
55.0	4.0	40.8	568	93	2 liq
52.0	6.0	42.0	565	40	Ag_2S + 2 liq
47.8	7.0	45.2	561	36	$Ag_2S + AgBiS_2$ + S liq
48.9	6.6	44.5	555	5	$Ag_2S + AgBiS_2$ + S liq

temperature the Ag_2S-rich liquid decomposes to $Ag_2S + AgBiS_2$ + S liquid (Table 7). The composition of this ternary monotectic is 54 ± 1 mol-% Ag, 6 ± 1 mol-% Bi, 40 ± 1 mol-% S.

Below 563 ± 2°C the stable condensed univariant assemblages are: Ag_2S + Ag + AgBi liquid, $Ag_2S + AgBiS_2$ + AgBi liquid, $Ag_2S + AgBiS_2$ + S liquid, $AgBiS_2 + AgBi_3S_5$ + AgBi liquid, $AgBiS_2 + AgBi_3S_5$ + S liquid, $AgBi_3S_5 + Bi_2S_3$ + AgBi liquid and $AgBi_3S_5 + Bi_2S_3$ + S liquid. At 563°C solid solution in the $AgBiS_2$ phase extends from 45 to 55 mol-% Bi_2S_3 along the Ag_2S–Bi_2S_3 join (*Van Hook*, 1960). Except for the gradual reduction in the size of the AgBi liquid field, these assemblages remain stable over the temperature range 563–343°C. The general phase relations within this tem-

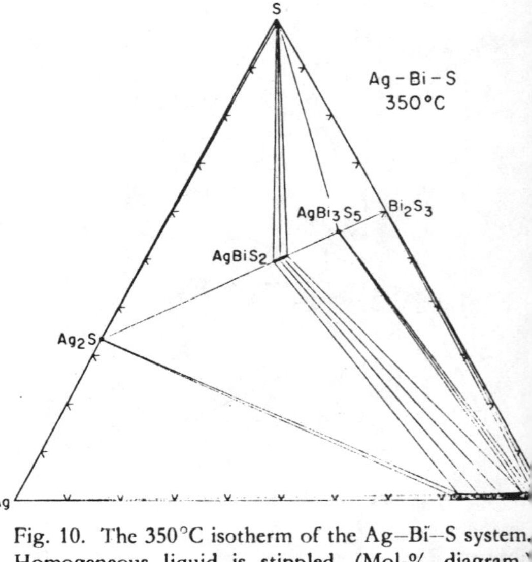

Fig. 10. The 350°C isotherm of the Ag–Bi–S system. Homogeneous liquid is stippled. (Mol-% diagram.)

with a small excess of Bi_2S_3 or an excess of $AgBiS_2$ all indicate identical reaction temperatures (Table 3).

Table 3. *Results of D.T.A. experiments*

Composition, mol-%				Thermal Effect		Interpretation
Ag	Bi	Pb	S	heating	cooling	
—	100	—	—	271		melting
—	—	—	—		262	freezing
3	8	—	89	115		S melting
—	—	—	—	676	678	$AgBi_3S_5 + S \rightleftarrows$ liquid
5	30	—	65	116		S melting
—	—	—	—	677	649	$AgBi_3S_5 + Bi_2S_3 + S \rightleftarrows$ Liq
4	7	—	89	115		S melting
—	—	—	—	678	676	$AgBiS_2 + AgBi_3S_5 + S \rightleftarrows$ Liq
47	6	—	47	116		S melting
—	—	—	—	561	511	$Ag_2S + AgBiS_2 + S \rightleftarrows$ Liq
50	15	—	35	345	318	$AgBiS_2 + Ag \rightleftarrows Ag_2S$ + Bi-Liq
—	—	—	—	525	516	Ag_2S inversion
—	—	—	—	596	598	$Ag_2S + AgBiS_2$ + Bi-Liq $\rightleftarrows$ Liq
45	—	8	47	116		S melting
—	—	—	—	176		Ag_2S inversion
—	—	—	—	457	450	Ag_2S inversion
—	—	—	—	528	529	$Ag_2S + PbS + S \rightleftarrows$ Liq
—	24	9	67	115		S melting
—	—	—	—	661	643	Bi_2S_3 + IV + S $\rightleftarrows$ Liq
—	—	—	—	691	682	Complete liquification

Continued lowering of the temperature leads to a decrease in the width of the liquid field in the metal-rich portion of the system and an increase in size of the $AgBiS_2$ + AgBi-liquid divariant region. At 669 ± 6°C intersection of the metal-rich two-liquid field and the $AgBiS_2$ + AgBi-liquid field occurs with the subsequent formation of the univariant assemblage $AgBiS_2$ + 2 liquids + vapor. The composition of intersection of these fields is 31 ± 2 mol-% Ag, 32 ± 2 mol-% Bi, 37 ± 2 mol-% S (Table 4).

Table 4. *Experimental data on intersection of metal-rich two-liquid field and $AgBiS_2$-liquid field (669 ± 6°C)*

Composition, mol-%			Temp.	Time	Products
Ag	Bi	S	[°C]	[hrs.]	
33.9	26.0	40.1	675	360	$AgBiS_2$ + liq
30.0	34.5	37.5	675	42	$AgBiS_2$ + liq
29.0	32.9	36.1	675	112	$AgBiS_2$ + liq
32.5	31.0	36.5	675	42	Liq
30.1	35.1	34.8	675	42	Liq
29.2	34.7	36.1	675	360	Liq
35.8	27.9	36.2	675	360	2 liq
35.2	27.1	37.6	662	112	$AgBiS_2$ + 2 liq
33.5	25.7	40.8	662	112	$AgBiS_2$ + 2 liq
36.0	28.0	36.0	662	112	2 liq

Table 5. *Experimental data at 625°C*

Composition, mol-%			Time	Products
Ag	Bi	S	[hrs.]	
65.0	5.0	30.0	116	Ag_2S + liq
57.7	7.9	34.4	98	Liq
50.0	44.9	5.1	23	Ag_2S + liq
50.0	15.0	35.0	116	2 liq
44.4	30.5	25.1	13	Ag_2S + 2 liq
44.2	15.1	40.7	98	$AgBiS_2$ + liq
35.0	29.4	35.6	98	$AgBiS_2$ + 2 liq
27.3	50.0	22.7	24	2 liq
20.0	53.0	27.0	23	$AgBiS_2$ + liq
20.0	60.0	20.1	23	Liq
10.4	49.8	39.8	69	$AgBiS_2 + AgBi_3S_5$ + liq
5.1	61.5	33.4	24	Liq
5.1	49.9	45.1	69	$AgBi_3S_5 + Bi_2S_3$ + liq

Phase relations as determined at 625°C (data in Table 5) are shown in Fig. 9. The relations in this figure remain with little change to 615 ± 2°C (*Van Hook*, 1960) at which temperature tie lines are established between Ag_2S and $AgBiS_2$. In a T–X section normal to the perature interval are shown in the 350 °C isotherm of Fig. 10 (data in Table 8). At 350 °C the metal-rich liquid is confined to a narrow zone containing less than 2 mol-% S except in the extremely Bi-rich portion where 4 mol-% S may be taken into solution.

Table 8. *Experimental data at 350 °C*

Composition, mol-% Ag	Bi	S	Time [hrs.]	Products
60.5	13.2	26.3	1325	Ag + Ag_2S + AgBi liq
21.0	69.9	9.1	675	Ag_2S + $AgBiS_2$ + AgBi liq
14.0	70.0	16.0	675	$AgBiS_2$ + AgBi liq

Developments below 350 °C. – At 343 ± 3 °C the assemblage Ag_2S + AgBi liquid becomes unstable and is replaced by the assemblage Ag + $AgBiS_2$ as indicated by the data in Table 9a. The changes in phase assemblages at this invariant point are illustrated in the schematic P–T diagram of Fig. 12.

In this diagram each line emanating from the invariant point represents a P–T curve of univariant (four-phase) equilibrium and is labeled by the phase which is absent. At the invariant point five phases coexist–Ag, Bi liquid, Ag_2S, $AgBiS_2$, and vapor. Change in either pressure or temperature necessitates elimination of one or more of these phases. The sequence of curves about this invariant point may be deduced from the principles described by *Morey* and *Williamson* (1918) and *Morey* (1957). The nearly vertical curve (V) defines the effect of confining pressure on the upper stability of the Ag + $AgBiS_2$ assemblage. Experiments conducted under a confining pressure of 2050 ± 20 bars (Table 9b) indicate that the stability of this assemblage is raised 6 °C per 1000 bars. The changes in phase assemblages occurring at

Table 9. *Experimental data on reaction 3 Ag + $AgBiS_2 \rightleftarrows 2\ Ag_2S$ + Liq*

Reactants	Temp. [°C]	Time [hrs.]	Products
(a) *In the presence of vapor*			
Ag + $AgBiS_2$	400	230	Ag + Ag_2S + AgBi liq
Ag + $AgBiS_2$	350	1325	Ag + Ag_2S + AgBi liq
Ag + $AgBiS_2$	346	888	Ag + Ag_2S + AgBi liq
Ag_2S + Bi	346	888	Ag_2S + $AgBiS_2$ + AgBi liq
Ag + $AgBiS_2$	343	460	Ag + $AgBiS_2$ + AgBi liq + Ag_2S
Ag + $AgBiS_2$	342	460	Ag + $AgBiS_2$
Ag + $AgBiS_2$	335	555	Ag + $AgBiS_2$
Ag + $AgBiS_2$	326	1996	Ag + $AgBiS_2$
Ag_2S + Bi	325	775	Ag + $AgBiS_2$ + AgBi liq
(b) *Under confining pressure of 2050 ± 20 bars*			
Ag + $AgBiS_2$	350	4	$AgBiS_2$ + Ag
Ag_2S + Bi	350	4	Ag_2S + Bi + Ag + $AgBiS_2$
Ag + $AgBiS_2$	355	3½	Ag + $AgBiS_2$
Ag_2S + Bi	355	3½	Ag_2S + Bi + Ag + $AgBiS_2$
Ag + $AgBiS_2$	360	3½	Ag + $AgBiS_2$ + Ag_2S + Bi
Ag_2S + Bi	360	3½	Ag_2S + Bi + Ag + $AgBiS_2$*

* Ag and $AgBiS_2$ only as rims formed on quenching.

This reaction may be written approximately as 3 Ag + $AgBiS_2 \rightleftarrows 2\ Ag_2S$ + Bi liquid. Evidence of this reaction may be observed in Fig. 11 in which are shown rims of Ag_2S which formed when Ag and $AgBiS_2$ were heated above 343 °C. temperatures above that of the invariant point are defined by the two curves (Ag) and ($AgBiS_2$). Changes occurring at temperatures below that of the invariant point are represented by the two curves (*L*) and (Ag_2S). Although not

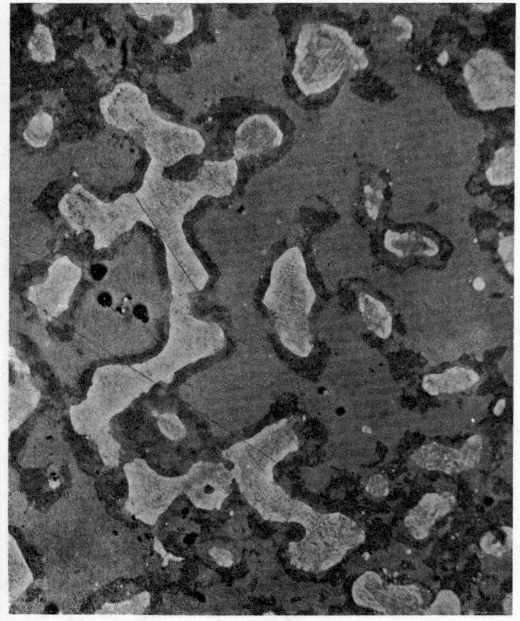

Fig. 11. Reaction rims of Ag_2S (dark grey) formed between $AgBiS_2$ (light grey) and Ag (white); light mottled grey within Ag-rims is Bi. Specimen heated 1 day at 344°C; previous heating at 342 C for 19 days formed only $AgBiS_2$, Ag, and Bi. (× 385 oil.)

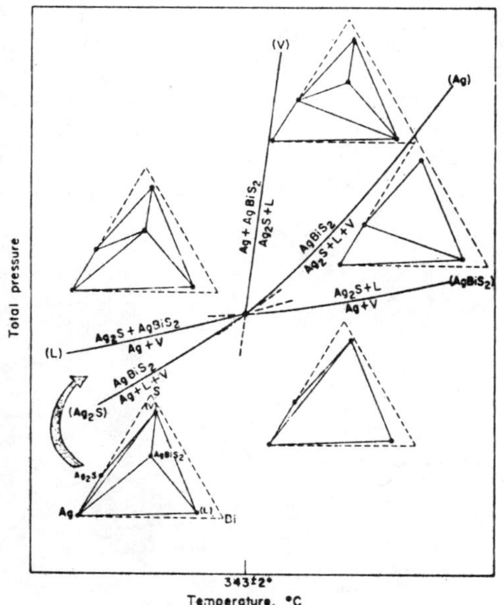

Fig. 12. Schematic pressure-temperature diagram showing phase relations around the invariant point which defines the maximum stability of the silver-matildite assemblage in the presence of vapor. The triangles depict the phase assemblages stable within the divariant regions in the relevant portion of the Ag–Bi–S system.

known with certainty, the composition of the vapor is assumed to be essentially sulfur.

Phase relations below the invariant point are shown in Fig. 13 (data in Table 10). Further decrease in temperature below 325°C results in reduction in area of the Bi-rich liquid field, appearance of Bi as a phase at 271°C, and final crystallization of Ag + Bi + $AgBiS_2$ from Bi-rich liquid at 259 ± 2°C. The composition of this ternary eutectic is 94 ± 1 mol-% Bi, 4 ± 1 mol-% Ag, 2 ± 1 mol-% S. The essential phase changes occurring within the Ag–Bi–S system between 259 and 25°C include decrease in the extent of solid solutions, and inversion of Ag_2S and $AgBiS_2$ to low temperature polymorphic forms.

Table 10. *Experimental data at 325°C*

Composition, mol-% Ag	Bi	S	Time [hrs.]	Products
39.9	18.8	41.3	775	Ag + Ag_2S + $AgBiS_2$
24.8	52.9	22.3	1275	Ag + $AgBiS_2$ + AgBi liq
14.0	69.9	16.1	775	$AgBiS_2$ + AgBi liq
6.0	70.0	24.0	775	$AgBi_3S_5$ + AgBi liq

Individual Phases

Ag and Bi. – Silver exists in only one modification below its melting point of 960.5°C (*Kracek*, 1946). Optical and physical properties

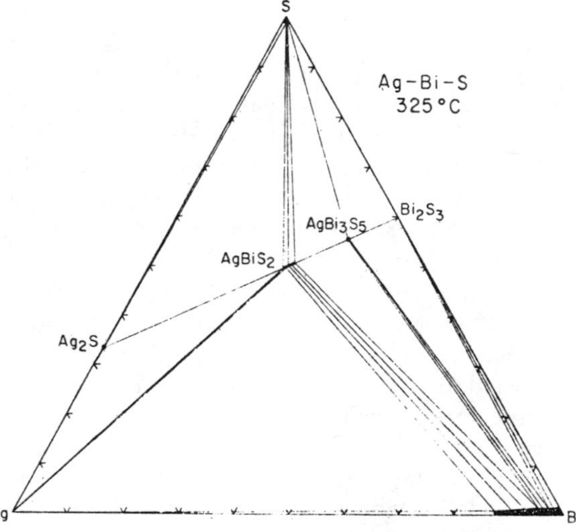

Fig. 13. The 325°C isotherm of the Ag–Bi–S system. Homogeneous liquid is stippled. (Mol-% diagram.)

remain essentially unaffected by the presence of other phases in the Ag–Bi–S system.

Native bismuth crystallized in the laboratory and observed in nature is characteristically twinned as shown in Figs. 14 and 15. This twinning has been ascribed to an inversion at 75°C (*Wurschmidt*, 1914) and has thus been considered as indicative of initial formation above the inversion temperature (*Bateman*, 1950; *Edwards*, 1954). Differential thermal analysis of bismuth has not revealed any thermal expression of such an inversion, and a high-temperature X-ray powder photograph at 200°C reveals only the lines of normal rhombohedral bismuth.

Fig. 14. Twinning in native bismuth (Nipissing Mine, Cobalt, Ontario). (Crossed nicols; × 425 oil.)

Fig. 15. Twinning in native bismuth (Nipissing mine, Cobalt, Ontario). Note two microfaults in center of photomicrograph. (Crossed nicols; × 425 oil.)

The characteristic twinning has been observed in all samples quenched from above liquidus temperatures. One twinned sample which was heated to 100°C in air while under microscopic observation retained its twin lamellae. Samples annealed at 250°C for 7 days contained many small recrystallized grains and twin lamellae boundaries which were irregular and less well defined than those in freshly quenched specimens.

Cooling of pure bismuth or of bismuth-rich compositions (>65 mol-% Bi) from the liquid state results in the formation of small beads of pure or nearly pure Bi on the upper surface of the experimental charges. Optical examination reveals that these beads result from the expulsion of bismuth at or near the temperature of final crystallization of the liquid. As the bismuth solidifies, it passes from a dense liquid phase to a less dense solid phase; this results in an increased volume which cannot be accommodated within the outer first crystallized shell of the charge. Thus, material is expelled from the interior of the charge through the outer shell onto the surface in the form of small beads. Since at the temperature of final crystallization the solubility of other phases in the Bi-rich liquid is small, the beads formed on the surface are nearly pure Bi. Stresses created within the bismuth on crystallization are apparently sufficient to produce twinning.

Intense twinning may also be induced by grinding. Sets of "herring bone" twin lamellae frequently parallel deep scratches in polished surfaces and result from the pressure exerted by plucked grains in the polishing process.

Ag_2S. — Ag_2S, analogous to the mineral acanthite exists in three crystalline modifications (*Kracek*, 1946; *Djurle*, 1958). Evidence of the high-temperature non-quenchable cubic forms exists in the presence of well developed crystals with cubic morphology in experiments above 176 °C.

Solid solution in this phase has been reported along the Ag–S join by *Kracek* (1946) and along the Ag_2S–Bi_2S_3 join by *Van Hook* (1960). The maximum solubility of Bi_2S_3 in Ag_2S is 3.0 ± 0.5 mol-% at 615 °C. On cooling the Ag_2S solid solution breaks down to form low-temperature Ag_2S with small (0.01 mm) exsolved lamellae of $AgBiS_2$.

Ramdohr (1960) states that the frequent twinning observed in natural Ag_2S results from inversion and thus is indicative of initial formation, or subsequent heating, above 176 °C, the lower transition temperature. Ag_2S synthesized from the elements above the upper transition (725 °C) and between the upper and lower transitions (325 °C) is highly twinned whereas Ag_2S synthesized below the lower transition (150 °C) is untwinned. Synthetic Ag_2S is optically identical with acanthite.

Bi_2S_3. — Bi_2S_3, analogous to the mineral bismuthinite, exists in only one modification below its melting point of 760 ± 3 °C. The solid solubility of Bi and $AgBi_3S_5$ in Bi_2S_3 is less than 1 wt-%.

$AgBiS_2$. — $AgBiS_2$ exists in two modifications below its congruent melting point of 801 ±4 °C. The high-temperature, or α, form is cubic ($Fm3m$, a_0 = 5.648 ± 0.002 Å). The low-temperature, or β, form is hexagonal, D_{3d} (5)–R_{3m}, a = 4.07 ± 0.02, c = 19.06 ± 0.05 Å (*Geller* and *Wernick*, 1959), and corresponds to the mineral matildite. Experiments on stoichiometric $AgBiS_2$ indicate that the α form is stable above 195 ± 5 °C, the α and β forms coexist between 195 ± 5 °C and 182 ± 3 °C, and the β form is stable below 182 ± 3 °C. Gold tube experiments on stoichiometric $AgBiS_2$ indicate that at a confining pressure of 2050 ± 20 bars the α and β forms coexist between 173 ± 5 °C and 205 ± 5 °C.

The coexistence of the two forms of $AgBiS_2$ over a temperature range indicates that near the inversion temperature the β form does not have stoichiometric composition. The absence of visible quantities of any other phase in polished sections of β-$AgBiS_2$ inverted from α-$AgBiS_2$ indicates, however, that this difference is small. When excess amounts of Ag_2S or $AgBi_3S_5$ are present only the α form of $AgBiS_2$ occurs above 182 ± 3 °C.

The maximum limits of solid solution of $AgBiS_2$ along the Ag_2S–$AgBiS_2$ join are 57 ± 1 mol-% Ag_2S at 615 ± 5 °C and 43 ± 1 mol-% Ag_2S at 732 ± 4 °C. The solid solution of Bi in $AgBiS_2$ is less than 1 mol-%. The lattice parameter of cooled α-$AgBiS_2$ ss is a function of composition along the Ag_2S–Bi_2S_3 join; a_0 varies from 5.648 ± 0.002 Å for stoichiometric $AgBiS_2$ to 5.700 ± 0.002 Å for a composition of 57 mol-% Ag_2S, 43 mol-% Bi_2S_3.

α-$AgBiS_2$ solid solutions containing more than 1 mol-% Bi_2S_3 in excess of the stoichiometric composition are not quenchable. At room temperature fine exsolution lamellae of a second phase are visible in the α-$AgBiS_2$; X-ray examination reveals these lamellae to be $AgBi_3S_5$.

$AgBi_3S_5$. — $AgBi_3S_5$, analogous to the sulfosalt pavonite, exists in only one structural form below its incongruent melting point at 732 ± 4 °C. Maximum limits of solid solution in this phase are less than 1 mol-%.

Bi–Pb–S System

Previous Work

Bi–S System. — Previous work in the Bi–S system was discussed above.

Pb–S System. — The Pb–S system has been investigated by *Bloem* and *Kroger* (1956), *Van Hook* (1960), and *Kullerud* (1965). The only binary phase is PbS, which melts congruently at 1115 ± 3 °C. *Kullerud* (1965) reports the presence of two two-liquid fields — one metal-rich, stable above 1042 ± 3 °C (77.2 ± 0.5 to 60.0 ± 0.5 mol-% Pb), and one sulfur-rich, stable above 799 ± 2 °C (28.8 ± 0.3 to <5 mol-% Pb).

Bi–Pb System. — A temperature-composition diagram of the Bi–Pb system compiled from many sources is presented by *Hansen* and *Anderko* (1958). The ε phase is stable below 184 °C and has a wide solid solution range (27 to 40 mol-% Bi at 100 °C) but is unknown as a mineral. *Bridgman* (1955) has suggested the existence of another phase, γ (63 % Bi, 37 % Pb), which he formed irreversibly when alloys were subjected to pressures of 10 kb or greater.

Bi–Pb–S System. — Fifteen mineral species have been reported having compositions within the Bi–Pb–S system. These minerals, which all lie on the PbS–Bi_2S_3 join, are listed with reported compositions in Table 11.

Beegerite (*Harcourt*, 1942) and goongarrite and warthaite (*Thompson*, 1949; *Rieder*, 1963) have been shown to be mixtures of galena and cosalite. Cannizzarite and bismutoplagionite have been shown to be identical with galenobismutite by *Berry* (1940) and *Palache* et al. (1944), respectively. Chiviatite, from the type locality was found by *Short* (1940) to be a mixture of bismuthinite and copper minerals. *Emmons* et al. (1926) have examined lillianite from the type locality and report it to be a mixture of galena, bismuthinite, and argentite; another sample from Sweden has been reported to be a mixture of galena and galenobismutite (*Berry*, 1940). Recently, the possible existence of lillianite has been suggested by *Syritso* and *Senderova* (1964). *Graeser* (1963) originally reported the composition of giessenite as 8 PbS · 3 Bi_2S_3, but more recently (*Graeser*, personal communication, 1964) has reported that giessenite contains Cu and Sb and that the accurate formula is $Pb_9CuBi_6Sb_{1.5}S_{30}$.

X-ray data for bonchevite (*Kostov*, 1958) and ustarasite (*Sakharova*, 1955) are very similar to that for bismuthinite and the X-ray powder diffraction pattern for bursaite (*Tolum*, 1956) has been noted by *Fleisher* (1956) to be suspiciously similar to that for cosalite.

Table 11. *Mineral species reported in the Bi–Pb–S system*

Mineral	Formula PbS/Bi_2S_3 ratio
Galena	PbS
Beegerite*	6/1
Goongarrite*	4/1
Warthaite*	4/1
Lillianite*	3/1
Giessenite**	8/3
Bursaite	5/2
Cosalite	2/1
Bismutoplagionite*	5/4
Cannizzarite*	6/5
Galenobismutite	1/1
Chiviatite*	3/4
Bonchevite	1/2
Ustarasite	1/3
Bismuthinite	Bi_2S_3

Several of these minerals originally designated as being in the B–Pb–S system have subsequently been either discredited (indicated by *) or shown to contain essential amount of elements other than Bi, Pb, and S (indicated by **).

Experimental studies within the Bi–Pb–S system have been conducted by *Schenck* et al. (1939), *Graham* (1950), *Ross* (1954), and *Van Hook* (1960). *Schenck* et al. (1939) and *Ross* (1954) report synthesis of 6 PbS · Bi_2S_3 ("beegerite"), 3 PbS · Bi_2S_3 ("lillianite"), and PbS · Bi_2S_3 (galenobismutite). *Graham* (1950) reports synthesis of 2 PbS · Bi_2S_3 (cosalite), PbS · Bi_2S_3 (galenobismutite), phase 2 ("cannizzarite") and phase 3 ("weibullite"). The compositions of the latter two phases were unknown. *Van Hook* (1960) reported three ternary phases designated as II, III, and IV of approximate compositions 5 PbS · Bi_2S_3, 8 PbS · 3 Bi_2S_3, and PbS · Bi_2S_3 respectively, but noted that "synthetic analogues of the known sulfosalt minerals were not found."

Experimental Data and Discussion

Phase relations within the Bi–Pb–S system were determined at 920, 400, and 250°C and on the PbS–Bi_2S_3 join from liquidus temperatures to 400°C.

The PbS–Bi_2S_3 Join. – Phase relations on the PbS–Bi_2S_3 join above 400°C are illustrated in Fig. 16 (data are given in Table 12). The binary and ternary phases encountered are listed in Table 13. Designation of phases is based upon the similarity of the compositions of phases in this study with those synthesized by *Van Hook* (1960). Since X-ray data on *Van Hook*'s phases are lacking, correlation cannot be definite. Phases II and III are probably identical with "beegerite" and "lillianite" synthesized by *Schenck* et al. (1939) and by *Ross* (1954). Phase IV corresponds to galenobismutite. Phases 1, 2, and 3 of *Graham* (1950) have not been encountered in the present study. Phase V was first reported by *Craig* (1965).

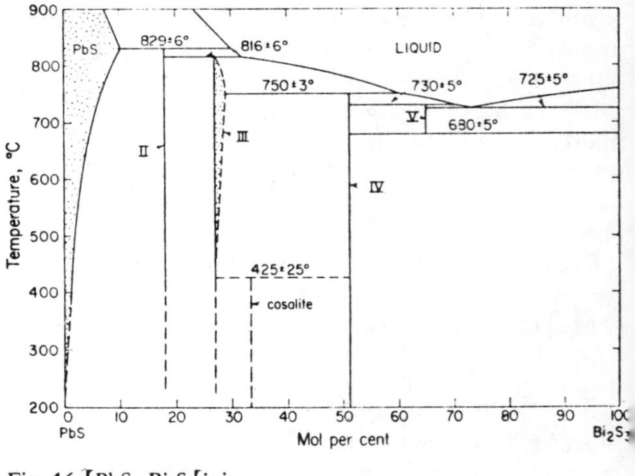

Fig. 16. PbS–Bi_2S_3 join.

Table 12. *Experimental data on the PbS–Bi₂S₃ join*

Composition, mol-% PbS	Bi$_2$S$_3$	Temp. [°C]	Time [hrs.]	Products
98.0	2.0	780	843	PbS ss
96.0	4.0	780	843	PbS ss
94.0	6.0	780	843	PbS ss
92.0	8.0	780	843	PbS ss
90.0	10.0	780	843	PbS ss + II
97.0	3.0	700	123	PbS ss
94.5	5.5	700	123	PbS ss + II
97.1	2.9	600	2273	PbS ss
95.4	4.6	600	2273	PbS ss + II
97.5	2.5	500	2272	PbS ss + II
82.0	18.0	835	12	PbS ss + liq
82.0	18.0	822	11	II
83.5	16.5	800	480	PbS ss + II
80.5	19.5	800	480	II + III
82.0	18.0	400	3105	II
72.7	27.3	835	12	PbS + liq
72.7	27.3	822	11	II + liq
72.7	27.3	811	45	III
77.7	22.3	700	123	II + III
73.7	26.3	700	207	II + III
72.7	27.3	700	123	III
71.8	28.2	700	318	III
70.9	29.1	700	207	III + IV
72.7	27.3	500	2272	III + IV
71.7	28.3	500	978	III + IV
72.7	27.3	200	3105	II + III
48.8	51.2	811	45	Liq
48.8	51.2	753	2	III + liq
48.8	51.2	747	2	IV
50.0	50.0	700	318	III + IV
48.8	51.2	700	318	IV
47.6	52.4	700	318	IV + V
49.0	51.0	500	2272	III + IV
47.8	52.2	500	954	IV + Bi$_2$S$_3$
33.0	67.0	742	4	Liq
35.4	64.6	742	4	IV + liq
33.0	67.0	735	16	IV + liq
33.0	67.0	725	3	V + Bi$_2$S$_3$
35.5	64.5	700	117	IV + V
34.5	65.5	700	120	V + Bi$_2$S$_3$
33.0	67.0	684	91	V + Bi$_2$S$_3$
33.0	67.0	675	379	IV + Bi$_2$S$_3$
29.7	70.3	400	3105	IV + Bi$_2$S$_3$
2.2	97.8	600	2773	IV + Bi$_2$S$_3$
1.2	98.8	500	2772	IV + Bi$_2$S$_3$

Subsolidus phase relations with the exception of those involving phase V are in general agreement with the findings of *Van Hook* (1960); liquidus relations are considerably different from those suggested by *Van Hook*.

Table 13. *Phases synthesized on the PbS–Bi₂S₃ join*

Phase	Composition
PbS ss	PbS–PbS ss containing 10 mol-% Bi$_2$S$_3$
II	82 ± 2 mol-% PbS, 18 ± 2 mol-% Bi$_2$S$_3$
III	73 ± 1 mol-% PbS, 27 ± 1 mol-% Bi$_2$S$_3$
IV	48.8 ± 0.5 mol-% PbS, 51.2 ± 0.5 mol-% Bi$_2$S$_3$
V	35 ± 0.5 mol-% PbS, 65 ± 0.5 mol-% Bi$_2$S$_3$
Bi$_2$S$_3$	Bi$_2$S$_3$ – no measurable solid solution toward PbS

An independent simultaneous investigation of the PbS–Bi$_2$S$_3$ join by *Salanci* (1965, 1966) confirms, in general, the relations in Fig. 16.

PbS Solid Solution. – The limit of solid solution of Bi$_2$S$_3$ in PbS has been determined as 9 ± 1 mol-% at 780 °C, 5 ± 1 mol-% at 700 °C, 3.5 ± 1 mol-% at 600 °C, and less than 2.5 mol-% at 500 °C (Table 12). These values extrapolate to a maximum solubility of 10 ± 1 mol-% Bi$_2$S$_3$ at 829 ± 6 °C, the incongruent melting point of phase II. The lattice parameter of PbS ss quenched from 780 °C varies linearly as function of composition from 5.936 ± 0.002 Å for stoichiometric PbS to 5.907 ± 0.002 Å for PbS containing 8% Bi$_2$S$_3$ in solid solution. The maximum solid solution of Bi in PbS is less than 1 wt-% at 600 and 300 °C. PbS ss containing less than 7 mol-% Bi$_2$S$_3$ is isotropic and optically indistinguishable from stoichiometric PbS; PbS ss containing more than 7 mol-% Bi$_2$S$_3$ is very slightly anisotropic.

Phase II. – Phase II, of composition 82 ± 2 mol-% PbS, 18 ± 2 mol-% Bi$_2$S$_3$ (Table 12), is not analogous to any known sulfosalt mineral. This phase melts incongruently at 829 ± 6 °C to PbS ss and liquid. No limit of lower stability has been evidenced; samples annealed at 325 °C for 3000 hours show no sign of breakdown. This phase is creamy white in polished section, weakly bireflectant, moderately anisotropic (tan, grey, lavender), and is slightly softer than PbS. The crystal structure of this phase has not yet been determined; however X-ray powder diffraction data are given in Table 14.

Phase III. – Phase III, of composition 73.0 ± 1 mol-% PbS, 27.0 ± 1 mol-% Bi$_2$S$_3$ (Table 12), is not analogous to any known sulfosalt mineral. This phase melts incongruently to phase II + liquid at 816 ± 6 °C; maximum solid solution is

2 ± 1 mol-% Bi_2S_3 at 750 ± 3°C. No lower limit of stability of this phase has been evidenced; samples annealed at 200°C for over 3000 hours showed no sign of breakdown. Phase III is creamy white in reflected light, moderately bireflectant, moderately anisotropic (lavender, gray, brown), and is very slightly softer than phase II. Although the crystal structure of phase III has not been determined, X-ray powder diffraction data are presented in Table 14.

Table 14. *X-ray powder diffraction data for phases II, III, and V*

Phase II		Phase III		Phase V	
d_{obs}	I_{obs}	d_{obs}	I_{obs}	d_{obs}	I_{obs}
3.75	4	4.10	2	5.99	3
3.53	3	3.92	½	4.96	1
3.41	10	3.67	2	3.99	½
3.35	4	3.52	10	3.77	10
3.11	2	3.42	4	3.60	½
3.07	½	3.39	3	3.31	1½
3.01	3	3.35	2	2.97	2
2.94	4	3.34	2	2.77	½
2.85	4	3.23	½	2.72	½
2.64	½	3.21	½	2.43	3
2.24	½	3.12	1	2.30	6
2.18	3	3.06	2	2.22	1
2.15	½	3.00	3	2.16	1
2.09	4	2.91	3½	2.12	½
2.07	3	2.85	1	2.02	½
2.04	2	2.77	2	1.994	1
1.96	1	2.70	1	1.969	1
1.89	1	2.58	½	1.937	½
1.85	1	2.36	1	1.895	1
1.78	2	2.15	3	1.873	½
1.77	2	2.06	3½		
		1.98	1		
		1.96	1		
		1.89	1		
		1.83	1		

Phase IV. – Phase IV, of composition 48.7 ± 0.5 mol-% PbS, 51.3 ± 0.5 mol-% Bi_2S_3 (Table 12), is analogous to the sulfosalt mineral galenobismutite. This phase melts incongruently to phase III and liquid at 750 ± 3°C; no evidence of solid solution has been observed. Phase IV is white in polished section and is moderately anisotropic (blue, tan, grey). The lattice parameters of synthetic single crystals as determined by Dr. *Nobuo Morimoto* (personal communication, 1965) are $a_0 = 11.73$, $b_0 = 14.47$, $c_0 = 4.08$ Å; these values are in good agreement with those determined on natural material by *Berry* (1940) ($a = 11.72$, $b = 14.52$, $c = 4.07$ Å).

Phase V. – Phase V, not previously reported, has a composition of 35.0 ± 0.5 mol-% PbS, 65.0 ± 0.5 mol-% Bi_2S_3 (Table 12); no evidence of solid solution beyond these limits has been observed. Phase V is stable only between 730 ± 5°C, the temperature at which it melts incongruently to phase IV and liquid, and 680 ± 5°C, the temperature of pseudobinary "eutectoidal" breakdown. Breakdown results in the formation of complex lamellar or myrmekitic intergrowths as shown in Fig. 26. Phase V is creamy white in polished section, but slightly darker than phase IV or Bi_2S_3, is moderately bireflectant, and moderately anisotropic (reddish brown, tan, lavender). The crystal structure of phase V has not as yet been determined; however, X-ray powder diffraction data are given in Table 14.

Bi_2S_3. – Bi_2S_3, analogous to the mineral bismuthinite, is the only bismuth sulfide; maximum solid solution is less than 1 mol-% toward Bi, S, or PbS (Table 12). Synthetic Bi_2S_3 is identical in physical and optical characteristics to natural samples. In reflected light it is creamy white, strongly pleochroic, and strongly anisotropic.

Cosalite. – The common Pb–Bi sulfosalt cosalite ($2\,PbS \cdot Bi_2S_3 = Pb_2Bi_2S_5$) was not encountered in the study on synthetic materials. The reported composition of cosalite (33.3 mol-% Bi_2S_3) lies between that of phase III (27.0 mol-% Bi_2S_3) and phase IV (51 mol-% Bi_2S_3). Experimental compositions of cosalite composition, prepared with the elements, with PbS and Bi_2S_3, and with phases III and IV, and heated at temperatures between 200 and 600°C for periods up to 3 months failed to produce cosalite. The absence of this phase in the study on synthetic materials may indicate (a) kinetic difficulties – PbS is very unreactive in the Bi–Pb–S system at temperatures below 500°C – or (b) cosalite is not a phase in the pure Bi–Pb–S system. Since chemical analyses of cosalites (e.g. *Palache* et al., 1944) generally contain only small amounts of extraneous elements, much of which can probably be attributed to mineral impurities, cosalite is tentatively considered to be a Bi–Pb–S phase.

This supposition is supported by the results of heating experiments (Table 15) conducted with natural cosalites, from Timiskaming and Brennan's Lake, Ontario, Boliden, Sweden, and the Cariboo Gold Quartz Mine, British Columbia. It must be noted, however, that the liquid field which spans the sulfur-rich portion of the system, and a small liquid field of nearly pure sulfur. Liquidus positions on the Pb–S join are from *Kullerud* (1965). The two-liquid field which exists between Pb and PbS at higher temperatures (above 1042 ± 3°C; *Kullerud*, 1965) is not present in the ternary system at 920°C.

Table 15. *Heating experiments with natural cosalite*

Specimen Locality	Temp. [°C]	Time [hrs.]	Analogous Synthetic Products
Timiskaming, Ont.	>900	5 min.	III*
Timiskaming, Ont.	805	15	III + tr. IV
Boliden, Sweden	805	23	PbS + III + tr. IV
Timiskaming, Ont.	700	120	III + tr. IV
Timiskaming, Ont.	600	160	III + tr. IV
Boliden, Sweden	600	274	II
Timiskaming, Ont.	600	274	III + tr. IV
Brennan's Lake, Ont.	500	166	III + tr. IV
Timiskaming, Ont.	500	166	III + tr. IV
Timiskaming, Ont.	450	155	Cosalite + III + tr. IV
Cariboo Gold Quartz Mine, B.C.	450	209	III + tr. cosalite
Brennan's Lake, Ont.	400	600	PbS + cosalite
Brennan's Lake, Ont.	325	1675	PbS + cosalite
Brennan's Lake, Ont.	200	2160	PbS + cosalite

* This sample was melted in oxygen-natural gas torch in an evacuated silica tube. The crystalline product was examined.

natural samples available contained variable amounts of galena as an impurity. In these experiments cosalites heated to 400°C underwent no change, whereas samples heated to 450°C or above showed signs of breakdown to phases analogous to the synthetic phases III and IV. On the basis of these experiments the stability limit of cosalite is tentatively given as 425 ± 25°C in the presence of vapor.

Ternary Phase Relations. – The phase relations in the Bi–Pb–S system were determined at 920, 400, and 250°C. Phase relations at intermediate temperatures have been deduced from these known isotherms, the bounding binary systems, and the PbS–Bi_2S_3 join.

Ternary phase relations at 920°C as determined by the experiments in Table 16 are illustrated in Fig. 17. At this temperature the system consists of a large divariant region of PbS ss and liquid, a large homogeneous liquid region, a two-

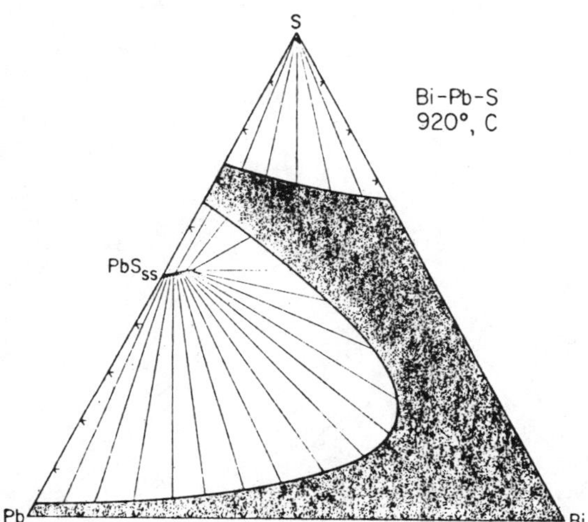

Fig. 17. The 920°C isotherm of the Bi–Pb–S system. Homogeneous liquid is stippled. The PbS ss extends toward Bi_2S_3 composition. (Mol-% diagram.)

Table 16. *Experimental data at 920°C*

Composition, mol-% Bi	Pb	S	Time [hrs.]	Products
65.0	20.3	14.7	3	Liq
54.9	19.8	25.3	4	PbS + liq
45.0	50.0	5.0	3	Liq
44.8	15.0	40.2	2.5	Liq
30.1	24.9	45.0	3	PbS + liq
19.4	25.8	54.8	1	Liq
14.2	32.3	53.5	1	PbS + liq

At lower temperatures phase relations in the Bi–Pb–S system are complicated by the appearance of phases II, III, IV and V in that order as indicated by liquidus relations in Fig. 16. The two-liquid field in the sulfur-rich portion of the system remains down to 661 ± 3°C (Table 3) at which temperature the more metal-rich member breaks down to Phase IV, Bi_2S_3 and S-liquid. In the metal-rich portion of the system a liquid exists at least to 125°C, the eutectic temperature on the Bi–Pb join.

The 400°C isotherm shown in Fig. 18 has been constructed with cosalite shown as a stable phase on the basis of the experiments conducted with natural cosalites and the experiments in Table 17. Phase II and III have not been reported as minerals and thus are believed to break down at low temperature. Therefore, although their lower stability limits were not established, these phases are omitted in Fig. 19, which illustrates low-temperature relations in the Bi–Pb–S system.

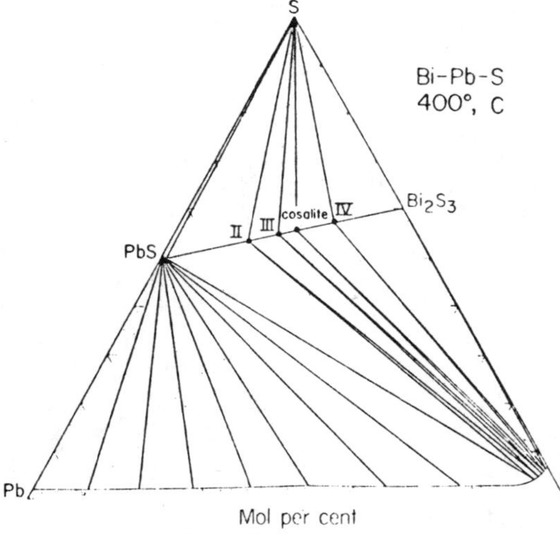

Fig. 18. The 400°C isotherm of the Bi–Pb–S system. Homogeneous liquid is stippled. (Mol-% diagram.)

Table 17. *Experimental data at 400°C*

Composition, mol-% Bi	Pb	S	Time [hrs.]	Products
70.0	25.0	5.0	336	PbS + PbBi liq
70.0	20.0	10.0	336	PbS + PbBi liq
70.0	15.0	15.0	336	PbS + PbBi liq
70.0	10.0	20.0	336	PbS + II + PbBi liq

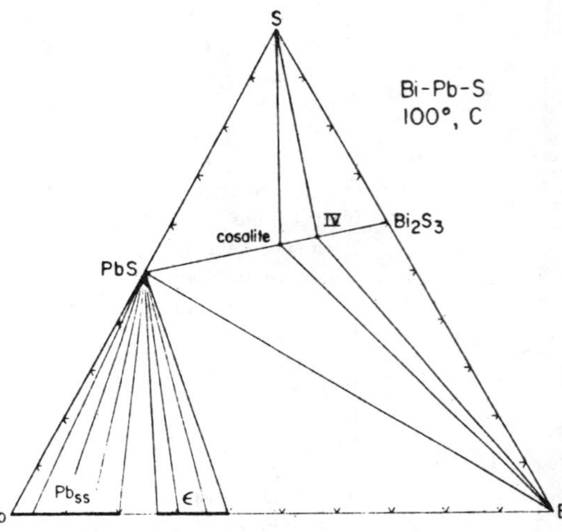

Fig. 19. Schematic 100°C isotherm of the Bi–Pb–S system. (Mol-% diagram.)

Ag–Pb–S System

Previous Work

Ag–S and Pb–S Systems. – Previous work on the Ag–S and Pb–S systems was discussed above.

Ag–Pb System. – Phase relations in the Ag–Pb system were summarized by *Hansen* and *Anderko* (1953). Maximum solubility of Pb in Ag is 2.8 mol-% at 650°C (*Raub* and *Polezec-Wittek*, 1942), whereas maximum solubility of Ag in Pb is 0.2 mol-% at 304°C (*Seith* and *Keil*, 1933); no intermediate compounds are reported.

Ag–Pb–S System. – *Friedrich* (1907) in a D.T.A. study of the PbS–Ag_2S join reports that Ag_2S and PbS are stable together below 630 ± 5°C. *Nissen* and *Hoyt* (1915) report the maximum solubility of Ag_2S in PbS as 0.6 wt.-% *Guertler* and *Luder* (1924) observed that mixtures of Ag, Pb, Ag_2S, and PbS heated at 1200°C display pronounced layering. They note that the upper layer, composed primarily of Ag_2S and PbS, contains about 12 wt-% S; the lower layer is primarily Pb and

Ag and contains only 3–5 wt-% S. *Vogel* (1953) in a D.T.A. study of the Ag–PbS and Ag_2S–Pb joins reports the maximum stability of the Ag–PbS assemblage as 732°C. *Van Hook* (1960) reexamined the Ag_2S–PbS join and found the maximum stability of this assemblage to be 605 ± 5°C and the maximum solubility of PbS in Ag_2S to be 3 mol-%.

Experimental Data and Discussion

Ag_2S–PbS and Ag–PbS Joins. – Experiments on the Ag_2S–PbS join confirmed the minimum melting temperature, 605 ± 5°C, as reported by *Van Hook* (1960). Investigation of the Ag–PbS join (Table 18) indicates that the Ag–PbS assemblage is stable below 784 ± 2°C in the presence of vapor; experimental data are given in Table 18. Above this temperature PbS and Ag react to form two liquids. Optical examination of the quench products of experiments above 784°C indicates that the composition of one of these liquids lies close to the Ag_2S–PbS join (approximately 75 mol-% Ag_2S, 25 mol-% PbS) and that of the other is near the Ag–Pb join (approximately 60 mol-% Ag, 40 mol-% Pb). Experiments prepared from Ag and PbS reacted between 700 and 784°C contain traces (<1%) of Ag_2S PbS liquid in addition to Ag and PbS, thus indicating that Ag in equilibrium with PbS above 700°C contains a small amount of Pb in solid solution.

Table 18. *Experimental data on the Ag–PbS join*

Composition wt-% Ag	PbS	Temp. [°C]	Time [hrs.]	Products
50.0	50.0	800	45	PbS + 2 liq
50.0	50.0	792	21	PbS + 2 liq
50.0	50.0	786	37	PbS + 2 liq
50.0	50.0	784	20	Ag + PbS + 2 liq*
50.0	50.0	782	93	Ag + PbS (+ Ag_2S liq)**
50.0	50.0	780	23	Ag + PbS (+ Ag_2S liq)**
42.0	58.0	750	4	Ag + PbS (+ Ag_2S liq)**

* Disequilibrium.
** Trace of Ag_2S-rich liquid; see text for explanation.

Ternary Phase Relations. – Ternary phase relations in the Ag–Pb–S system as determined in the temperature range 925–900°C by experiments listed in Table 19 are shown in Fig. 20; knowledge of the bounding binary systems and of the Ag_2S–PbS and PbS–Ag joins permits extrapolation of phase relations to lower temperatures.

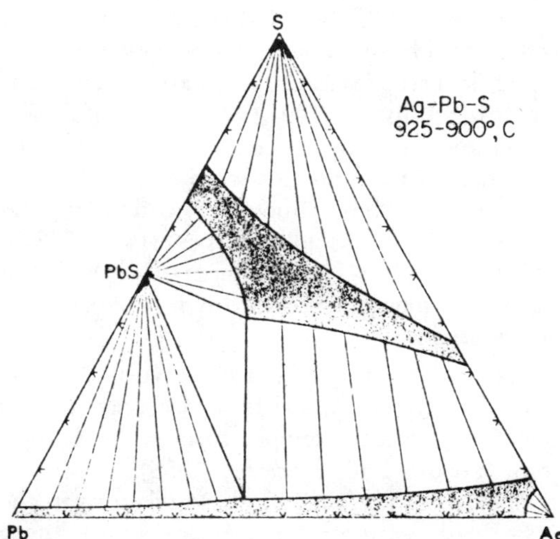

Fig. 20. Ag–Pb–S system at 925–900°C. Homogeneous liquid regions are stippled. (Mol-% diagram.)

Table 19. *Experimental data at 900–925°C*

Composition, mol-% Ag	Pb	S	Temp. [°C]	Time [hrs.]	Products
65.0	20.1	14.9	925	9.5	2 liq
59.9	34.9	5.2	900	2.5	2 liq
42.7	46.8	10.5	920	2.5	2 liq
36.8	25.1	38.2	920	2.5	2 liq
30.1	65.2	4.7	900	2.5	PbS + liq
25.1	59.9	15.0	925	9.5	PbS + liq
25.1	44.9	29.9	925	9.5	PbS + 2 liq

In the 925–900°C range the Ag–Pb–S system contains a small liquid field of nearly pure sulfur, a sulfur-rich two-liquid field which spans the entire system, a central liquid region, a metal-rich two-liquid field, a homogeneous Ag–Pb liquid which nearly spans the entire system, two PbS + liquid fields and a PbS + two-liquid region. The metal-rich boundary of the central liquid field almost coincides with the Ag_2S PbS join; the Ag–Pb liquid contains less than 5 mol-% S.

At higher temperatures (above 1042 ± 3°C, the minimum temperature at which a two-liquid field exists between PbS and Pb) it is probable that the metal-rich two-liquid field spans the entire system. This two-liquid field in the ternary system persists to 784 ± 2 °C, below which temperature the join Ag and PbS is stable. The sulfur-rich two-liquid field spans the system above 799 ± 2 °C; below this temperature PbS and S liquid are stable together and the two-liquid field recedes into the ternary system from the Pb–S join. At 740 ± 2°C the sulfur-rich two-liquid field also recedes into the ternary system as Ag_2S and S liquid become stable together. In the sulfur-rich portion of the system the two-liquid field persists to 528 ± 3 °C (Table 3) at which temperature the more metal-rich member breaks down to Ag_2S, PbS and S liquid.

Phase relations at 400 °C are shown in Fig. 21. The Pb-rich liquid exists at least as low as 304 °C, the eutectic temperature on the Ag–Pb join (*Hansen* and *Anderko*, 1958). At lower temperatures the general phase relations in the Ag–Pb–S system are identical with those shown in Fig. 21, except that the Pb-rich liquid is replaced by solid Pb.

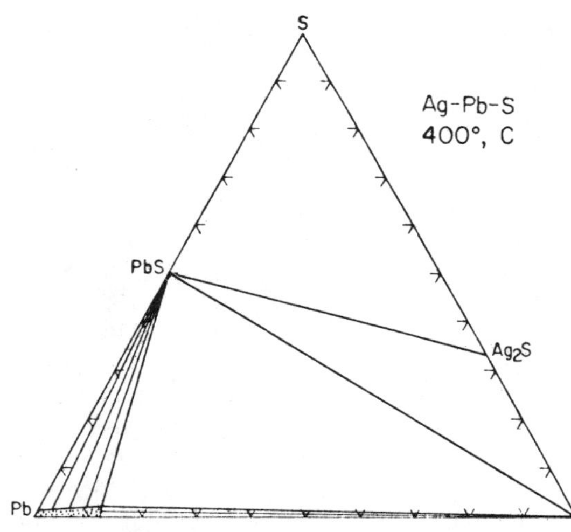

Fig. 21. The 400 °C isotherm of the Ag–Pb–S system. (Mol-% diagram.)

Ag–Bi–Pb–S System

Previous Work

Previous studies in the binary and ternary systems that make up the Ag–Bi–Pb–S quaternary system have been discussed with the exception of the Ag–Bi–Pb. This latter system comprises the fourth face of the quaternary compositional tetrahedron and contains no ternary compounds.

Extensive or complete solid solution between $AgBiS_2$ and PbS at elevated temperatures was first suggested by *Ramdohr* (1938) to account for complex intergrowths of the mineralogical equivalents (matildite and galena, respectively). He speculates that at 250 °C 30% PbS dissolves in $AgBiS_2$ and 15% $AgBiS_2$ dissolves in PbS and that solid solution is complete at temperatures above 350 °C. The same conclusion was reached by *Leutwein* and *Herrmann* (1954). Complete solid solution near liquidus temperatures was demonstrated by *Wernick* (1960) and *Van Hook* (1960). *Van Hook* (1960) further investigated the Ag_2S–$AgBiS_2$–PbS portion of the Ag_2S–PbS–Bi_2S_3 plane and reports the presence of a liquid as low as 592 °C.

Two minerals have been reported as occurring within the Ag–Bi–Pb–S system, schirmerite ($PbAg_4Bi_4S_9$) and schapbachite ($PbAg_2Bi_2S_5$); the compositions of both phases lie on the PbS–$AgBiS_2$ join. Schapbachite has been shown to be a mixture of matildite and galena (*Ramdohr*, 1938). *Van Hook* (1960) points out that schirmerite is probably also a mixture of galena and matildite.

Experimental Data and Discussion

Phase relations within the quaternary system must be built upon the known phase relations in bounding binary and ternary systems and upon additional data obtained from systematic investigation of certain planes in the quaternary system. This investigation has been conducted on the Ag_2S–PbS–Bi_2S plane, which separates the quaternary system into sulfur- and metal-rich portions, and on the Ag–PbS–$AgBiS_2$ plane.

$AgBiS_2$–PbS Join. — At all temperatures below the liquidus, the $AgBiS_2$–PbS join serves conveniently to divide the Ag_2S–Bi_2S_3–PbS plane into two portions. The unit cell dimensions of the continuous solid solution along the $AgBiS_2$–PbS join vary uniformly from 5.648 ± 0.002 Å for stoichiometric $AgBiS_2$ to 5.936 ± 0.002 Å for PbS (Table 20); these values are in close agreement with those determined by *Van Hook* (1960).

Ramdohr (1938) inferred the crest of the solvus between PbS and $AgBiS_2$ to be about 350°C. *Van Hook* (1960), though unable to determine the position of the solvus, reported breakdown of one specimen of the $AgBiS_2$ PbS solid solution (3.9 mol-% $AgBiS_2$) at 170°C.

In the present study synthetic samples originally homogenized at 400 °C did not break down when annealed at 250 °C for 57 days. In ex-

Table 20. *Experimental Data on $AgBiS_2$–PbS Join*

Composition at. % AgBiS$_2$	PbS	Temp. [°C]	Time [hrs.]	a_0 of homogeneous solid solution ±0.002
9.9*	90.1	250	1365	5.916
10.1*	89.9	400	1275	5.915
30.0*	70.0	250	1365	5.832
30.0*	70.0	400	1275	5.835
50.0*	50.0	250	1365	5.765
50.0*	50.0	400	1275	5.765
69.9*	30.1	250	1365	5.712
70.0*	30.0	400	1275	5.718
90.0*	10.0	250	1365	5.672
90.0*	10.0	400	1275	5.666
58**	42	300	213	5.746
59**	41	250	260	5.743
64**	36	225	167	5.730
66***	34	200	200	5.725
8	92			5.916

* Experiments prepared by melting appropriate mixtures of PbS and $AgBiS_2$; these were melted in an oxygen-natural gas torch and annealed at the indicated temperature.

** Experiments conducted with samples of natural galena-matildite intergrowths from Leadville, Utah; the compositions were determined by the application of the spacing curve to the unit cell of the homogenized product.

*** Experiment conducted with natural galena-matildite intergrowth from Leadville, Utah, in which homogenization did not take place.

Effect of PbS on the Stability of the Ag–$AgBiS_2$ Assemblage. – The maximum stability of the silver-matildite assemblage in the Ag–Bi–S system is 343 ± 2°C. Solid solution of galena in matildite raises the maximum temperature of coexistence with silver. The T–X curve defining the stability limit of silver and matildite-galena solid solutions passes through the points indicated in Table 21.

Assemblages of silver and matildite-galena solid solution heated to temperatures above the curve passing through the points in Table 21 react to form argentite + silver-bismuth liquid + a matildite-galena solid solution which is more PbS-rich than the original sample. At higher temperatures (above 597 ± 3°C on the Ag–Bi–S face and above 784 ± 2°C on the Ag–Pb–S face) the Ag_2S + AgBi liquid + ($AgBiS_2$–PbS) solid solution assemblage reacts to form two liquids. The composition-temperature variation of this latter reaction with composition in the quaternary system has not been determined.

Quaternary Phase Relations. – On the basis of known phase relations in the bounding binary and ternary systems and in selected planes within the quaternary system, it is now possible to construct the tentative phase relations within the Ag–Bi–Pb–S quaternary system. These relations are schematically presented in three quaternary figures – 1. high temperature, 1050°C; 2. intermediate temperature, 400°C; 3. low temperature, 100°C – and are briefly discussed below.

Table 21. *Maximum stability of Ag + ($AgBiS_2$–PbS) solid solution assemblages*

$AgBiS_2$–Pbs comp. in mol-% PbS	0	21	44	70	87	100
Maximum T of stability in °C	343 ± 2	359 ± 16	387 ± 13	450 ± 50	550 ± 50	784 ± 2

periments conducted with natural intergrowths of matildite and galena from Leadville, Colorado, complete homogenization was observed at 225°C whereas the two phases remained distinct at 200°C (Table 20). Traces of chalcopyrite and tetrahedrite present as impurities in the natural samples were not visibly altered after being heated at 200 and 225°C. The composition of the homogenized sample, as determined by measurement of the resulting lattice dimensions, was about 36 mol-% PbS. Assuming a relatively symmetrical solvus, it is probable that the crest of the solvus lies at 215 ± 15°C.

1050°C Isotherm. – The phase relations at 1050°C, although of limited geologic interest, are informative in the development and understanding of phase relations at lower temperatures. The system, as shown schematically in Fig. 22, contains two large two-liquid fields, a large homogeneous liquid field and a field of PbS + liquid. Tie-lines in the sulfur-rich two-liquid field have been omitted for clarity. The metal-rich two-liquid field occupies a principal volume of the metal-rich portion of the quaternary system. This two-liquid field spans the Ag–Pb–S face of the tetrahedron, and extends about 54 mol-% Bi across the Ag–Bi–S face. The extent of this field on the Bi–Pb–S face is unknown. The large field of homogeneous liquid extends through the central portion of the quaternary system separating the two regions of liquid immiscibility.

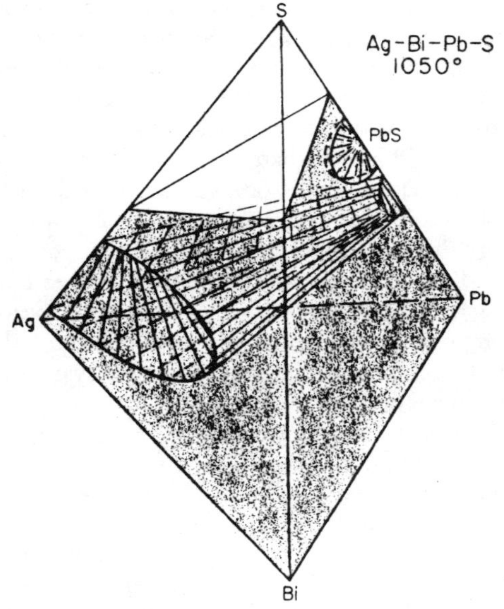

Fig. 22. Schematic phase relations of the Ag–Bi–Pb–S system at 1050°C. (Mol-% diagram.)

Developments between 1050 and 400°C. — The two-liquid field which occupies the entire sulfur-rich volume of the quaternary system at 1050°C begins to withdraw into the quaternary below 799 ± 2°C (*Kullerud*, 1965) at which temperature PbS becomes stable with S liquid. Withdrawal from the Ag–S and Bi–S joins occurs at 740 ± 2°C and 723 ± 3°C, respectively. At the latter temperature tie-lines are established between S liquid and the entire $AgBiS_2$–PbS solid solution, resulting in division of the remaining two-liquid field within the quaternary system into two separate two-liquid fields. One of these sulfur-rich two-liquid fields is stable on the Ag–Pb–S face to 528 ± 3°C; the other is present on the Bi–Pb–S face to 661 ± 3°C. It is not known if these two-liquid fields persist to lower temperatures within the quaternary system.

The metal-rich two-liquid field shown in the 1050°C isothermal tetrahedron withdraws into the quaternary system from the Pb–S join below 1042 ± 3°C (*Kullerud*, 1965). This field persists on the Ag–Pb–S face to 784 ± 2°C, on the Ag–Bi–S face to 597 ± 3°C, and within the quaternary system to 589 ± 4°C. *Van Hook* (1960) reports liquid on the Ag_2S–PbS–Bi_2S_3 plane at 592 ± 5°C, at a composition of 81 Ag_2S–5 Bi_2S_3–14 PbS. An experiment conducted with Ag_2S, Bi_2S_3, and PbS in the same ratio but with an additional 5 wt-% Bi contained no evidence of liquid at 585°C.

400°C Isotherm. — Reduction and disappearance of the sulfur-rich and metal-rich two-liquid fields, appearance of binary and ternary phases as previously noted, and diminution in size of the metal-rich liquid result in phase relations at 400°C as illustrated in Fig. 23.

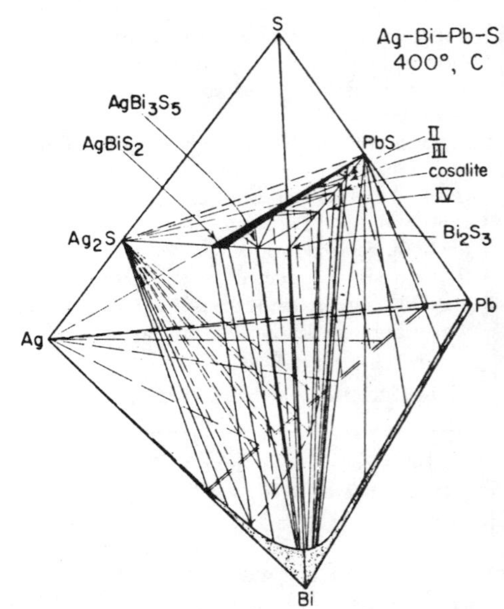

Fig. 23. Schematic phase relations of the Ag–Bi–Pb–S system at 400°C. (Mol-% diagram.)

All phases, with the exception of phase V in the Bi–Pb–S system, which appear in the constituent ternary systems at higher temperatures persist to 400°C. Phase V, stable in the Bi–Pb–S system only in the interval 730 ± 5°C to 680 ± 5°C, is stabilized to 625 ± 5°C by the presence of 2 wt-% Ag_2S (Table 22).

Phase relations at 400°C are illustrated in Fig. 23; tie-lines between ternary phases and liquid sulfur have been omitted for clarity. At this temperature the PbS–$AgBiS_2$ solid solution exists in equilibrium with all liquidus compositions of the metal-rich liquid except for a narrow zone near the Ag–Bi–S face of the tetrahedron and a wedge which coexists with Ag_2S. Phase relations of the $AgBiS_2$–PbS–Bi_2S_3 portion of the Ag_2S–PbS–Bi_2S_3 plane are deduced from data on the $AgBi_3S_5$–PbS join (Table 22).

the system into four-phase volumes. These volumes are Ag + Ag_2S + PbS + $AgBiS_2$, Ag + Bi + PbS + $AgBiS_2$, $AgBiS_2$ + PbS + Bi + cosalite, $AgBiS_2$ + cosalite + $AgBi_3S_5$ + Bi, $AgBi_3S_5$ + cosalite + IV + Bi, and $AgBi_3S_5$ + IV + Bi_2S_3 + Bi. The first three correspond to observed natural mineral associations. The latter three assemblages may be observed when additional occurrences of the rare sulfosalt pavonite are reported.

Table 22. *Experimental data on Ag_2S–Bi_2S_3–PbS plane*

Composition, mol-%			Temp.	Time	Products
Ag_2S	Bi_2S_3	PbS	[°C]	[hrs.]	
2.2*	6.5	91.3	400	4680	PbS ss**
4.7*	14.2	81.1	400	4680	PbS ss + II
7.3*	22.0	70.7	400	1272	PbS ss + II
11.3*	34.0	54.7	400	4680	PbS ss + III
15.6*	46.7	37.8	400	4680	PbS ss + III + IV
21.3*	64.1	14.6	400	4680	$AgBi_3S_5$ + PbS ss + IV
23.2*	59.8	7.0	400	3480	$AgBi_3S_5$ + PbS ss + IV
24.1*	72.3	3.7	400	3480	$AgBi_3S_5$ + PbS ss + IV
2.0***	65.4	32.6	670	134	V
2.0***	65.4	32.6	650	117	V
2.0***	65.4	32.6	630	117	V
2.0***	65.4	32.6	620	240	$AgBi_3S_5$ + Bi_2S_3 + IV

* Prepared with PbS and $AgBi_3S_5$.
** PbS ss designates PbS–$AgBiS_2$ solid solution.
*** Prepared with V and Ag_2S.

Developments below 400°C. The metal-rich liquid phase persists within the quaternary system at least as low as 125°C, the eutectic temperature on the Bi–Pb join.

The $AgBiS_2$–PbS solvus is intersected at 215 ± 15°C.

Phases II and III of the Bi–Pb–S system stable at 400°C are not known as minerals and therefore are presumed to break down at some temperature below 400°C but above that of the diagram of Fig. 24. The phase relations within the quaternary system at low temperature are shown schematically in this figure; tie-lines between the ternary phases and sulfur have been omitted for clarity.

Examination of Fig. 24 reveals that three-phase planes Ag + $AgBiS_2$ + PbS, Bi + $AgBiS_2$ + PbS, Bi + $AgBiS_2$ + cosalite, Bi + $AgBi_3S_5$ + cosalite, and Bi + $AgBi_3S_5$ + IV – divide the geologically significant portion of

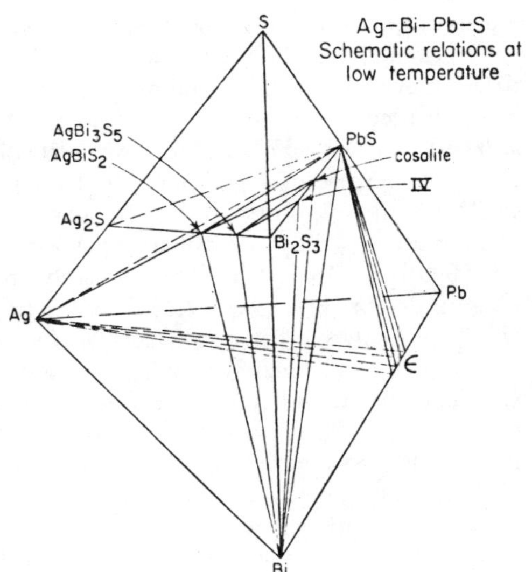

Fig. 24. Schematic relations of the Ag–Bi–Pb–S system at low temperature. (Mol-% diagram.)

It is noted that the Ag + PbS + Bi plane separates the geological from the metallurgical portion of the system. The Ag + ε + PbS plane subdivides the metallurgical portion and the Ag_2S + PbS + Bi_2S_3 plane separates the geological portion from the sulfur-rich portion.

Geological Implications

Mineral assemblages of the Ag–Bi–Pb–S system analogous to synthetic phase assemblages prepared in the laboratory occur in many and varied deposits, including volcanic fumaroles (Lipari Islands), massive replacement bodies (Leadville, Colorado), and fissure veins (Cobalt, Ontario). With the exception of galena, the Ag–Bi–Pb–S minerals generally occur as accessories. Certain of the naturally occurring assemblages are discussed and interpreted below in light of experimental studies. Most experimental data presented in this study were obtained from rigid silica tube experiments in which a vapor phase is inherently present. Therefore, in the strictest sense the majority of conclusions reached in this study are applicable only to natural situations in which a vapor was also present. Since some ores may have formed in the absence of vapor, it is important to investigate the phase relations in regions of P–T space where vapor is not a phase; such investigation is conducted by means of gold tube experiments as described by *Kullerud* and *Yoder* (1959). To date experimental data are available only on the melting point of bismuth [decrease of 4°C per 1000 bars (*Kennedy* and *Newton*, 1963)], on the lower transition in Ag_2S [increase of 4.5°C per 1000 bars (*Roy* et al., 1959)], and on the maximum stability of the Ag and $AgBiS_2$ assemblage (increase 6°C per 1000 bars, Table 9b). The relatively small magnitudes of these pressure effects are probably representative of such effects on the system as a whole. Therefore it appears safe to conclude that the effects of pressure, of a few kilobars on the phase relations in the Ag–Bi–Pb–S system are minor. The effect of partial pressures $P_s < P_{total}$, especially when P_{total} is not sufficient to condense the vapor phase, on phase relations in sulfide systems has been discussed in detail by *Kullerud* and *Yoder* (1959).

Native Elements

Mineral assemblages containing native silver and/or bismuth are characteristic of the "cobalt-nickel-native silver ore type" (*Bastin*, 1939), which occurs in important deposits such as Joachimsthal, Czechoslovakia, Freiberg, Germany, and Cobalt and Gowganda, Ontario. The presence of a silver-bismuth assemblage, as has been observed at Cobalt, Ontario, indicates that crystallization could only have been completed below 262°C, the temperature of the eutectic between the two phases. The effects of other phases, such as arsenides, with which silver and bismuth are commonly associated, on the silver-bismuth eutectic are not known.

The characteristic twinning commonly observed in native bismuth (see Figs. 15 and 16) and previously ascribed to passage through an inversion at 75°C (*Wurschmidt*, 1914; *Edwards*, 1954) apparently results from stresses created by the expansion of liquid bismuth during crystallization. Twinning of an identical nature is readily induced in bismuth by deformative pressures and even by scratching during the polishing process. Accordingly, twinning in native bismuth is not a valid indication of conditions under which mineralization occurred.

Sulfides

Galena, PbS, occurs nearly ubiquitously in base metal and precious metal deposits. The solid solution of Bi_2S_3 in this phase (maximum of 10 mol-% = 19 wt-%) may vary sufficiently with temperature to provide a potential geologic indicator. Slow cooling, or annealing of synthetic members of this solid solution at temperatures of 400 and 500°C, results in exsolution of phase II in oriented laths. Continued exsolution at lower temperatures or for longer periods of time results in deterioration of the exsolution textures to coarse myrmekitic intergrowths, segregations, and granular aggregates. Similar textures of galena and the Pb–Bi sulfosalt mineral cosalite have been observed in ores from the Cariboo Gold Quartz Mine, British Columbia, Timiskaming and Brennan's Lake, Ontario, and Boliden, Sweden, and are reported in the literature from many other localities (*Ramdohr*, 1960). Phase II has not been reported as a mineral.

Bismuthinite, Bi_2S_3, is noted by *Palache* et al. (1944) as typically found in hydrothermal veins (Cerro de Pasco, Peru; Potosi and Llallogua, Bolivia; Cornwall, England) and in pegmatites (Kingsgate, New South Wales), and has been reported as a volcanic sublimate from the Lipari Islands. Its high melting point and existence in only one polymorphic form negate its use as a geologic indicator when occurring alone.

Silver sulfide, Ag_2S, though occurring naturally in only one structural form (monoclinic), has frequently been described with cubic morphology, argentite (*Ellsworth*, 1916; *Ramsdell*, 1943; *Palache* et al., 1944). The cubic morphology is inherited from one of the high-temperature cubic forms and is indicative of formation above 176°C, the lower inversion temperature. Profuse polysynthetic twinning is produced in argentite on passage through the inversion at 176°C and thus may be indicative of initial formation or subsequent heating above this temperature. Pressure raises the lower inversion temperature 4.5°C per 1000 bars (*Roy* et al., 1959). Caution should be exercised, however, in usage of such twinning as a geologic indicator, in that similar twinning may be induced through deformation. The effect of annealing at low temperatures on the removal of inversion twinning, and the effects of most chemical elements on the Ag_2S inversions are unknown.

Silver-Matildite Assemblages

The assemblage silver and matildite is stable only below 343 ± 2°C. Above this temperature these phases react to form argentite and bismuth-rich liquid. Incorporation of galena into the matildite raises the maximum temperature of coexistence of the silver and matildite solid solution. The effect of galena in amounts up to 50 mol-% of the matildite-galena solid solution is relatively small, raising the maximum temperature of coexistence with silver only to 347 ± 4°C for 25 mol-% galena and 380 ± 25°C for 50 mol-% galena. Increased amounts of galena raise the temperature of coexistence to a 475 ± 50°C for 75 mol-% galena and to 784 ± 2°C for pure galena.

Confining pressure raises the upper stability limit of the silver + matildite assemblage approximately 6°C per 1000 bars. Although the effect of pressure on the stability of silver with other compositions of the matildite-galena solid solution has not been determined, it is assumed to be similar in magnitude to that on the matildite end member.

The effect of pressure on the silver-matildite assemblage demonstrates that the alternative mineral assemblage — crystalline bismuth + argentite — is unstable under all geologic conditions. This conclusion is compatible with the observation that, although both bismuth and argentite are common in ores of the "cobalt-nickel-native silver type" (*Bastin*, 1939), these minerals do not coexist in polished section. In the laboratory, reaction between argentite and bismuth liquid to form silver and matildite is initially rapid; however, once an intermediate reaction rim has formed, further reaction is retarded. The matildite-silver assemblage has been reported from Cobalt, Ontario, by *Ramdohr* (1938). Preservation of remnants of the high-temperature assemblage in ores has not been reported.

Lead-Bismuth Sulfosalts

Cosalite is the most common of the lead-bismuth sulfosalts and occurs in hydrothermal deposits (Cobalt, Ontario; Cariboo Gold Quartz Mine, British Columbia; Ouray, Colorado) and in contact metamorphic deposits (Falun and Nordmark, Sweden). *Banás* (1965) has noted that cosalite is a characteristic mineral in skarn deposits. Experimental investigations of natural cosalites suggest that this mineral is stable only below 425 ± 25°C, thus pointing out the value of this mineral as an indicator of the limiting thermal conditions at the time of mineralization.

Galenobismutite (reportedly $PbS \cdot Bi_2S_3$), a less common lead-bismuth sulfosalt than cosalite, occurs in the deposits at Nordmark and Gladhammar, Sweden, and has been observed in the high-temperature areas (550–610°C) of volcanic fumaroles on the Lipari Islands (*Palache* et al., 1944). Its high melting temperature (750°C) and existence in only one form unfortunately negate its value as a geologic indicator when occurring alone. The maximum temperature of coexistence of galenobismutite with most other minerals is unknown. The laboratory study in the Bi–Pb–S system indicates that above 300°C galenobismutite is not stoichiometric but contains approximately 1 mol-% Bi_2S_3 in excess on the reported composition.

The most characteristic occurrence of the lead-bismuth sulfosalts is that of complex lamellar intergrowths suggestive of exsolution and/or breakdown of high temperature intermediate phases. Especially common is the occurrence of galena-cosalite intergrowths; a typical example from Boliden, Sweden, is shown in Fig. 25. These intergrowths may be envisioned as resulting from (a) exsolution of sulfosalts from the extensive PbS solid solution towards Bi_2S_3 and (b) breakdown of natural phases analogous to synthetic phases II and III.

Fig. 25. Lamellar intergrowth of cosalite (black and white) and galena (grey) from Boliden, Sweden. (Crossed nicols; × 365 oil.)

Particular note of textures of bismuthinite and galenobismutite has been made by *Ramdohr* (1960, p. 719), who states, "the intergrowth [of galenobismutite] with bismuthinite is often skeletal or dendritic, indicating the possible breakdown of a complex sulfosalt." Experimental studies prove the existence of such a complex compound, designated as phase V (65 mol-% Bi_2S_3, 35 mol-% PbS), which is stable above 680 ± 5°C in the pure Bi–Pb–S system and which breaks down to form complex lamellar intergrowths (Fig. 26) such as those described by *Ramdohr* (1960).

Galena-Matildite Assemblage

Galena is ubiquitous whenever matildite occurs, and the two minerals characteristically form Widmanstätten-like structures, two typical examples of which are shown in Figs. 27 and 28. These textures develop from breakdown of the solid solution, which is complete between the two phases at high temperature. Heating experiments with natural matildite-galena intergrowths indicate that in the presence of vapor the solid solution becomes incomplete below 215 ± 15°C and that at 200°C the $AgBiS_2$ solid solution will accept approximately 34% PbS, whereas the PbS solid solution will accept only about 8% $AgBiS_2$. The

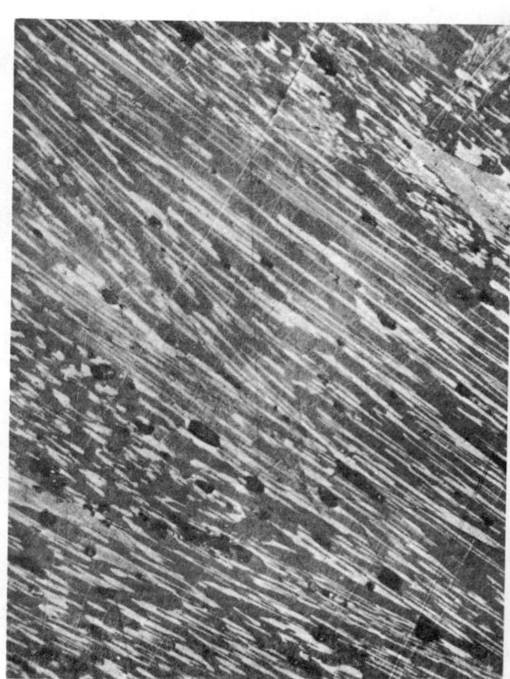

Fig. 26. Lamellar intergrowth of galenobismutite (dark) and bismuthinite (light) resulting from eutectoidal breakdown of phase V. Sample initially synthesized at 700°C, annealed at 575°C for 7 days. (Crossed nicols; × 490 oil.)

effect of pressure on the $AgBiS_2$–PbS solvus is unknown; however, on the basis of other sulfide systems it is probable that increasing confining pressure decreases mutual solubility of the phases and raises the solvus 10–15 °C per 1000 bars.

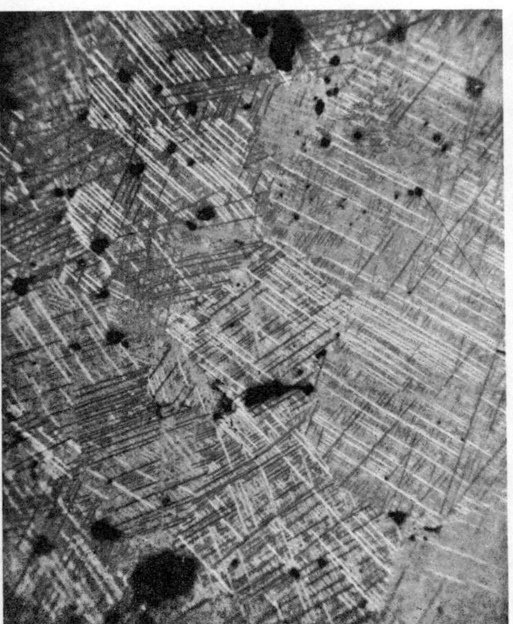

Fig. 27. Fine intergrowths formed through exsolution of matildite (white and dark grey) and galena (grey) from previously homogeneous phase (Leadville, Utah). (Crossed nicols; × 365 oil.)

Suggestions for Future Study

Although the general phase relations are now known, several facets of the Ag–Bi–Pb–S system merit further study.

The effects of confining pressure on the subsolidus phase relations in the Ag–Bi–Pb–S system have been assumed to be small. The actual effects of pressure can, however, be determined only through experimentation.

The phase relations presented on the PbS–Bi_2S_3 join are considered to represent equilibrium conditions only above 400°C. In order to obtain complete understanding of phase relations in the Bi–Pb–S system, it will be necessary to investigate this join at lower temperature. Of special importance is the problem of cosalite stability. This phase, though common in ore deposits, has not been synthesized in the pure Bi–Pb–S system, and its position in the system is accordingly uncertain. Also unknown are the low-temperature stability relations of synthetic phases II and III.

Fig. 28. Coarse intergrowths of matildite (black and white) and galena (grey) formed through exsolution from a previously homogeneous phase (Ouray, Colorado). (Crossed nicols; × 365 oil.)

The present study has considered only the pure Ag–Bi–Pb–S system. Since chemical analyses of phases in this system generally indicate the presence of only small amounts of impurities, the effects of other elements on phase equilibria in the system are probably small. Nevertheless, the effects of elements such as Cu, Fe, Zn which are present in the common sulfides and of Se and Te, which may substitute for S, should be systematically investigated.

Acknowledgements

I am indebted to Drs. *G. Kullerud* and *A. J. Naldrett*, who critically read this manuscript and provided many valuable suggestions. Dr. *Kullerud* generously permitted use of his laboratory for a portion of this study and gave much of his time to serve as thesis advisor. I am also grateful to fellow students at Lehigh University for many fruitful discussions. Samples were donated by Dr. *A. J. Naldrett* of the Geophysical Laboratory, Dr. *E. Grip* of Bolidens Gruvaktiebolag, Sweden, and Mr. *G. W. Gilbert* of the Cariboo Gold Quartz Mining Company. Dr. *N. Morimoto* of Osaka University, Japan provided single-crystal data on synthetic galenobismutite. Dr. *W. Schreyer* kindly translated the Germa summary. Financial support was obtained throug National Science Foundation Fellowships.

References

Aten, A. H. W.: Über Phasengleichgewichte im System Wismut und Schwefel. Anorg. Chem. **47**, 387–398 (1905).

Banas, M.: Polymetamorphe Skarngesteine mit Eisen- und Zinkvererzung in den Sudeten. Freiberger Forschungshefte C 186 (1965).

Bastin, E. S.: The nickel-cobalt-native silver ore type. Econ. Geol. **34**, 1–19 (1939).

Bateman, A. M.: Economic Mineral Deposits, 916 p. New York: Wiley 1950.

Berry, L. G.: Studies of mineral sulpho-salts: IV galenobismutite and lillianite. Amer. Min. **25**, 726–734 (1940).

— Recent advances in sulfide mineralogy. Amer. Min. **50**, 301–313 (1965).

Bloem, J., and *F. A. Kroger:* The P–T–X phase diagram of the lead-sulfur system. Zeit. Physik. Chem. **7**, 1–14 (1956).

Bridgman, P. W.: Effects of pressure on binary alloys: IV, six alloys of bismuth. Proc. Amer. Acad. Arts Sci. **84**, 43–109 (1955).

Craig, J. R.: A systematic study of phase equilibrium in the Ag–Bi–S system and exploration of the geologically significant portion of the Ag–Bi–Pb–S system. Unpub. Ph. D. Thesis, 272 p. Lehigh Univ., 1965.

Djurle, S.: An X-ray study of the system Ag–Cu–S. Acta Chem. Scand. **12**, 1427–1436 (1958).

Edwards, A. B.: Textures of the Ore Minerals and Their Significance. Austr. Inst. Min. and Met. Melbourne, 242 p., 1954.

Ellsworth, H. V.: A study of certain minerals from Cobalt, Ontario, 25th Ann. Rept. Ont. Bur. Mines, p. 200–243 (1916).

Emmons, S. F., C. H. Stockwell, and *R. H. B. Jones:* Argentite and acanthite. Amer. Min. **11**, 326–328 (1926).

Fleisher, M.: New mineral names: bursaite. Amer. Min. **41**, 671 (1956).

Friedrich, K.: Die Schmelzdiagramme der binären Systeme Bleiglanz–Magnetkies und Bleiglanz–Schwefelsilber. Metallurgie **4**, 477–485 (1907).

Gaudin, A. M., and *D. W. McGlashan:* Sulfide silver minerals – a contribution to their pyrosynthesis and to their identification by selective irridescent filming. Econ. Geol. **33**, 143–193 (1938).

Geller, S., and *J. H. Wernick:* Ternary semiconducting compounds with sodium chloride-like structure $AgSbS_2$, $AgSbTe_2$, $AgBiS_2$, $AgBiSe_2$. Acta Cryst **12**, 46–54 (1959).

Graeser, S.: Giessenite – ein neues Pb–Bi Sulfosalz au dem Dolomit des Binnatales. Schweiz. Mineral. u Petro. Mitteil. **43**, 471–478 (1963).

Graham, A. R.: Synthesis and X-ray study of compound in the system Pb–Bi–S, Ag–Bi–S, and $Ag(Bi,Sb)S_2$ Unpub. Ph.D. Thesis, Univ. of Toronto, 1950.

— Matildite, Aramyoite, Miargyrite. Amer. Min. **36** 436–449 (1951).

Guertler, W., and *E. Luder:* Gleichgewichte zwische Metallpaaren und Schwefel: Das ternäre Syster Silber–Blei–Schwefel. Metall und Erz **21**, 173–17 (1924).

Hansen, M., and *K. Anderko:* Constitution of Binar Alloys, 2nd Ed., 1305 p. New York: McGraw-Hi 1958.

Harcourt, G. A.: Tables for the identification of or minerals by X-ray powder patterns. Amer. Min. **2** 63–113 (1942).

Kennedy, G. C., and *R. C. Newton:* Solid-liquid and soli solid phase transitions in some pure metals at hig temperature and pressure. In: Solids under Pressur p. 163–178. Ed. by *W. Paul* and *D. M. Warschauer*. New York: McGraw-Hill 1963.

Kostov, I.: Bonchevite, $PbBi_4S_7$, a new mineral. Minera Mag. **31**, 821–828 (1958).

Kracek, F. C.: Phase relations in the system sulfur–silv and the transitions in silver sulfide. Trans. Ame Geophys. Union **27**, 364–374 (1946).

Kullerud, G.: Thermal stability of pentlandite. Cana Min. **7**, 353–366 (1963).

— The lead–sulfur system. Carnegie Inst. Wash Year Book **64**, 195–197 (1965).

—, and *H. S. Yoder:* Pyrite stability relations in th Fe–S system. Econ. Geol. **54**, 533–572 (1959).

Leutwein, F., u. *A. G. Herrmann:* Kristallchemische ur geochemische Untersuchungen über Vorkomme und Verteilung des Wismuts im Bleiglanz der ki sig-blendigen Formation des Freiberger Gan reviers. Geologie **3**, 1039–1056 (1954).

Morey, G. W.: The system nepheline–albite: a theore cal discussion. Amer. Jour. Sci. **255**, 461–4 (1957).

—, and *E. D. Williamson:* Pressure–temperature curves in univariant systems. J. Amer. Chem. Soc. **40**, 59–84 (1918).

Nissen, A. E., and *S. L. Hoyt:* On the occurrences of silver in argentiferous galena ores. Econ. Geol **10**, 172–179 (1915).

Nuffield, E. W.: Pavonite, a new mineral. Amer. Min. **39**, 409–415 (1954).

Palache, C., H. Berman, and *C. Frondel:* The System of Mineralogy. Vol. I, 834 p. New York: Wiley 1944.

Petrenko, G. I.: Über die Legierungen des Silbers mit Thallium, Wismut und Antimon. Zeit. Anorg. Chem. **50**, 133–144 (1906).

Rahlfs, P.: Über die kubischen Hochtemperaturmodifikationen der Sulfide, Selenide, und Telluride des Silbers und des einwertigen Kupfers. Zeit. Physik. Chem. **31**, 157–194 (1936).

Ramdohr, P.: Über Schapbachite, Matildite und den Silber- und Wismutgehalt mancher Bleiglanze. Sitzungsberichte der Preuss. Akad. Wiss., Phys.-Math. Kl., p. 71–91 (1938).

Die Erzmineralien und ihre Verwachsungen, 2nd Ed., 1089 p. Berlin: Akademie-Verlag 1960.

Ramsdell, L. S.: The crystallography of acanthite, Ag_2S. Amer. Min. **28**, 401–425 (1943).

Raub, E., and *A. Polezec-Wittek:* The temperature dependence of the solubility of lead in solid silver above the eutectic temperature. Zeit. für Metallkunde **34**, 93–96 (1942).

Rieder, M.: X-ray powder data for two discredited minerals: "warthaite" and "goongarrite." Acta Univ. Carolinae **2**, 115–119 (1963).

Roessler, F.: Synthese einiger Erzmineralien und analoger Metallverbindungen durch Auflösen und Kristallisierenlassen derselben in geschmolzenen Metallen. Zeit. Anorg. Chem. **9**, 31–77 (1895).

Ross, V.: The formation of intermediate sulfide phases in the solid state. Econ. Geol. **47**, 734–752 (1954).

Roy, R., A. J. Majumdar, and *C. W. Hulbe:* The Ag_2S and Ag_2Se transitions as geologic thermometers. Econ. Geol. **54**, 1278–1280 (1959).

Salanci, B.: Untersuchungen am System Bi_2S_3–PbS. N. Jb. Miner. Monatshefte p. 384–388 (1965).

— Experimentelle Untersuchung des binären Systems Bi_2S_3–PbS. Unpublished Ph.D. Thesis, 116 p. University of Heidelberg, 1966.

Sakharova, M. S.: On bismuth sulfosalts of Ustarasaisk Deposit. Trans. Miner. Mus. Acad. Sci. USSR **7**, 112–126 (1955).

Schenck, R., I. Hoffmann, W. Knepper u. *H. Vogler:* Gleichgewichtsstudien über erzbildende Sulfide. I. Zeit. Anorg. u. Allg. Chem. **240**, 173–197 (1939).

Seith, W., u. *A. Keil:* Diffusion in Au–Pb- und Ag–Pb-Legierungen. III. Mitteilung über Diffusion von Metallen im festen Blei. Zeit. Physik. Chem. **B 22**, 350–358 (1933).

Short, M. N.: Microscopic Determination of the Ore Minerals. U.S. Geol. Surv. Bull. **914** (1940).

Syritso, L. F., and *V. Senderova:* The problem of the existence of lillianite. Zapiski Vses. Mineralog. Obshch. **93**, 468–471 (1964).

Thompson, R. M.: Goongarrite and warthaite discredited. Amer. Min. **34**, 459–460 (1949).

Tolum, R.: A study on the concentration tests and beneficiation of the Uludag tungsten ore. Bull Min. Research and Explor. Inst. Turkey (Foreign Ed.) **46–47**, 106–127 (1956).

Van Hook, H. J.: The ternary system Ag_2S–Bi_2S_3–PbS. Econ. Geol. **55**, 759–788 (1960).

Vogel, R.: Über das System Blei–Silber–Schwefel. Zeit. Metallkunde **44**, 133–135 (1953).

Wagner, C.: Investigations on silver sulfide. J. Chem. Phys. **21**, 1819–1827 (1953).

Wernick, J. H.: Constitution of the $AgSbS_2$–PbS, $AgBiS_2$–PbS, and $AgBiS_2$–$AgBiSe_2$ Systems. Amer. Min. **45**, 591–598 (1960).

Wurschmidt, J.: Volume changes in amalgams. Berichte der deutschen physik. Ges. **16**, 799 (1914).

Dr. *James R. Craig*, Geophysical Laboratory, Carnegie Institution of Washington, 2801 Upton Street, N.W., Washington, D.C. 20008

Part V

ECONOMIC GEOLOLGY—ORE DEPOSITS

Editors' Comments on Papers 16 Through 22

16 **MINTSER**
The Geochemical Properties of the Behavior of Bismuth in Hypogenic Processes

17 **MALAKHOV**
Bismuth and Antimony in Galena as Indicators of Certain Conditions of Ore Deposit Formation

18 **CHUKHROV, SENDEROVA, and ERMILOVA**
The Mineralogy of Bismuth in the Zone of Oxidation

19 **NAUMEV, PACHADZHANOV, and BURICHENKE**
The Behavior of Bismuth, Silver, Lead, and Zinc in a Process of Secondary Sulfide Enrichment

20 **DZHANDZHGAVA**
The Problem of the Correlation of Selenium and Tellurium with Bismuth Using Georgia as an Example

21 **KOLONIN**
Native Bismuth as a Geological Thermometer IV. The Existence of a Stability Field of Metallic Bismuth at Temperatures Above Its Boiling Point

22 **VASILENKO, EFIMOVA, and KOLESNIKOV**
Native Bismuth as a Geological Thermometer in the Ore of the Smirnovsk Deposits (Southern Maritime Territory)

It has long been common knowledge that bismuth is present in many minerals and rocks from known ore deposits as a dispersed or camouflaged element. If reported, it is usually present as a trace element. While discussions of its chemical and mineralogical vagaries have been given by several authors, our search of the literature clearly indicated that Russian geochemists have possibly studied the geochemistry of bismuth more than any other group. This fact becomes apparent in this section that includes several key papers from the Russian literature—all of which, to the best of our knowledge, appear in English for the first time.

Paper 16 by Mintser leads off this section. It is the most complete review of the Russian literature that we know of. It covers the geochemistry of bismuth and the part this element plays in the ore-forming processes. As a review of the Russian viewpoint toward this subject, it is excellent. While in some respects the article might be better placed in the section on magmatic processes, we feel it is more appropriately placed here. It provides a good introduction to the papers that follow and includes discussions of the migration pathways of bismuth in most rocks and many minerals of economic importance as well as its correlation with other selected trace elements such as thallium.

Mintser begins his discussion with a brief review of bismuth chemistry. He then provides a review of the crystallo-chemical properties of bismuth and its compounds (minerals). Mintser points out that following the pathways of bismuth in the magmatic process is difficult due to fragmentary data. In general, bismuth concentration is very low or absent in the early crystallization products and is sporadically encountered and enriched only in the late ones. The concentration of bismuth in rocks is apparently related to that present in accessory minerals contained in the rock. Bismuth is relatively common in many ores formed in post-magmatic processes, although the parageneses are sometimes unclear. Mintser points out that although bismuth mineralization is not associated with specific magmatic bodies, its association with the acid and moderate-acid intrusive and effusive rocks is clearly seen.

If bismuth is to have extended use as an indicator element in the search for sulfide ore deposits, we will need a clearer understanding of the concentration mechanisms for bismuth in galena. This topic is discussed in considerable detail in Paper 17 by Malakov. He notes that the geologic factors considered to affect the variation of bismuth concentration in galena are the element concentrations in and the reactivity of the enclosing rock. Physical and chemical factors affecting concentration include the rate of temperature and pressure decrease, the volume of the solutions and of dissolved substances, and the alkalinty and acidity of the micro-inclusions in the minerals. Data from Malakov's study indicates that the amount of bismuth found in galenas increases with depth. Study of bismuth minerals isomorphous with galena suggests that these two compounds are miscible under a wide range of natural conditions. He also feels that the ratio of antimony to bismuth in galena indicates the temperatures and pressures that existed at the time of crystallization. Malakov provides a good discussion of these controls in Paper 17, which is a good companion article to that by Rose (1967) on trace elements in sulfide minerals from Utah and New Mexico.

Among the least studied minerals of the oxidation zone are those of bismuth. Several reasons for this state and lack of knowledge are known. One is the similarity of external appearance of the bismuth metals when their bismuth content is quite different. Chukhrov et al. (Paper 18) discuss the results of a rather comprehensive study of the minerals of bismuth from the zone of oxidation and how these minerals related to the primary ore minerals present in certain deposits. This paper is the only one we found in our literature search that treats this problem strictly for the bismuth minerals. Additionally, it presents considerable chemical and physical data on selected bismuth minerals. Extended discussions of the mode of occurrence, associations, and mineralogy of some of the more common bismuth minerals are presented. Much of this data has not been presented elsewhere.

Serving as a companion paper to that on Bi reactions in the oxidation zone is the one by Naumev et al. discussing the behavior of bismuth in the zone of secondary sulfide enrichment (Paper 19). They present the results of studies of the copper-bismuth deposits in the Karamazar area of Russia. A straightforward discussion of the mineral reactions occuring for Bi, Ag, Pb, and Zn in a classic Russian mineral deposit is presented. They emphasize that during the secondary sulfide enrichment process, the behavior of bismuth is controlled by the migration mobility of the element during the oxidation of its minerals and the elemental affinity for sulfur. The approach needs to be tested elsewhere to assess its applicability on a wider scale. The basic controls on the concentration of bismuth and related elements are the exchange reactions between the sulfates of natural waters and the sulfides of the ore bodies.

The use of bismuth as a guide element in exploration geochemistry is normally tied to its correlation with the presence of other elements such as tellurium, selenium, and thallium. The brief article by Dzhandzhgava (Paper 20) discusses this problem using the Georgia deposits of Russia as examples. The correlation is shown to be controlled by the isomorphic capacity of galena for Bi. An interesting aspect of this study was the inability of the author to induce the inclusion of tellurides and bismuthic minerals in experimental galenas. From the data presented the author points out that bismuth is usually present in those ore deposits that contain selenium and tellurium. They conclude that on the basis of their study of the Georgian deposits a direct correlation exists between bismuth, tellurium, and selenium concentrations. This paper should be read as a companion to the brief article by Brooks (1961) on the apparent geochemical association between bismuth and thallium in which lead is proposed as the common link between Bi and Tl.

The use of bismuth concentration as a geological thermometer in ore deposits is mentioned in papers on sulfide mineralization. We complete this section by presenting two papers (Papers 21 and 22) dealing with this topic. The one by Kolonin is a mathematical and theoretical approach. It presents a series of calculations made as a means to evaluate the influence "of the melting of bismuth on the equilibrium constants of those reactions which lead to its formation as well as the thermodynamic possibility of their occurrence." This paper suggests on a mathematical basis that natural bismuth, notwithstanding its transformation to the melted state (under certain conditions), remains a thermodynamically stable phase.

The article by Vasilenko (Paper 22) gives as an example the use of bismuth as a geological thermometer in working out the ore sequence of the Smirnov sulfide deposits. Three generations of natural bismuth separations are recorded in the Smirnov ore deposit. The first two separate at temperatures above 271°C—the melting point of bismuth. Colloidal inclusions of natural bismuth separate below 271°C. These authors make good use of Kolonin's experimental and numerical data, along with their own thermobaric data to explain the occurrence of bismuth in the Smirnov deposits. As a result, the two papers complement each other and should be read as a pair.

REFERENCES

Brooks, R. R. 1961. Apparent geochemical association of bismuth and thallium. *Nature* **4768**:910–11.

Rose, A. W. 1967. Trace elements in sulfide minerals from the Central District, New Mexico and Bingham District, Utah. *Geochim. et Cosmochim. Acta* **31**:547–85.

16

THE GEOCHEMICAL PROPERTIES OF THE BEHAVIOR OF BISMUTH IN HYPOGENIC PROCESSES

E. F. Mintser

This article was translated expressly for this Benchmark volume by Kathleen Ottinger, University of Kansas, Lawrence, from the original Russian article in Forma Nakhozhdeniia i Osobennosti Raspredeleniia Vismuta v Gidrotemal'nykh Mestorozhdeniiakh, *V. V. Ivanov, ed., Moscow: "Nauka," 1969, pages 6–51.*

Due to a series of historical and economic reasons, bismuth has remained until recently one of the least studied elements. Well known in its natural state since ancient times, bismuth was long considered a species of antimony, lead, or tin. The first information about it as a specific metal is found in Georg Agrikol's "Dvenadtsat knig metallurgi" (Twelve Books of Metallurgy) of 1529 where the extraction and processing of bismuth ores are described. A sufficiently complete description of bismuth as a chemical element was compiled only in the eighteenth century.

The history of the name of the element is inconsistent. The majority of researchers consider it to be derived from the German word "Wismut," which, according to one version, is derived from the Old German "Wismuth"--a white metal; according to another, it derives from two words--"Wiessen" (meadow) and "Muten" (to work a mine), since bismuth was extracted in districts around Schneiburg that were overgrown with meadows. V. I. Vernadskij, quoting A. E. Krymskij, considers this word to be derived from the Arab "bi ismid"--that is, "a possessor of the properties of antimony."

In recent years, the rapidly growing interest in bismuth has prompted a turn primarily to clarification of the most general regularities of the concentration of this element in the processes of post-magmatic ore-formation. The basic features of the geochemical behavior of bismuth in the process of the evolution of the matter of the earth's crust, and the properties of its association with a series of heavy metals have been elucidated in the works of Vernadskij

(1915, 1916, 1922), Nenadkevica (1922), Fersman (1939), Zuckert (1926), Ahlfeld (1933), Goldschmidt (1954), and other scholars; however, new information accumulated in recent years makes it possible to specify and add to several of the earlier ideas.

PHYSICAL AND CHEMICAL PROPERTIES OF BISMUTH AND ITS COMPOUNDS

Before proceeding to a discussion of the questions to be examined, let us consider the most general features of bismuth in order to compare it with other, better studied elements and to clarify several characteristics governing the nature of its behavior in post-magmatic process.

Bismuth is the heaviest element in the main sub-group (the nitrogen sub-group) of group V of the periodic table. Its closest analogs, according to the sub-group are antimony and arsenic, and its neighbors in the periodic table are lead and polonium.

Beyond bismuth (atomic number 83 and atomic weight 209)[1] the periodic table indicates a series of elements with an absence of stable isotopes, or a series of so-called natural emitters. At the present time, 19 isotopes are known for bismuth, 13 of these (Bi^{198}, Bi^{199}, Bi^{200}, Bi^{201}, Bi^{202}, Bi^{203}, Bi^{204}, Bi^{205}, Bi^{206}, Bi^{207}, Bi^{208}, and Bi^{215}) are obtained only in artificially induced reactions. Five (Bi^{210}, Bi^{211}, Bi^{212}, Bi^{213}, Bi^{214}) are members of the radioactive families of urnaium, thorium, actinium and neptunium. Their maximum half-life rate does not exceed a few days. As known (Rankama, 1956), the only isotope encountered in nature, Bi^{209}, also represents a final quasi-stable product of the neptunium series with a half-life of more than $2 \cdot 10^{18}$ years. The relative stability of this isotope, that is the unusually low energy of decay, is the result of the closed state of its neutron shell (A-Z = 126), and also of the fact that the transition of $Bi^{209} . Tl^{205}$ is connected with a change of mechanical momemt of 4 units (i.e. it is forbidden). The characteristics of the natural isotopes of bismuth are shown in Table 1.

The electron configuration of the bismuth atom is characterized

1. See notes at end of article

Table 1. Characterization Pattern and Properties of the Natural Isotopes of Bismuth.

Isotope	Natural Radioactive Family	Nature of the decay Decomposition	Half-life	Energy Emission/ Radation MEV	Products of Decom- position
Bi^{209}	Neptunium		2.10^{18} yrs	3.0	Tl^{205}
Bi^{210}	Uranium	(>99%)	5 days	1.17	Po^{210}(RaF), Tl^{206}
Bi^{211}	Actinium	(99.68%)	2.16 minutes	6.62(82.6%)	$Po^{211}(AcC^{I})$, $Tl^{207}(AcC^{II})$
Bi^{212}	Thorium	(63.8%) (36.2%)	60.5 minutes	2.25 6.086(27.2%)	$Po^{212}(ThC^{I})$, $Tl^{208}(ThC^{II})$
Bi^{213}	Neptunium	(98%)	47 minutes	1.39 5.86	Po^{213}, Tl^{209}
Bi^{214}	Uranium	(>99%)	20 minutes	5.52	$Po^{214}(RaC^{I})$, $Tl^{210}(RaC^{II})$

Note: The content of Bi^{209} in a natural mixture is 100%.

by the presence of level P of 2s and 3p electrons and the unfilled f-shell in the 0 level.

The atom of bismuth is analogous to N, P, As and Sb; however, the 18-electron penultimate level (of the "cupro" type) that is characteristic of chalcophiles and absent with N and P, restricts the chemical effects within the limits of the sub-group only with As and Sb, and links bismuth with a whole series of heavy chalcophiles of other groups (Pb, Tl, Au, Ag, and others).

Bismuth is a white metal with a reddish tinge; it is sparkling and brittle. At 20°C, its density is 9.8, and at its melting point it is 10.067 g/cm^3--i.e., the expansion of bismuth at solidification is 3.32% of its volume at its melting point. The density of molten bismuth decreases proportionately with increasing temperature (from 10.062 at 275°C to 9.611 g/cm^3 at 650°C). The hardness of bismuth, according to Mohs scale is 2.5. Bismuth has a low heat conductivity; it is also the most diamagnetic of all the metals. A few of the physical and crystallochemical properties of bismuth and of elements similar to it are given in Table 2.

Like all the elements that relate to E sub-group V6, bismuth can possess a maximum valence of 5+, but in view of the fact that in going from nitrogen to bismuth, the tendency of the elements toward a trivalent condition increases, the basic valence for bismuth in compounds is 3+.

In accordance with the general chemical regularity within the limits of the sub-group, bismuth possesses the most clearly expressed metallic properties. Whereas trivalent nitrogen and phosphorus are typical non-metals (acid-formers), and whereas arsenic and antimony are amphoteric substances, forming both acids and bases, trivalent bismuth possesses only basic properties. This situation is illustrated by regular changes in the character of the hydroxides of the elements.

Nitrogen	HNO_2		Antimony	$Sb(OH)_3$	H_3SbO_3
Phosphorus	H_3PO_3		Bismuth	$Bi(OH)_3$	
Arsenic	$As(OH)_3$	H_3AsO_3			

Furthermore, it should be noted that in compounds with such an electro-positive element as hydrogen, the elements of the Bismuth sub-group emerge in the capacity of negative trivalents. This

Table 2. The Physical and Crystallo-Chemical Properties of Bismuth and Some Elements Related to it.

Property	Bi	Sb	As	Pb	Ag	Cu
Atomic No.	83	51	33	82	47	29
Atomic Weight						
On basis of Oxygen	209.00	121.76	74.91	207.21	107.88	63.54
On basis of Carbon	208.98	121.75	74.9216	207.19	107.87	63.54
Density at 20°C	9.80	6.68	5.73	11.34	10.5	8.92
Atomic Volume, $cm^3/g \cdot atom$	21.31	18.19	12.98	18.27	10.27	7.11
Radii, Å						
Atomic	1.55	1.45	1.21	1.75	1.44	1.28
Ionic, acc. to Arsen, 1952	Bi^{3+}--0.96	Sb^{3+}--0.76		Pb^{2+}--1.20	Ag^{+}--1.26	Cu^{+}--0.96
	Bi^{5+}--0.74	Sb^{5+}--0.62	As^{5+}--0.46			Cu^{2+}--0.72
Acc. to Belov and Bokij, 1954	Bi^{3+}--1.20	Sb^{3+}--0.90	As^{3+}--0.69	Pb^{2+}--1.26	Ag^{+}--1.13	Cu^{+}--0.98
	Bi^{5+}--0.74	Sb^{5+}--0.62	As^{5+}(0.47)			Cu^{2+}--0.80

Table 2. (Continued)

Property	Bi	Sb	As	Pb	Ag	Cu
Coordination Number.	3.6	3.6	3.6	6	4	4
Electronegativity, acc. to Porarennyx, 1951.	195(III) 265(V)	195(III) 265(V)	220(III)	170(II)	175(I)	177(I)
Potential of ionization, ev. . .						
I_1	7.277	8.64	9.81	7.415	7.574	7.724
I_2	19.3	16.7±0.5	18.7±0.1	15.03	21.48	20.29
I_3	25.6	24.8	28.3	31.93	36.10	36.83
I_4	45.3	44.1	50.1	39.0		
I_5	56.0	63.8±0.5	62.9±0.1	69.7		
I_6	94.4	119	127.5			
Temperature, ^{o}C						
Melting	271	631	814	327.3	960.8	1083
Boiling	1560	1635	615	1751	2163	2580

ability, however, decreases from nitrogen to bismuth, and the compound BiH_3 (bismuth hydride) is remarkably unstable.

At normal air temperature, bismuth is stable; at red-heat temperature it burns, converting Bi_2O_3. On heating, bismuth reacts also with bromine, iodine, sulphur, selenium and tellurium. At normal temperature, water has no effect on bismuth, but at calcination in an atmosphere of steam it slowly oxidizes. The oxidation-reduction potential (E_0) of the system $Bi^{3+}/Bi = +0.226$ v.--i.e., the atom of bismuth possesses a fairly high electron affinity, transforming comparatively easily into a neutral atom. In connection with this, the ions of trivalent bismuth are reduced by solutions of Cr^{2+}, Ti^{3+}, and by a series of metals (Mg, Zn, Cd, Fe, Sn). On the other hand, under certain conditions, metallic bismuth is oxidized by solutions of trivalent iron, bivalent copper, iodine, and others.

As is evident from the significance of the normal oxidation potential, bismuth is "more precious" than hydrogen and does not displace the latter in non-oxidizing acids (i.e. it is insoluble in them). Nitric acid and hot concentrated sulfuric acid dissolve bismuth easily.

Bismuth easily forms alloys with metals but rarely compounds. With oxygen, bismuth forms the oxides Bi_2O_3, Bi_2O_4, Bi_2O_5. The trioxide of bismuth, Bi_2O_3, is readily soluble in acids and slightly soluble in concentrated alkalis. In correspondence with the distinctly basic properties of the trioxide of bismuth, its compounds with the more electronegative atoms have the character of salts. The pentoxide, Bi_2O_5, and its corresponding bismuthic acid are not known as individual compounds. The salts of bismuthic acid--bismuthates--are produced only by the action of strong oxidants (chlorine, potassium, permanganate and others) on the hydroxide of bismuth in a concentrated alkali. These salts are strong oxidants even in an alkali medium and their existence in nature is difficult to prove. For this reason, only compounds of trivalent bismuth, as the only oxidized condition of this element in natural compounds, will be examined further. The oxide Bi_2O_4, not typical of bismuth, apparently contains one atom of bismuth in a trivalent state and one in a five-valent state (Busev, 1953).

The hydrolysis reactions that lead to the forming of slightly soluble basic salts, for the most part of a variable composition, are quite typical of compounds of bismuth. The salts of the strong acids (chloride, sulfate, nitrate) are readily soluble in water only with an excess of the corresponding acid. With this, bismuth can be present in the form of the free hydrated ions of Bi^3 only in a highly acid medium. In insufficiently acidic solutions bismuth is found in the form of the polymerized ions $Bi_4O_4^{4+}$; the ions of bismuthile BiO^+ have not been discovered (Busev 1953; Remi 1963).

Slightly soluble basic salts are formed by increasing the solution pH with dilution by water or the addition of corresponding reagents. At reduction, the solution pH of the basic salt is evidently explained by the formation of complexes with the ions Cl^-, $SO_4 2^-$, NO_3^- or the molecules HCl, H_2SO_4, HNO_3. The nature of the basic salts of bismuth have received little investigation; however, it is suggested that in the basic salts, the atoms of bismuth are connected by oxygen bridges (Busev et. al., 1953). For example the following structure is assigned to the basic nitrate $2Bi(NO_3)_3Bi_2O_3$:

$$\begin{array}{ccccccccccccc} NO_3 & - & Bi & - & O & - & Bi & - & O & - & Bi & - & O & - & Bi & - & NO_3 \\ & & \uparrow & & & & \uparrow & & & & \uparrow & & & & \uparrow & & \\ & & NO_3 & & & & NO_3 & & & & NO_3 & & & & NO_3 & & \end{array}$$

The magnitude of the quantitative precipitation of bismuth in the form of a basic salt depends on many factors (the nature and concentration of the anion, the initial concentration of the bismuth ion, the concentration of neutral salts, the presence of complex-formers, the temperature, the means of increasing the solution pH, etc.) and presently, for the majority of cases is not determined.

Tananaev and others (1962) consider that if the solution has a $pH > 2$, then the presence of the ion, Bi^{3+}, in it is practically excluded. Of the solutions containing salts of bismuth and of alkali metals, dual salts (acidosalts) of the following type are often crystallized:

$$M[BiX_4],\ M_2[BiX_5],\ M_3[BiX_6]$$

where X = F, Cl, Br, I; M is an ion of a univalent metal. This indicates that the salts of bismuth are inclined to combine the acid ions found in excess with the formation of the acido-ions, for example, $BiCl_3+Cl^- = BiCl_4^-$.

Very characteristic of bismuth are the compounds with sulphur and its analogs--selenium and tellurium. Within the limits of the sub-group bismuth possesses maximum stability. The sulphide of Bi bears a close resemblance chemically, to an oxide. Just as with Bi_2O_3, the Bi sulphide rarely reacts with the sulphides of alkalis, whereas the sulphides of the amphoteric substances, As and Sb form salts of the corresponding acids, with soluble sulphurous metals according to the reaction:

$$Sb_2S_3 + Na_2S = 2NaSbS_2.$$

The sodium salt of the sulphoantimonous acid.

Oxidizing acids easily decompose the sulphide of bismuth. In diluted HNO_3 it easily dissolves even at room temperature. The sulphide of bismuth is quite similar to its selenide which is also insoluble in KOH and K_2S.

In connection with the fact that the capacity of the element for complexing is one of its most important geochemical characteristics, we will now examine bismuth as a complex-forming element.

In the limits of the sub-group the capacity for complexing increases with the transition from nitrogen and phosphorus towards arsenic and especially towards antimony and bismuth.

Thus, for As^{5+}, the only known halide complexes are the fluorides $[AsF_6]^-$ and $[AsF_7]^{2-}$ and for As^{3+}, only the slightly stable complex chlorides $[AsCl_4]^-$, $[AsCl_5]^{2-}$ and $[AsCl_6]^{3-}$ are known. For Sb^{5+}, the relatively stable chloride and bromide complexes $[SbCl_6]^-$ and $[SbBr_6]^-$ are characteristic. Similarly, the complex chlorides and bromides of trivalent antimony are well known.

The complex halides of five-valent bismuth are unstable in a water solution because of its very strong tendency toward reduction. Bi^{3+} is a good complex-former. Chloride, bromide, and iodide complexes are well known for it. As was demonstrated by Babko and Bolub (1953), the stability of the halide complexes of trivalent bismuth increases from the chlorides to the iodides (the instability constants for $[BiCl_6]^{3-}$, $[BiBr_6]^{3-}$, $[BiI_6]^{3-}$ are respectively $3.8 \cdot 10^{-7}$; $2 \cdot 10^{-10}$; $3.1 \cdot 10^{-12}$. Also well known are the complex compounds of Bi^{3+} with sulphur-containing addends, for example, Thiourea.

The complex-formers can be divided into 2 groups according to their capacity for forming halide complexes: The stability of the complexes of one group decreases in the series $F^- \gg Cl^- > Br^- > I^-$, and the second changes in the reverse order, i.e., $I^- > Br^- > Cl^- >> F^-$ (Babko, 1955; Scerbina, 1962).

The variation of the stability of the complex halides of bismuth, shown above, indicates that this element belongs to the second group. The relative stability of the halide complexes permits an approximate evaluation of the nature of the interaction of the central atom with the addends. Thus, for the first group, the electrostatic (ionogenic) type of chemical bound is primarily assumed, and for the second--primarily the covalent type with partially dual bonds (Leden and Chatt, 1955). Together with Bi^{3+}-Cu^+, Tl^+, Ag^+, Cd^{2+}, Hg^{2+}, and Pb^{2+} also relate to the second group. The pronounced tendency of these elements toward complexing is due to the fact that the energy of formation of the corresponding ions (the sum of the sublimation energy and the ionization energy) is very great and in view of this tends to form, to a large degree, covalent bonds with the corresponding addends. The stability of the compounds of bismuth, as that of many other heavy metals, in the molecular state must increase in the series $F < O < Cl < I << S$, because in this sequence, the fraction of the covalent bond increases (Figure 1). In fact, Grinberg (1951) notes that the sulphur-containing addends possess complexes which are stable in solution only with elements which form bonds with a significant degree of covalence. Since the stability of the containment of the acido-ion in the internal coordination sphere is proportional to its affinity with the proton (Smits-Dju Mont 1961), the halide ions in the complexes of heavy metals possessing a large bond energy must be displaced by sulphur.

These facts, therefore, give reason to assume that in a structurally complicated hydrothermal solution, bismuth and a series of other metals more than likely must form sulphide (generally chalcogenide) complex compounds in stable and transient systems. At the present time it is a generally accepted fact that in a solution system where the stability consists of the cation of the trivalent metal (Bi^{3+}) and the divalent anion (S^{2-}), it follows that the presence of the groups $[BiS]^+$, $[BiS_2]^-$, $[BiS_3]^{3-}$ is anticipated. The existence or predominance in a solution of one or another group will be determined by specific physico-chemical conditions, however it is not possible at the present time to determine them for each group.

Table 3. The Distribution of Bismuth Minerals According to Classes.

Class	Sub-class and group	Number of Minerals
Natural elements and intermetallic compounds		4
Sulphides	Simple	2
	Complex Pb-Bi	12
	Cu-Bi	5
	Ag-Bi	3
	Pb-Cu-Bi	5
	Pb-Ag-Bi	2
	Ni(CO)-Bi	4
Tellurides	Simple	4
	Complex	9
Selenides	Simple	4
	Complex	3
Oxides, Oxyhalides		5
Carbonates		3
Arsenates		4
Vanadates		1
Molybdates		1
Tellurates		1
Silicates		2
Tantalo-niobates		1

At the present time, 74 minerals of bismuth are known. The presentation of their distribution according to compound classes and the number of mineral forms within the latter is given in Table 3. A few of the bismuth minerals (for example, chiviathite, klaprotholite, chammarite, lindstromite, and others) have not received general acceptance as independent minerals and are considered by different researchers to be mechanical mixtures. However, the absence of objective criteria supporting or disproving the existence of these minerals as independent forms, and the extremely limited information about them, allow us to accept, with an equal degree of probability, one or the other point of view. On the other hand, in a group of bismuth minerals there are minerals of other metals that have not been examined and that sometimes contain from one to ten percent of bismuth. It is more correct to consider such minerals as bismuth-containing varieties of the basic mineral form. These include: alloclas--bismuth-containing dyalitte, rionitte--bismuth-containing tennanthite, bismuth-scutterudite, bismuth-microlite.

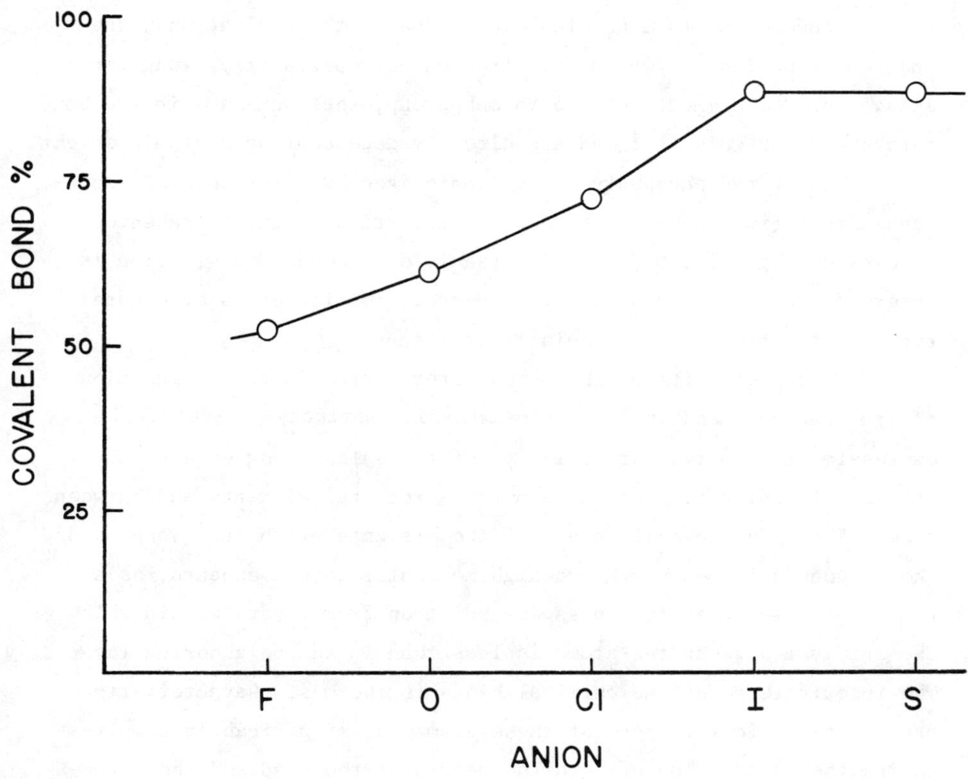

Figure 1. Covalent character of bismuth associations with the halogens, oxygen, and sulfur (Taken from C. C. Badanova, 1962).

THE CRYSTALLO-CHEMICAL PROPERTIES OF BISMUTH AND ITS COMPOUNDS

As is known, the most general (although far from adequate) principle determining the crystallo-chemical properties of the b sub-groups, is the Hume-Rothery rule, in accordance with which the coordination number (c.n.) of the element is equal to 8-N (where N is the number of the group of the periodic system). The regularity is explained by the capacity of the elements of the b sub-groups to form stable covalent bonds owing to the unpaired s and p electrons, the number of which is equal to 8-N. Therefore, for the elements of sub-group Vb the c.n. must be equal to 3.

Krebs (1961) demonstrated that with the transition from the light atoms at the top of the periodic table (which tend to form an sp^3 electron configuration with tetrahedrally directed bonds) towards the heavy elements at the lower end of the table, there is an increase in the tendency to form chemical bonds due to the p-electrons, the inherent functions of which are directed perpendicularly along three axes. For the elements of the Vb sub-group, these changes in the bond nature are very distinct and are directly dependent upon atomic weight.

Thus, if red phosphorus is characterized by a tetrahedral short-range order (in black phosphorus the distortion of the tetrahedron appears only partially), then in arsenic a pronounced distortion is observed and a layered structure appears. The latter is even more typical of antimony and especially of bismuth.

The strengthening of the p-character of the intrinsic function of the chemical bond in the series arsenic--antimony--bismuth, finds expression in the regular variation of the valance angles and the interatomic intervals both within the structural elements and between them. Thus, in the solid state of the elements of Vb sub-group, and the compounds formed by them a high-molecular nature appears, as a result of which the structures are built up from layers within which the interval between the atoms is less than in the neighboring layers. The inequivalence of the chemical bonds is the most characteristic crystallo-chemical feature of these elements, significantly complicating the description of both the nature of the bond and the structures of the compounds formed.

In the same series, arsenic--antimony--bismuth, the transition temperatures of low-molecular forms of the elements into high-molecular ones, and the high-molecular amorphous ones to crystalline, change regularly (they drop rapidly). See Table 4. The variation of the character of the chemical bond is expressed in the magnification of the degree of its metallicity with which such properties of the elements as electrical conductivity, color, etc., are directly linked.

From Table 5, it can be seen that in the crystalline phases arsenic, antimony, and bismuth, in accordance with the Hume-Rothery rule, have three equivalent bonds with minimal interatomic intervals oriented along three almost perpendicular axes. Besides this, there are three additional and also equivalent bonds going in directions opposite to the first.

Generally, when bismuth combines with atoms of various size, the electrons of which belong to different energy levels, the picture becomes even more complicated and the degree of anisotrophism increases.

Table 4. Low and High-Molecular Forms of the Elements of the Vb Subgroup. (Acc. to G. Krebs, 1961).

Elements	Low Molecular Form	Temp. of Conversion. °C.	High Molecular form		
			Amorphous	Temp. of conversion. °C.	Crystal
Phosphorus	(P_4) gaseous			400-500 ⟶	Phosphorus, red crystalline
Phosphorus	Liquid	180 →	red vitreous		↑ 550
	Hard			300 → (with Hg)	Phosphorus, black rhombic
Arsenic	(As_4) gaseous	? →		270	Arsenic, rhombohedral ↑270 (with Hg ↑200)
Arsenic	Hard	<20 →	black, vitreous	125 → (with Hg)	Arsenic, rhombic
Antimony	(Sb_4) gaseous	—180 →	Black	20 →	Antimony rhombohedral
Bismuth	(Bi_2) gaseous	—269 →		—253 ⟶	Bismuth rhombohedral

* The table shows the minimum temperatures of formation.

This gives rise to a series of difficulties with the development of the regularity of the stereochemistry of the compounds of bismuth, since, connected with the inconstancy of the distance Bi-X, where X is Cl, O, S, Se, etc., a certain indeterminacy appears in the separation of the coordination polyhedrons.

In nature the chalcogenides of bismuth are widespread. For this reason, we examine the crystallo-chemistry of these compounds first.

In a series of studies directed toward the stereochemistry of the elements of sub-group Vb, it was noted that arrangement according to the type of macromolecule is the most tenable principle for the classification and description of the structures of their compounds

Table 5. The interatomic intervals within the structural elements (chains, lattices) (r_1) and between them (r_2) the number of the neighbors and valent angles (ϕ) of the structural elements for different conditions of As, Sb and Bi.

	Liquid Phase		Amorphous Phase			Crystal (rhombohedral) Phase				
Elements	Inter-Atomic Interval Å	Number of the Neighbors	Inter-Atomic Interval Å		Number of the Neighbors	Inter-Atomic Interval		Number of the Neighbors	r_2/r_1	ϕ
			r_1	r_2		r_1	r_2			
			2.5		3.1	2.51		3		
As				3.75			3.15	3	1.25	97
			2.87		3	2.87		3		
Sb				3.75	2.6		3.37	3	1.17	96
	3.32	7--8				3.10		3		
Bi	(6.6)						3.47	3	1.12	94

(Tarasov 1958; Krebs 1961; Polynova and Poraj-Kosits 1966; and others). Following this principle, it is possible to divide the sulphides of bismuth (using their most general features) according to polymeric chain type into:

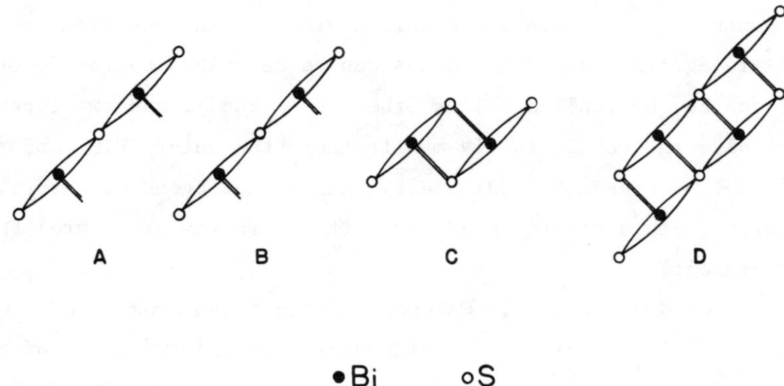

Figure 2. Patterns of polymer chains and bands, sulfides of bismuth a) simple chains, b & c) binary chains [simple bands], d) binary bands.

1. A simple chain with the content of then $[BiS]_n^{n+}$ (Figure 2a); similar chains are characteristic of the thiohalides of bismuth.

2. Double chains (simple bands) of two forms of the content $[Bi_2S_5]_n^{n4-}$ and $[Bi_2S_4]_n^{n2-}$ (see Figures 2 b and c); cosalite is an example of a mineral with a double chain of the first form where the binary chains (bands) are surrounded on all sides by atoms of lead in an octahedral coordination; emplectite (an atom of bismuth surrounded by six atoms of sulphur giving a distorted octahedron with three atoms of sulphur of this band and three of the neighboring--has the double link of the second form.

3. The four chained double bands of the form $[Bi_4S_6]_\infty$ arise in the structure of bismuthine and aikinite (see Figure 2 d). In these bands, bismuth has two different types of atomic coordination.

The atoms of Bi_I have a coordination in the form of a trigonal pyramid with the interval Bi_I-S of the order 2.8-2.9°A: The distorted octahedron have added 3 atoms of sulphur from the neighboring band at significantly large intervals (of the order 3.1-3.2°A). The atoms of Bi_{II} are characterized by a coordination polyhedron in the form of a tetragonal pyramid, the atom of bismuth being somewhat deflected outwards from the plane of the pyramid base. The minimal length of the bond is the interval from Bi_{II} to the S atom at the top of the pyramid. In this structure, the distinction between the short-range order "terminals" and "bridges" of the ligands is especially clear. In compounds with a simple and double chain, all the positions of bismuth are equivalent and the bonds can be described by the formula 1+2+(2) to the quadruplicate link (the double band), and two types of atoms of bismuth are distinctly manifested: the outer (Bi_I) and the inner (Bi_{II}). Guanajuatite $Bi_2(Se,S)_3$ has an analogous structure. In aikinite (with a similar band, but with Bi in its structure) Pb also is present.

4. Layered structures. Galenobismutite is an example of a mineral with a similar design. The alternation of the two types of coordination polyhedrons, analogous to the polyhedrons of Bi_2S_3, forms undulating layers which touch upon each other. In the gaps between the layers the paired atoms of Pb assume an octahedral arrangement (Figure 3).

In the transition to the tellurides (sulphotellurides) the structural patterns are sharply altered. If in the structures of the sulphides of bismuth, the inequivalence of the atoms of Bi appears, then in the tellurides, the structural pattern is determined by the inequivalent atoms of Te. The layered structure of tellurobismuthite is formed as a result of the sequential application of the five-layered groups Te_I--Bi--Te_{II}--Bi--Te_I.

The atoms Te_{II} are almost exactly surrounded octahedrally by the atoms of Bi; the distorted octahedral environment is observed also in the atoms of Bi which, on the one hand, are linked by the 3 atoms of Te_{II} (the length of the bond is $3.22A^o$), and the other, by the three atoms of Te^I (the length of the bond is 3.12^oA).

The crystalline structures of the other tellurides of bismuth represent the structures of tellurobismuthite, which are complicated due to the interchanging of the one or the other type of atoms of Te and S (tetradymite, ciclovaite) or the magnification of the layers in groups.

It is necessary to consider the important but small group of com-

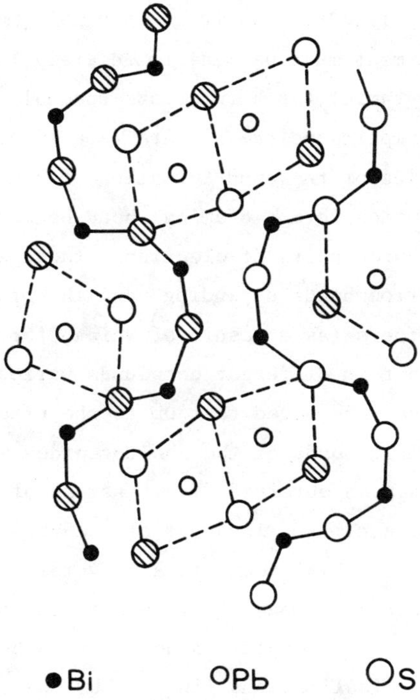

Figure 3. Structural pattern of galeno bismutite.

pounds of bismuth with the transition metals and platinoides. These possess the structural type of NiAs. At the present time only a few minerals of this group are known, but judging by the stability of the parageneses of the bismuth minerals with the minerals of iron, the crystallo-chemical bond of bismuth with the elements of the VIIIth group of the periodic table is much closer. This question is discussed later in greater detail. Let it be noted here that those structures that may be deduced from the NiAs type structure (with the filling of a part of the octahedral voids by the atoms of the transition metals) are possessed by moncheite (Pt, Pd) $(Te, Bi)_2$ and kotulskite Pd(Te, $Bi)_{1-2}$. Michenerite (Pt, Pd) (Bi, Te) has a structure of the pyrite type.

As is evident from the diagrams given, in all instances, despite the complication of the pattern, the short-range order of bismuth is characterized by the presence of three atoms of sulphur at minimal intervals--i.e., the trigonal pyramid with bismuth at the top is the basic coordination polyhedron, as it is in simple molecules (see Table 5). The statement must be made immediately that the bond lengths (the edges of the pyramid), as a rule, are unequal. The cause of this phenomenon, which complicates the picture to a significant degree, lies in the properties of the bond-formation of the ligands. As is known, the chalcogenides form a chemical bond because of the p-orbits. However, having unshared pairs of electrons, these elements can form additional coordination bonds depending upon the properties of the electropositive partners (as a result of which, for example, the valence angle for sulphur in different compounds varies from 75 to 138°). With this, it must be remembered that up to the present no compound containing the multiple bonds of the chalcogenides with an electropositive element has been obtained (i.e. instead of -P=S, -As=S, etc., we have >P-S, >As-S, etc.; Andrianov et al., 1965).[2]

The fact that bismuth silicate is a very rare mineral and the comparatively low concentrations of bismuth in the oxygen compounds of elements close to it, is explained not so much by crystallochemical as by geochemical reasons (in part by the comparatively large ionogenicity of the bonds Bi-O). In the isostructural compounds $CePO_4$ (monazite) and $BiPO_4$ the constants of the elementary cell vary up to a maximum of 1%.

In concluding the examination of the basic crystallo-chemical properties of bismuth and its compounds, it is imperative once more to emphasize the following:

1. In chalcogenides, an irregular octahedral environment is characteristic of Bi. Because of this and the large variance of the bond lengths, it is often fairly difficult to separate the bond-forming atoms of short-range order, which leads to the indeterminacy of the coordination number of bismuth. The inequivalence of the atoms of Bi in the lattice appears more distinctly in those sulphides where there are three strong pyramidally directed bonds or five bonds along the quadratic pyramid (with the atom of Bi in the center of the base);

2. A series of structural complications arise as a result of the inequivalence of the ligands. Polynova and Poraj-Koscits (1966)

2. See notes at end of article.

explain the large variance of the intervals in the structures of the compounds of trivalent antimony in part by the specific properties of the ligands (especially sulphur). This is apparently also correct for Bi compounds.

The noted crystallo-chemical properties of Bi hinder the determination of the range of elements which it can replace in natural formations. Only the limited isomorphism of bismuth with antimony and lead is generally acknowledged, and an isomorphism with tellurium, calcium, rare earths and palladium is assumed.

THE EXTENT AND FORMS OF THE OCCURRENCE OF BISMUTH IN NATURE.

Despite the often-noted presence of bismuth in ores of the most varied deposits, this element is comparatively uncommon. The unrelated and contradictory information about bismuth concentration in various types of rocks in the earth's crust and in cosmic material does not, for the present, give any reliable criteria for the establishment of the degree of its occurrence. In the first place, this affects the inadequate reliability of the Clark of bismuth, which is different according to the findings of various researchers. Values are illustrated here (in g/t):

Clark	Author
0.01--0.1	Clarke and Washington (1924)
0.034	Berg (1925; 1933)
0.07	Noddack (1934)
0.2	Goldschmidt (1937)
0.1	Fersman (1939)
0.002	Green (1954)
0.n	Vinogradov (1956)
0.17	Brooks and Ahrens (1961)
0.009	Vinogradov (1962)

The Clark of bismuth, according to Vinogradov (1962), is lower than the Clarks of such elements as Tl, In, Cd, Ag, Ga, Ge, and only a little higher than the Clarks of Re, Te, and Au. On this basis according to the degree of occurrence, bismuth is a typically rare element. The data given in Table 6 on the content of bismuth in cosmic materials show that the sulphide and metallic phases are enriched in this element. Brooks and Ahrens (1961), considering the chondrites equivalent in composition to the cosmic substance from which the Earth was formed, utilize the ratio given here for the determination of the degree of enrichment of the earth's core by a specific element.

$$\frac{\text{Content of element in the earth's core (clark)}}{\text{Content in the earth as a whole equal to the content in chondrites}}$$

For bismuth, in connection with the low reliability of the Clark, this ratio is not significant because the factor of enrichment varies between 100 (if we take the figures of Fersman (1939), or Brooks and Ahrens (1961) for the Clark) and 0.5 (if we take the figures of Vinogradov, 1962).

If we add to this the fact that the Bi content in rare carbonate chondrites is higher than in the normal ones by 50 times (Reed, 1960), then it becomes clear that it is not possible to evaluate the degree of enrichment of the earth's core for this element.

Table 6. The Content of Bismuth in Meteorites.

Type of meteorite	Content of Bi, g/t	Literary Source
Troilite	2.0	Noddack (1930)
"	0.2	Noddack (1934)
Ferrous-nickel	0.5	Noddack (1930)
"	0.3	Noddack (1934)
Meteorite-iron	0.5	Braun (1949)
Chondrites including the sulphides and the ferrous-nickel phase.	0.02	Noddack (1934)
Chondrites	0.02	Urey (1952)
"	0.0022	Ehmann and Huizenga (1959)

BISMUTH IN THE MAGMATIC PROCESS.

Only fragmentary, unsystematic data exist concerning the geochemical history of bismuth in the processes of the formation of igneous rocks. The content of bismuth in rocks is lower than the sensitivity limit of the normal semi-quantitative spectral analytical methods. Chemical techniques for the determination of small quantities of bismuth in silicate material are inadequately developed.

As is known, the behavior of the element in the processes of differentiation and crystallization of magmas together with geologico-tectonic and physico-chemical conditions is determined by crystallo-chemical factors. Due to these properties, the element is dispersed by its isomorphous entry into the crystalline lattices of rock-forming minerals, or its accumulation in the detrital products of magmatic activity. With this, the crystallo-chemical criteria have a more general, universal significance with the development of the rules of distribution of bismuth in the particular minerals of igneous rocks--i.e., in the processes of magmatic crystallization.

An analysis of the data in the literature has shown that bismuth is absent (its content is lower than the analysis of the limit of sensitivity) in the early crystallization products (granitic magma, feldspars, and quartz) and is sporadically encountered only in the late ones (e.g., in the accessory minerals and biotite). Goldschmidt (1954) indicated the possibility of the replacement of $Bi^{3+} \rightarrow Ca^{2+}$ in early minerals of magmatic rocks (partly in apatite) and the replacement of $Bi^{3+} \rightarrow TR^{3+}$ in rare-earth minerals with the crystallization of the detrital fusions. However, Brooks and Ahrens (1961) give attention to the fact that, in calcium rich rocks (basic), as well as in rocks which are relatively poor in calcium (acid and alkali), the content of bismuth is of low order; no enrichment by bismuth of the calcium minerals (wollastonite, actinolite, apatite, and the carbonates) is observed. Thus, these authors consider the assumption of a significant geochemical relationship between Ca and Bi in the process of differentiation and crystallization of magmas to be untenable.

At the same time Brooks (1961) noted a correlating dependence between the content of thallium and bismuth in the rocks. Brooks, on the basis of Goldschmidt's data, hypothetically considers that the lead present in rocks is partly replaced by thallium + bismuth. To some extent, this association may be explained by the formation of the stable isotopes of these elements as a result of the nuclear transmutations of $Bi^{207-215} \rightleftarrows Pb^{195-214} \rightleftarrows Tl^{173-186}$ (Krasnikov et al. 1965). However, there is no completely satisfactory explanation of this dependence.

Bismuth (of the order of a few grams to the ton) is sometimes present in biotite (Koptev-Dvornikov et al. 1960; Brooks and Ahrens 1961; Ezov 1964). Since biotite often contains micro-inclusions of the accessory minerals, then the development of a method for the location of bismuth in this mineral is a special task. There are practically no data on the evaluation of the content of bismuth in the accessory rare-earth minerals. It is known, however, that monacite is rich in bismuth.

The basic concentration of bismuth in rocks is apparently related to that present in the accessory minerals, by native bismuth and bismuthine in the course of weathering and conversion into bismuthite and bayerite. We will discuss the characteristics of the accessory bismuth minerals later.

Information about the behavior of bismuth in the process of formation of various types of magmatic rocks is very scanty. On this basis, for example, bismuth was excluded from examination in the survey work of Anikeeva (1964), which examined the association of chemical elements in various rocks. The data of various researchers regarding the content of bismuth in the main types of rocks differs greatly and, evidently, are not reliable (Table 7). Nevertheless, all the authors note that the content of bismuth is somewhat higher in acid and moderately acid rock types in contrast to the basic rock.[3]

Table 8 shows the content of bismuth in several igneous rocks of S. Africa according to the findings of Brooks and Ahrens (1961). As seen in this table, the content of bismuth in acid and ultrabasic-alkali rocks is of one order of magnitude and its variation is insignificant. In acid and basic rocks the diversity of values exceeds one order of magnitude. The reasons for this are not clear. It can be postulated that this is related to the properties of the distribution of the intrinsic accessory minerals of bismuth.

3. See notes at end of article.

Table 7. The Content of Bismuth in the Main Types of Igneous Rocks (in g/t)

Acid	Average	Basic	Ultrabasic	Literary Source
2	--	--	--	Preuss (1941)
0.18	--	0.15	--	Brooks, Ahrens (1961)
0.01	0.01	0.007	0.001	Vinogradov (1962)

Indications of the presence of accessory bismuth minerals in granitoides are fairly numerous, but as a rule, a quantitative appraisal of these minerals is not given. The contamination of intrusive rocks by bismuth on a provincial basis is clearly apparent.

Bismuth minerals (bismuthine, native bismuth, and the products of their oxidation) have been discovered in the granitoides of Kalba in the Rudnij and Gornij Alta regions (Nikolskij 1952; Ljaxovic and Zolotarev et al. 1959), in the granites of the circumpolar Urals (Preobrazenskij 1940), in the biotite granites of the Megrinskij pluton in the Caucasus (Meliksetjan 1960)[4], in the variously aged granitoide complexes of Eastern Mongolia (Bobrov 1962), and in a series of granitoide massifs in the North-Eastern U.S.S.R. (Ipatjev 1960).

Accessory bismuthine is known in the aegirine-riebeckite granites in the North-East U.S.S.R. (Pecerskij 1961). However, accessory minerals of bismuth have not been discovered, for example, in the granitoides of Eastern Sayan (Ljaxovic, Nonesnikova et al. 1959).

A quantitative characterization of the accessory bismuth minerals in granitoides is found only in the works of V. V. Ljaxovic. Table 9 shows the statistical appraisals of the content of bismuth minerals in the granitoides. As seen from this table, the coefficient of variation (v) is exceptionally high.

Calculations based on the data of this table have shown that the content of bismuth due only to the accessory bismuth minerals, in the biotite granites can reach 0.02, in the leukocratic granites 0.06, and in the alaskite 0.8 g/t--i.e., in all cases it significantly

4. See notes at end of article.

Table 8. The Content of Bismuth in Different Igneous Rocks (Brooks, Ahrens, 1961)

Rocks, deposit	Bi, g/t	Rocks, deposit	Bi, g/t
ALKALI		BASIC	
Acid-alkali		Dolerite, Silly, Jagersfontain, S. Africa	0.08
Nepheline syenite, S. W. Africa	0.03	" "	0.15
" "	0.02	" "	0.80
Ultrabasic alkali		" "	0.15
Sovite, S. W. Africa	0.04	" "	0.28
Freueite, S. W. Africa, Parezis	0.04	" "	0.48
Acids		" "	0.17
Ryolite, S. W. Africa, Parezis	0.03	" "	0.76
Ryolite sodium, S. W. Africa, Parezis	0.02	" "	0.16
Ryolite riebeckite, S. W. Africa, Parezis	0.22	" "	0.02
Ryolite Dike, S. W. Africa, Parezis	0.21	" "	0.13
		" "	0.04
		" "	0.04
		" "	0.01
		" "	0.03
		" "	0.23
		Eclogite, S. Africa	0.02
		" "	0.10
		Eclogite	0.03
		Basalt	0.02

Table 9. Statistical Estimate of Accessory Bismuth Mineral Content in Granitoids (according to V. V. Ljaxovic, 1963).

Mineral	Biotite Granites (107)				Leukocratic Granites (16)				Alaskites (20)			
Bismuth	0.003	1	1035	0.003	0.05	6	402	0.27	--	-	--	--
Bismuthine	0.01	5	622	0.09	--	-	--	--	0.81	5	477	4.43
Bismuthite	0.01	7	855	0.12	0.01	13	314	0.03	0.21	20	377	1.0
Tetradymite	0.02	1	1054	0.03	--	-	--	--	--	-	--	--

exceeds the content of bismuth in acid rocks according to the findings of Vinogradov (1962).

If, with this, the inevitability of some loss of the accessories at separation is taken into account, then the conclusion can be drawn that the primary source of bismuth in the granitoides is found in the form of accessory minerals of bismuth. As Ljaxovic demonstrated in the example of the Urals (Table 10), the content of bismuth minerals in the granitoides increases in the later phases of the intrusive complexes. This researcher notes that bismuthine, bismuthite, and tetradymite have been discovered only in the biotite granites of the geosynclines and are not found in the biotite granites of the platforms.

There is practically no information about the contents of bismuth in basic rock-formers and in the several accessory minerals of the granitoides, including apatite, the tantalo-niobates, and monacite, which, as Goldschmidt postulated, must concentrate bismuth. Fragmentary information, based on data from spectra analyses, shows that bismuth is in fact often encountered in these minerals, but its content does not usually exceed 50-100 g/t. It was established partly by Xrjukin and Gavrilova (IMGRE) that the F-apatite of the interval facies of the porphyro-like biotite granites of South-East Tuva does not contain bismuth (~5b/t), while the F, OH-apatite from the boundary facies of these granites contains 50 g/t of Bi.

There is relatively no data in the literature on the regularity of the distrubition of bismuth and the forms of its location in alkali and basic rocks. From an examination of the extremely limited information given above on the behavior of bismuth in the magmatic process, several conclusions can be drawn:

1. A relative enrichment in bismuth of the products of the final stages of magma crystallization is observed--i.e., an enrichment of bismuth takes place in detrital (or late) differentiates. To some extent, this may explain the fact that bismuth gives commercially important concentrations only in connection with the granitoides.

2. In the general picture of bismuth present in the granitoides, the role of its (i.e., bismuth) assumed mineral-concentrators (monacite, apatite, and others) is evidently insignificant. This is evident in the low degree of concentration of bismuth in them, and also with the accessory character of these minerals. An overwhelming

Table 10. The Average Content of the Accessory Bismuth Minerals in Granitoids of Various Phases of the Hercynian Multiphase Intrusives in the Urals. (In g/t). (Acc. to Ljaxovic, 1966).

Mineral	Phase I quartz-diorites, granodorites (16)*	Phase II granites biotite *32(	Phase III granites leucocratic and alaskite (17)	Average content in the Hercynian granitoids (65)
Bismuth	Undisc.	--	0.003	0.001
Bismuthine	"	0.003	--	0.0002
Bismuthite	"	0.06	0.38	0.13
Bayerite	"	0.0006	--	0.0003

* The number of samples from which the mean is deduced.

amount of bismuth is concentrated in the intrinsic minerals (native bismuth, bismuthine, tetradymite), which are accessories. In considering the basic geochemical factor facilitating the concentration of bismuth in the intrinsic minerals, it is necessary to acknowledge the absence of its dispersion in the common rock-forming minerals.

It is necessary to consider the behavior of bismuth in the magmatic liquefaction of sulphide ores.[5] In accordance with the data of Rankama and Sahama (1950), an analysis of 35 pyrrhotites, 10 pyrites, 8 pentlandites and 4 nickelous pyrites demonstrated the presence in them of 2 g/t of Bi on the average.

Even with such relatively low contents of Bi in ores the intrinsic minerals of bismuth can be established. Thus, in the boundary regions of the Sudbury rocks native bismuth, bismuthine, tetradymite and parkerite ($Ni_3Bi_2S_2$) are found in association with galena, nickeline and cubanite. Parkerite is found also in the Insizv bed (S. Africa) in association with pyrrhotite, pentlandite, chalcopyrite, cubanite and nickeline.

In the surface magnetite veins of the Moncegorskij deposit, the telluro-bismuthide palladiums and platinums--moncheite (Pt, Pd) $(Te, Bi)_2$, kotulskite $Pd(Te, Bi)_{1-2}$, and michenerite $(Pd_{0.75}Pt_{0.25})$ BiTe were discovered by Genkin et al. (1963). As a result of mineragraphic research by the indicated authors, it was established that the telluro-bismuthides are coordinated with regions of replacement of the pyrrhotite-pentlandite-chalco-pyritic veins by magnetite. Chalcopyrite was redeposited with this forming small veins in the magnetite. Related to these veins are moncheite, kotulskite and michenerite. In addition, in later magnetite associations, violarite, cubanite, and bornite are observed. In this manner, the development of these bismuth minerals is related to the effect on the early sulphides of the detrital solutions which are enriched in oxygen. Michenerite (froodite) has been discovered also in the ores of the Froude deposit (Sudbury, Canada).

5. See notes at end of article.

E. F. Mintser

BISMUTH IN THE PEGMATITIC PROCESS

Although bismuth minerals are commonly noted in pegmatites, especially the pegmatites connected with granites, they are never of any commercial interest. This evidently explains the insufficient attention given to the properties of the behavior of bismuth in pegmatitic processes, which, as will be shown, are sufficiently distinctive and certainly warrant special examination. According to the findings of Fersman (1939), bismuth is fixed mainly in the geo-phases D-F (i.e., in the transitions from the fluid-gaseous to the fluid-hydrothermal phase according to Niggli.

Sufficiently large, although narrowly localized concentrations, provide a basis for discussing the distinct enrichment by bismuth in detrital solutions. An analyses of the geologico-geochemical properties of the formation of bismuth mineralization shows that it is formed in the final phases of the pegmatitic process (i.e., the fluid-hydrothermal, the pneumo-hydrothermal, and the hydrothermal phases). Basically, in pegmatites, the same bismuth minerals are observed which are present as accessories in the corresponding mother rocks; namely, native bismuth, bismuthine, and more rarely tetradymite. Consequently, the factors controlling the formation of the intrinsic minerals of bismuth in the process of rock crystallization have probably continued to have a decisive control in the pegmatitic process as well--bismuth has not dispersed in the rock-forming minerals but has accumulated in the detrital solution mixtures.

However, in 1937 Bjorlikke (1937), established the presence of bismuth in pyrochlores from the Archean pegmatites of Setesdalen (S. Norway). A subsequent study (Rankama and Sahama 1950) showed the connection of bismuth with the complex oxides of Ta, Nb, and Ti. In this manner, bismuth is present in the pegmatites in the form of a natural metal of sulphide (chalcogenide). It also enters into the formation of oxygen compounds, which is not very characteristic of the hypogenic mineral-formation of bismuth.

The granite pegmatites, as already indicated, commonly show bismuth mineralization. Bismuth mineralization in pegmatites has been repeatedly noted in the literature, even in those cases where, in massif rocks to which they are related, bismuth has not been established (Fersman 1940; Ljaxovic and Nonesnikova et al. 1959; Ljaxovic and Zolotarev et al. 1959; Ljaxovic and Nonesnikova 1961).

Large deposits (to a width of several tens of centimeters) of bismuthine and native bismuth (sometimes in association with tetradymite) are related, as a rule, to late sulphide parageneses--pyrrhotite, chalcopyrite, arsenopyrite, and sphalerite. The presence of some quantity of bismuth is noted in galena (e.g., in the pegmatites of Fabulosa, Bolivia; Hindholmen, Tisfjord, Norway; Norr, Sweden; the pegmatites of the Urals, where native bismuth with inclusions of tetradymite (owing to which the content of tellurium in bismuth occasionally reaches 2%) is coordinated to plagioclase nuclei, closely associated with a green-colored plagioclase. In the case where the nuclei are composed of fluorite and muscovite in association with beryl, bismuth is present in the form of individual crystals enclosed in the cavities of beryl. Despite some peculiarity of the associations and forms of separation of the bismuth minerals examined, their formation is controlled by the action of the hydrothermal process and is not specific for pegmatites.

A problem of particular geochemical interest is presented by the bismuth mineralization in granite pegmatites. This problem is related to the action of alkali auto-metasomatism (albitisation of the mineralization process). Separate grains or irregular pocket-shaped deposits of bismuthotantalite are often encounted in the substituting complex, together with lepidolite, lithium muscovite, topaz, beryl, microlites, and tourmaline (e.g., the pegmatites of Brazil, the Mongol Altai, Africa, and China). They are also found in a series of bismuth-containing tantaloniobates (Table 11). Spectral analysis of the pegmatite minerals has shown that monacite, xenotime, pyrochlore, euxenite, and apatite are characterized by a constant concentration of bismuth (Table 12).

As mentioned above, Goldschmidt (1954) indicated the presence of bismuth in apatite and rare-earth minerals, the bismuth replacing the sterically close ions of Ca or Sr. However, there is still no information about the high bismuth contents of the minerals containing these elements. As is evident from the data given, this supposition of Goldschmidt's has some validity. Of great scientific interest, this question demands special examination.

Thus, in the granite pegmatites, we have two independent paragenetic series of bismuth minerals:

Table 11. Bismuth Content in the Tantalo-niobates of the Granite Pegmatites.

Mineral	Formula	Content	Deposit	Literary Source
Uranmicrolite (Djalmaite)	$(Ca,U)_{2-x}Ta_2O_{6-x}(OH,F)_{1+x}$	8800	Brazil	Dena et al. (1951)
Stibiotantalite	$Sb(Ta,Nb)O_4$	2700-7300	Varutersk, Sweden	Hintze (1933) Dena et al. (1951)
Stibiocolumbite	$Sb(Nb,Ta)O_4$	4700	W. Africa	Penfield and Ford (1906)
Stibiobismuth-otantalite	$(Sb,Bi)(Ta,Nb)O_4$	35700	W. Africa	Bandy (1951)
Bismuthom-icrolite	$(Na,Ca,Bi)_2Ta_2O_6(F,OH,O)$	30500	Mongolian Altai	Zalaskova and Kuxarcik (1957)
Betaphite (blomstran-dite)	$(U,Ca)_{2-x}(Ta,Nb,Ti)_2\cdot\cdot O_{6-x}IOH)_{1+x}$	3600	Madagascar	Lacroix (1912)

Table 12. The Content of Bismuth in some Minerals of the Granite Pegmatites. (Acc. to the data of spectral analyses).

Mineral	Content of Bi, g/t	Deposit	Literary Source
Monazite Xenotime Pyrochlore Euxenite Orthite	10-50 10 50 10 10	Kolskij p-ov	Data of A. P. Karlita, IMGRE
Xenotime Pyrochlore Euxenite Orthite	50 10-50 10-50 10	Priladozie	Data of A. P. Karlita, IMGRE
Apatite	50	N. Karelia	Data of A. P. Karlita, IMGRE
Apatite Microlite	5 10-50	Primorie	Data of G. B. Melentiev, IMGRE
Columbite Columbite-tantalite	25 50	The emerald mines, Urals, the pegmatites of "the pure line"	Vlasov, Kutukova, (1960)
Molybdenite Sphalerite Apatite Molybdenite Chrysoberyl Tin-tantalite Thalenite Gadolinite	50 25 25 200-350 500 100 1-20	The desilicised pegmatites	Matias (1961)[1,2] J. a. W. Noddack (1931)
Allemontite	200-2400	Varutresk, Sweden	Dana et al. (1951)
Galena	30,000	Norway	Goldschmidt (1954)
Chevkinite Igtrobritholite Fergusonite Euxenite Hatchettolite	50 5-100 10 10 10	The pegmatites of the alkali granites of W. Kejva	Data of A. P. Kalita, IMGRE

1. Native bismuth and its chalcogenides associated with the sulphides of Mo, Fe, Zn, Pb.

2. Bismuth tantalite. The bismuth-containing tantalo-niobates, the rare-earth, and calcium minerals in the replaced pegmatites are predominantly of an albite composition.

Bismuth minerals are noted also in pegmatitic migmatites where native bismuth and bismuthine are observed in the central parts of the pegmatite (Jurk 1939).

Intrinsic minerals of bismuth have not been discovered in the pegmatites connected with alkali rocks. Galena, however, is characterized by a fairly high content of this element. Bismuth is noted in amounts of 500-10,000 g/t in galena from the pegmatites of the Lovezerskij alkali massif (Vlasov et al. 1959). Galena is related here with the areas showing evidence of hydrothermal processes and in association with fluorite, sphalerite, molybdenite, and sodalite. It is also localized in natro-lithium zones. According to the findings of Goldschmidt (1954), the average content of bismuth in galenas from pegmatites in nepheline-syenites is 600 g/t. Although no data exists on the bismuth content in the niobotantalates and the rare-earth minerals from the pegmatites from the nepheline-syenite system, it can be postulated that in the niobo-tantalates the content of bismuth is lower than in the minerals of the granite pegmatites which are in turn markedly richer in tantalum. There is no information in the literature on the behavior of bismuth in the pegmatites of basic and ultrabasic rocks.

BISMUTH IN THE POST-MAGMATIC PROCESSES

Information on the behavior of bismuth in high-temperature metasomatic processes is exceptionally meager. The complete absence of accessory sulphides in granitoides which have undergone early sodium metasomatism (Liakovic and Nonesnikova 1960) and the increased content in them of tantalo-niobates give reason to postulate that bismuth is present in the latter in amounts which are related to its contents in the analogous minerals of the pegmatites. The greisenisation, following in time the process of albitisation (i.e., acid metasomatism), is characterized by a distinct increase in the bismuth content.

The presence of the accessory bismuth minerals (native bismuth and bismuthine) in the greisenised varieties and their absence in the original granites has been noted by Koptev-Dvornikov et al. (1960) for a series of massifs in Central Kazakstan, by Leontiev, Boiko (1959) and by Liakovic and Nonesnikova (1960) for the Altai. These same minerals are also known in the greisen portions of high-temperature veins where their maximum occurrence is correlated to the quartz-fluorite (Nakovnik 1954) and the quartz-siderophyllite (Levitskii 1964) facies.

Bismuth mineralization is noted in all types of scarnite deposits and in a series of deposits connected with secondary quartzites. Here, it is wholly connected with the evidence of the later hydrothermal phases of mineral formation.

In the albitized nepheline-syenites, the alkali syenites, and the alkali granites no bismuth mineralization has been discovered. Bismuth is present, however, in a series of late metasomatic minerals of quartz-microcline composition along with aegerine and riebeckite in alkali syenite massifs of Burpal (the Pribaikal). In these rocks gagarinite and bapherticite contain up to 10 g/t Pb pyrochlore u to 50, and galena up to 11,500 g/t of Bi (according to the finds of A. F. Efimov and A. A. Ganzeev, IMGRE).

In the capacity of an accessory mineral of the carbonatization phase of the post-magmatic stage, galena rich in bismuth (containing up to 27,000 g/t of Bi) has been reported (Zabin et al. 1960) in the exo-contact phenitized rocks of the nepheline syenites of the Visnevy Mountains. Galena occurs with quartz, pyrite, calcite, and sphalerite. In carbonatites connected with the ultrabasics--alkali intrusive complexes--bismuth is reported only rarely, at the threshold sensitivity of spectral analysis (~5 g/t).

A particular complexity distinguishes the behavior of bismuth in the hydrothermal process. Despite the fact that it is with this process that the larger concentrations of commercial interest are connected, the diversity of geological and geochemical conditions of mineral formation is so great that it is not possible to give a generalized characterization of the behavior of bismuth. To a significant degree, this is also inhibited by insufficient knowledge of the geochemistry of bismuth in the metalliferous process.

In high temperature deposits of skarnite formation bismuth is common,[6] although its concentration, as a rule, is not great. Of greatest interest are skarno-polymetallic ores where bismuth either enters into galena isomorphously, or forms a particular mineralization which combines intimately with galena, sphalerite and chalcopyrite. Galenobismuthite (Ketcikov 1958), aikinite, wittichenite, more rarely emplectite, native bismuth, cosalite, and very rarely tetradymite and tellurobismuthite (Yanulova 1962) have been noted in association with these sulphides.

In skarno-magnetite ores bismuth is rarely noted and, if present, is in low concentration (Karasik 1959; Kakkarov 1959). In the pyroxene-granite skarns of Western Moravia, bismuth mineralization indicated native bismuth and bismuthine is present in the magnetite scarns along with arsenopyrite. It is isolated from the poly-metallic ores (Nemets 1964). In the Cokadam-Bulak deposit (Central Asia), bismuthine and cosalite are found in close association with pyrite, chalcopyrite and magnetite (Naumova and Sidorenko 1965). The skarn cupric-magnetite deposits, in which bismuth mineralization is on occasion extensively developed (Bocegiano, Italy; Kazakstan, U.S.S.R.) warrant special attention.

Bismuth is also constantly noted in the ores of skarn-scheelite deposits. In the largest skarn-scheelite deposit Sangdong (southern Korea) the main occurrence of bismuth (in the form of bismuthine) is associated with the skarn scheelite-pyrrhotite ores. The latter containing molybdenite and chalcopyrite, (Klepper, 1947). The association of bismuth minerals (bismuthine and native bismuth) with pyrrhotite and chalcopyrite is extremely characteristic of a series of skarno-scheelite deposits of Central Asia (Primore and Jakutja). However, in several uniform deposits (the Caucasus, Central Asia), a very close association of these bismuth minerals with sphalerite is observed. As a result, the bismuth content in the latter exceeds its content in pyrrhotite and chalcopyrite by more than two orders of magnitude.

Bismuth mineralization is widespread in the distinctive pyritic

6. See notes at end of article.

metasomatic high-temperature deposits of shallow depth which are deposited in the basement rocks (the Scandinavian type). In them bismuthine, cosalite and galenobismuthite combine with arsenopyrite, pyrite and chalcopyrite, (the deposits of Boliden, Falun, Bielke et al., Sweden). Occassionally, in later parageneses tetradymite, tellurobismuthite and stibiotelluro-bismuthite are present in association with chalcopyrite and gold tellurides.

In high-temperature greisen deposits of tungsten, tin, and molybdenum, bismuth mineralization is almost universally noted[7] (e.g., Cornwall, England; Southern China; Queensland, Australia; Zabaikale and Central Kazakstan, U.S.S.R., etc.). High concentrations of bismuth are observed, however, only in deposits with an intensely developed sulphide phase.

For example, in the metalliferous veins of the Serlovogorsky deposit, bismuth is associated with arsenopyrite and berberite and in the Bukuka deposit--with sphalerite and chalcopyrite (Levitskii, 1964). In the ores of the tin tungsten deposits of Central Kazakstan, the nature of the parageneses is somewhat different. The large separations of bismuthine and native bismuth are correlated with the prizalband parts of the quartz-cassiterite-wolframite veins in the nodule cavities of which are observed bonchevite, galeno-bismuthite and cosalite (combining with pyrite, chalcopyrite, gilbertite, and fluorite).

As a result of the study of the Kara-Oba deposit by Popova et al. (1966), the regular complexity of the composition of bismuth minerals was established in time within each phase of the mineral-formation process as well as the juxtaposition of separate phases. These regularities are, evidently, general for deposits of the greisen type, since an analogous complexity of the composition of bismuth minerals in the ore-formation process of the Bukukinskii deposit has been noted by Barabanov (1961), and Ontoev (1964). Ontoev, in part, observed the shift from the early bismuthine-galenobismuthite paragenesis to the late galena-lillianite-cosalites right up to the formation of Bi- and Ag-containing galena.

7. See notes at end of article.

Bismuth minerals are often present in ores and cassiterite-sulphide deposits (Ivanov 1964). In early parageneses, native bismuth and bismuthine associate with arsenopyrite (due to which the content of Bi in it sometimes exceeds 0.5%), pyrrhotite, sphalerite, and chalcopyrite. Bismuthine, native bismuth, and emplectite are encountered in the later carbonate-sulphide paragenesis, where the ferrous carbonates (siderite, magnanosiderite) combine with late generations of sphalerite, stannite, galena, pyrite, chalcopyrite, and with the minerals of silver and the copper ores.[8]

Bismuth mineralization is characteristic also of the arsenopyrite high-mean-temperature deposits and occasionally reaches concentrations which are of commercial interest. Bismuthine, native bismuth, more rarely the complex sulphides of lead and bismuth, the bismuth-containing antimony minerals (kobellite, bournonite, boulangerite and jamesonite) are the main minerals of bismuth which commonly occur with arsenopyrite and pyrrhotite. As a rule, tetradymite is present. The mineral composition of ores and the nature of the parageneses is fairly uniform for deposits from various areas (Central Asia, The Caucasus, Eastern Zabaikale). Their mineralogico-geochemical properties are determined by the uniqueness of the genesis of high-mean-temperature formations with the localized manifestation of the processes of pneumatolysis which are developed under conditions of shallow and moderate depths. A high-temperature silicate metasomatism (quartzification, tremolitazation, tourmalinization et al.), the presence in the ores of the typical high-temperature minerals (tourmaline, cassiterite, tungsten, scheelite) together with galena, the sulpho-salts of antimony and sometimes with the sulphides of arsenic and mercury, are all characteristics of these deposits (Saxarova 1955; Dunin-Barkovskaja 1966; Malakov and Navirova 1966).

In the processes of ore-formation, there is constantly observed a paragenesis of bismuth minerals with gold; however, bismuth mineralization is widely developed particularly in the mean-temperature deposits (the deposits of Canada, Australia, the Urals, and Eastern Zabaikale). With this, the true gold-bismuth association is seen in

8. See notes at end of article.

the terminal phases of ore-formation. Cosalite and galenobismuthite (the deposit of Caribou, Canada; Warren 1936), aikinite (the Berezovskoe deposit, the Urals), and bismuthine (the deposits of Eastern Zabaikale and Kuznetskii Alatav) have all been noted in association with gold.

It is especially appropriate to discuss here the nature of bismuth's behavior in the ores of polymetallic deposits.

There exists a vast, specialized literature directed toward the great practical importance of bismuth in polymetallic ores. Galena is most often the basic concentrator of bismuth in lead-zinc deposits. The content of bismuth in other ore-forming sulphides, as a rule, is one to two orders of magnitude lower than its content in galena. At the present time it is considered to be firmly established that in most cases, the presence of bismuth in galena is explained by heterovalent isomorphism according to $Bi^{3+} + Ag^{+} \rightarrow 2\ Pb^{2+}$, and more precisely, by the solution of -$AgBiS_2$ (schapbachite) in galena at a temperature higher than 300°C. In the work of Godovikov (1966), where statistical mathematical conformation of this approach is given, it is also shown that for several galenas, the content of bismuth clearly exceeds that of silver, and does not correspond to the formula $AgBiS_2$. Godovikov (1966) considers this a consequence of the initial formation of solid solutions of PbS-Bi_2S_3, which disintegrate when the temperature is lowered. Actually, inclusions of complex sulphides of lead and bismuth are occasionally present in some galena; but more often, the inclusions are complex sulphides containing copper (aikinite, wittichenite). Recent work clarifies the presence of bismuth in galena and the presence of other impurities: Se, Te (Oftedal 1959; Neceljustov et al. 1962) and thallium (Malevskii 1966). Oftedal (1940), studying galenas of high-temperature sulphide deposits, showed that the amount of bismuth in them increases with an increase in temperature. However, within a group of deposits and even within one deposit, large variations are observed.

L. N. Ketcikov (1958) confirmed these regularities with one of the lead-zinc skarn-ore deposits. He explained the high bismuth concentrations in galenas from the skarn deposits (Janulova and Potok 1956; Neceljustov et al. 1961). In addition the bismuth content in galena varies considerably in different ore provinces. It is now clear how difficult it is to carry out an appraisal and comparative analysis of these findings. What has been said is illustrated to some degree in Table 13 in which data are presented on the content of bismuth in the galena from ore occurrences and deposits of various kinds.

Among the mid-temperature deposits (very unique and important in relation to bismuth) are those of the so-called five-element formations (Co--Ni--Bi--Ag--U), well known in the Rudnii mountains of Czechoslovakia, Canada, the U.S.A., and several other areas of the world. Native bismuth in close association with the arsenides of cobalt and nickel is practically the only mineral of bismuth in the ores of these deposits. Nickelite, rammelsbergite, scutterudite, and sphalerite occur in the siderite veins of the Rudnii mountains deposit (Dymkov 1960), and cobaltite, lollingite, scutterudite, dolomite, chalcopyrite, and hematite in the Eldorado deposit in Canada (Haycock 1935). Callilite [Ni(Sb, Bi) S], is found only in the siderite veins of Schonstein (Westphalia, F.D.R.) in association with gersdorfite and nickelite. According to the finds of Mrnja (1963), the solutions that formed the arsenide paragenesis in the Joachimstal deposit (Rudnii mountains) were neutral or weak in alkaline conditions--determined on basis of the pH of hydrous suspensions. The constant presence of carbonates indicates this.

In ores of the cupra-molybdenum deposits, bismuth is also quite common; however, its concentration is low. The minerals of bismuth in these ores, as a rule, are determined only by microscopic research and are represented by bismuthine, emplectite, and wittichenite (Gorokova 1960; Mamedov and Efendiev 1963). As shown by Gorokova, there is almost always a Bi content of 5-50 g/t in the molybdenites of the molybdenite-chalcopyrite phases.

The large and diverse group of pyritic deposits has been poorly studied in relation to bismuth. This is, evidently, explained by the fact that, present in these or other ore-forming sulphides, bismuth in most cases does not form segregations where its concentration is greatly increased. This hinders the study of the types of location as well as the regularities of its distribution. Bismuth has been qualitatively identified by Hawley and Nicol (1961) in the pyrites, pyrrhotites, and chalcopyrites of all the hydrothermal

Table 13. The Content of Bismuth in Galena.

Average content of Bi, g/t	Type of Deposit	Literary Source
To 30,000	Granite pegmatites	Goldschmidt (1954)
600	Pegmatites of the nepheline syenites	Goldschmidt (1954)
To 10,000	Lovozerskij alkali massif, pegmatites	Vlasov, Kuzmenko (1959)
To 11,500	Late metasomatisms of the alkali syenites	Acc. to materials of A. F. Efimov and A. A. Genzeev, IMGRE
To 27,000	Hydrothermalites of the nepheline syenites	Zabin et al. (1960)
To 76,100	Greisen, Bukuka	Ontoev et al. (1960)
To 8,300	Greisen, E. Conrad	Cuxroev (1960)
18,000-44,300	Greisen, Kara-Oba	Popova, Neceljustov et al. (1966)
8,000-12,000	Sulphides, mid-temperature	Hoehne (1934-1935)
160-32,500	Scarn and Polymetallic	Xetcikov (1958)
9,900-34,500	Scarn-lead-zinc-copper Akcagyl	Nesterova (1958)
3,900-16,800	Scarn-lead-zinc-copper Kyzyl-Espe	Nesterova (1958)
To 1,000	Sulphide-Cassiterite, Jakutija	Ivanov (1964)
<1-1,000	Mean low-temperature hydrothermal	Goldschmidt (1937)
To 114,700	Polymetallic, mid-temperature, Leadville, U.S.A.	Chapman, Stevens (1933)

cupric and pyritic-auric-ore deposits studied. Bismuth content of the order of tens of grams per ton is characteristic of the basic ore-forming sulphides of the cupropyritic and the pyritic-polymetallic deposits. With the exception of galena, chalcopyrite and pyrrhotite are characterized by a higher content of bismuth (>100 g/t). This is explained by the commonly observed association of the latter with native bismuth, bismuthine and wittechenite (?). Sometimes galeno-bismuthite and bismuthine also occur in regions of polymetallic mineralization (Gorzevskii 1955) associated with the bismuth-containing copper ores.

True bismuth mineralization is frequently present in ores of cupro-zinc deposits, where its distribution is not dependent upon the abundance of galena (the Ikumo and Tadzima deposits, Japan).

Bismuth is not characteristic of low-temperature, especially tele-thermal deposits. Under these conditions, the association of bismuth with tellurium, antimony, and gold is very clearly seen. Examples of low-temperature bismuth mineralization are the occurrence of telluro-bismuth mineralization in the Zakarpate (Maleev 1963) where the veins with cinnabar, arsenopyrite, pyrrhotite and galena contain wehrlite and native bismuth, and the gold-bismuth-tellurium mineralization of the Zodskii deposits.

In the quartz-carbonate veins of this deposit tetradymite, bismuthine, cuprabismuthite and zodite ($Bi_{10}Sb\ Te_{17}$) are encountered with pyrite, chalcopyrite, gold, antimonite, and cinnabar (Magakian 1957; Tvalcrelidze 1959).

In conclusion, it is imperative to discuss the nature of several hydrothermal bismuth-containing deposits. They are of special interest because of the unusually high degree of concentration of bismuth with significant degrees of mineralization. We have in mind here the tin-silver-bismuth deposits of Central Bolivia. Due to the specificity of the genesis of these deposits (clearly connected with the processes of vulcanism) their relation (by temperature) to some formation group is quite intricate. The rapid change in all the physico-chemical parameters of the solutions of ore minerals, which are usually divided in time and space, is known to occur.

Uncia-Lialiagua and Potosi are the best known and largest of the deposits of this type. In the Uncia-Lialiagua deposit, early quartz-bismuthine-cassiterite and wolframite are related to extrusive quartz-porphyries of the basic phase (Terner 1964). In the Potosi deposit it is connected with the extrusive etmolite of the quartz-monozite-porphyry. To this association arsenopyrite is added. It is interesting that close to these deposits, a little farther south, the Tansa deposit occurs. The high-temperature mineralization of the Tansa deposit was formed at relatively shallow depths in connection with a hypothetical unconcealed granitoid intrusion.

The nature of the bismuth association here is different. Bismuthine is found in the paragenesis with chalcopyrite which contains sphalerite. Jamesonite and copper ore are encountered in the ores of wolframite, phyrrhotite, arsenopyrite, and stannine.

The total assembly of the geologico-structural and mineralogico-geochemical properties of these deposits is proof of their formation from relatively high-temperature solutions at surface conditions (or at shallow depths). These conditions are, evidently, optimal, to the extent that, in the overwhelming majority of cases in those regions of evident bismuth mineralization (and interestingly, at a maximum high level of bismuth concentration), investigators are developing facts indicative of the shallow depths and comparatively high temperatures of the formation of these deposits (Ashfeld 1933; Warren 1936; Terner 1964; Janulova 1962; Levitskii 1964; and others). The genetic properties of the cupra-bismuth deposits of Eastern Karamazar also indicate the decisive role played by these two factors in affecting the bismuth concentration.

As emphasized above, the tendency of bismuth to form particular mineral phases is characteristic in the hydrothermal process. However, as an impurity, bismuth is reported in a whole series of ore minerals (Table 14). There is little data on the form or the location of bismuth in these ore minerals. It may be postulated that in many instances, the presence of bismuth in them is controlled by micro-inclusions of particular minerals. It is interesting that a range of minerals in which bismuth occurs at higher concentrations is basically limited by the complex sulphides of antimony, the arsenides and the tellurides. This once again emphasizes the geochemical relationship of these elements at various stages of the hydrothermal process.

A FEW REGULARITIES OF THE BEHAVIOR OR BISMUTH IN THE POST-MAGMATIC PROCESS

Having examined the basic paragenetic association of bismuth in the main types of hypogenic bismuth-containing deposits, it is now possible to attempt to clarify the role of the physico-chemical factors controlling the concentration of this element and in determining the forms of its occurrence.

The capacity of bismuth to form complex compounds, essentially with the covalent component of the bond (i.e., the very stable), affects its basic geochemical nature (i.e., its facility to form specific mineral phases in the most diverse processes of endogenic

Table 14. The Content of Bismuth in Various Minerals.*

Mineral	Formula	Average Content of Bi, g/t
Natural gold	Au	1,300-29,200
Natural arsenic	As	To 4,000
Antimonite	Sb_2S_3	80
Allemontite	AsSb	200-2,400
Ullmanite	NiSbS	6,800-117,600
Lollingite	$FeAs_2$	50-7,000
Arsenopyrite	FeAsS	100-41,300
Glaucodot	(Co,Fe)AsS	To 1,000
Scutterudite	$(Co,Ni)As_3$	600-20,500
Ni-Scutterudite	$(Ni,Co)As_3$	To 1,600
Smaltine (Bismutho-smaltine)	$(Co,Ni)As_{3-x}$	201,700
Maucherite	$Ni_{11}As_8$	To 5,500
Nickeline	NiAs	1,000-5,400
Sapphorite	$(Co,Fe)As_2$	200-8,000
Rammelsbergite	NiAs	To 2,100
Corynite	Ni(As,Sb)A	To 4,000
Luzonite	Cu_3AsS_4	To 15,000
Famatinite	Cu_3SbS_4	1,000-17,900
Jamesonite	$Pb_4FeSb_6S_{14}$	To 45,900
Polybazite	$(Ag,Cu)_{16}Sb_2S_{11}$	To 8,000
Tetrahedrite	$(Cu,Fe)_{12}Sb_4S_{13}$	3,400-45,500
Tennantite	$(Cu,Fe)_{12}AS_4S_{13}$	To 130,700
Ramdohrite	$Pb_3Ag_2Sb_6S_{13}$	To 25,000
Boulangerite	$Pb_5Sb_4S_{11}$	20-300
Hessite	Ag_2Te	To 3,000
Melonite	$NiTe_2$	To 400
Montbreitite	Au_3Te_3	To 28,100
Thalenite	$Y_2Si_2O_7$	1-20
Gadolinite	$Y_2FeBe_2[O(SiO_4)]_2$	

*The reference manuals and findings of the members of the IMGRE staff, V. V. Ivanov and T. N. Cvileva were utilised in the composing of the table.

mineral-formation). Localized, exceptionally high concentrations of bismuth, which exceed its Clark by millions of times, are also controlled by this. It is quite difficult to indicate specifically the nature of bismuth transfer in solutions. From the work of Craner et al. (1956) and Olin (1959), showing that in hydrous mixtures, bismuth is found in the form of multinuclear complexes with a degree of polymerization of 5--6, it is possible to postulate the existence of similar macromolecules in nature also. It is all the more probable that, as shown above, there exist separated polymeric chains, bands, and layers in the structures of almost all bismuth minerals.

In the course of evolution of post-magmatic solutions and the alteration of their physico-chemical parameters, the sites of bismuth location in them must also vary. As a result of this, the nature of the bismuth mineralization also varies.

As a result of the distinctly expressed chalcophilicity of bismuth in the hypogenic process, its oxygen compounds do not play a major role. At the time of early alkali metasomatism (albitization), bismuth forms complex oxides (bismuth-tantalite, bismuth-microlite, and the bismuth-containing minerals of the group of pyrochlore, etc.). According to the electron shell structure of bismuth atoms, the element must sharply manifest chalcophile properties in geochemical processes. This characteristic of bismuth was emphasized by Vernadskii (1916), Fersman (1939), Rankama and Sahama (1950), and Goldschmidt (1954). In fact, the overwhelming majority of the minerals of bismuth are chalcogenides. However, in the hydrothermal process, along with the chalcophile properties of bismuth, its siderophile tendencies are also markedly evident. The siderophilic nature of arsenic and to a lesser extent of antimony, is well known, and up to the present, the siderophilicity of bismuth has not received proper study, although the coordination of bismuth mineralization with pyrrhotite, arsenopyrite, cobalt-nickel, chalcopyrite, magnetite, and other high-iron ores in nature is clearly known. It has already been indicated that this relation is not only geochemical, but is also partially crystallo-chemical. As is evident from Table 15, arsenic forms a significant number of minerals with each of the metals of the iron group (Fe, Co, Ni), one mineral with palladium (arsenopalladinite), and one with platinum (sperrylite). Antimony has a smaller

Table 15. The quantity of Minerals of As, Sb and Bi with the elements of the VIIth group of the periodic table.

	Quantity of common minerals				
Elements	Fe	Co	Ni	Pd	Pt
As	4	5	6	1	1
As--Sb	-	-	2	-	-
Sb	3	1	2	1	-
Sb--Bi	2	-	3	-	-
Bi	-	-	1	4	1

number of common minerals, most of which are compounds with Fe and Ni. Bismuth forms minerals with iron only in the presence of antimony. With this condition, it also forms a series of minerals with nickel and also one non-antimonous sulphide of nickel and bismuth (parkerite--$Ni_3Bi_2S_2$). Minerals of bismuth with cobalt are completely unknown. However, the number of natural compounds that bismuth forms with the platinoides is sharply increased.

Thus, a crystallo-chemical bond with the latter members of the triad of group VIII of the periodic table--Ni, Pd, Pt, is characteristic of bismuth. The information given above on the paragenetic associations of bismuth indicate also its geochemical bond with the elements of the iron group.

It is interesting to examine how the parageneses of bismuth minerals vary in the course of the evolution of post-magmatic mixtures in accordance with the model of this process postulated by D. S. Korzinskii.

In conditions of increasing acidity ("the early alkali phase"), with sodium (sometimes lithium) metasomatism, there is a clearly indicated tendency of bismuth to enter into oxygen compounds--the complex oxides of Ta, Nb, Ti. An essential concentration of bismuth with this tendency is not observed.

Deposits or ore-manifestations of bismuth connected with the early stages of the "acid phase," are not established, but bismuth mineralization is extremely characteristic of greisen deposits that have been formed from a mixture of increased acidity. However, the coordination of bismuthine and native bismuth to the quartz-sericite, the quartz-siderophilite, and the quartz-fluroite facies of the griesens indicates the formation of bismuth minerals from the process of a rapid lowering of the acidity of the mixtures.

If, in the early stages, the paragenesis of these two minerals is observed only at a different degree of oxidation from that in which they are found, then the increasing alkalinity of the mixtures would quickly displace the equilibrium in favor of Bi^{3+}. Later bismuth mineralization is shown then only by the sulphides of bismuth.

Figure 4 shows the variation in time of the composition of the minerals of bismuth formed with ore-precipitation in the greisen tin-tungstic deposit of Kara-Oba. From Figure 4, the consequent complication of the composition of the bismuth minerals (the growth of the ratio $PbS:Bi_2S_3$) for each of the following phases, as well as within them is evident. Native bismuth, not shown in the figure, associates only with the bismuthine of the first two phases.

As is known, with the formation of the sulphide mineralization of hydrothermal deposits connected with the late phase of the acidic stage of the evolution of the post-magmatic mixtures, the alkalinity increases with time (i.e., with falling temperature). The association of bismuth with iron minerals (which are good indicators of the chemistry of the environment) makes it possible to trace the influence of oxidizing potential and the pH of the mixtures on the nature of the bismuth minerals formed.

On the basis of thermodynamic calculations and experimental data, it was shown that the presence of bismuthine in mineral associations indicates the weak-acid-neutral character of the mixtures, and the presence of native bismuth, the neutral-alkali nature. This is well illustrated by the predominance of these minerals in ores of relatively high-temperature deposits which have been formed under conditions of increased acidity of the mixture. An increase in alkalinity leads to complications in the composition of bismuth minerals within the separate phases, from phase to phase, and entirely in the hydrothermal process. In general, this is expressed in the alteration of the nature of the ore mineralization which has been formed in separate stages of the process, and perhaps is shown by the following series: bismuthine → native bismuth → Pb-Bi complex sulphides →

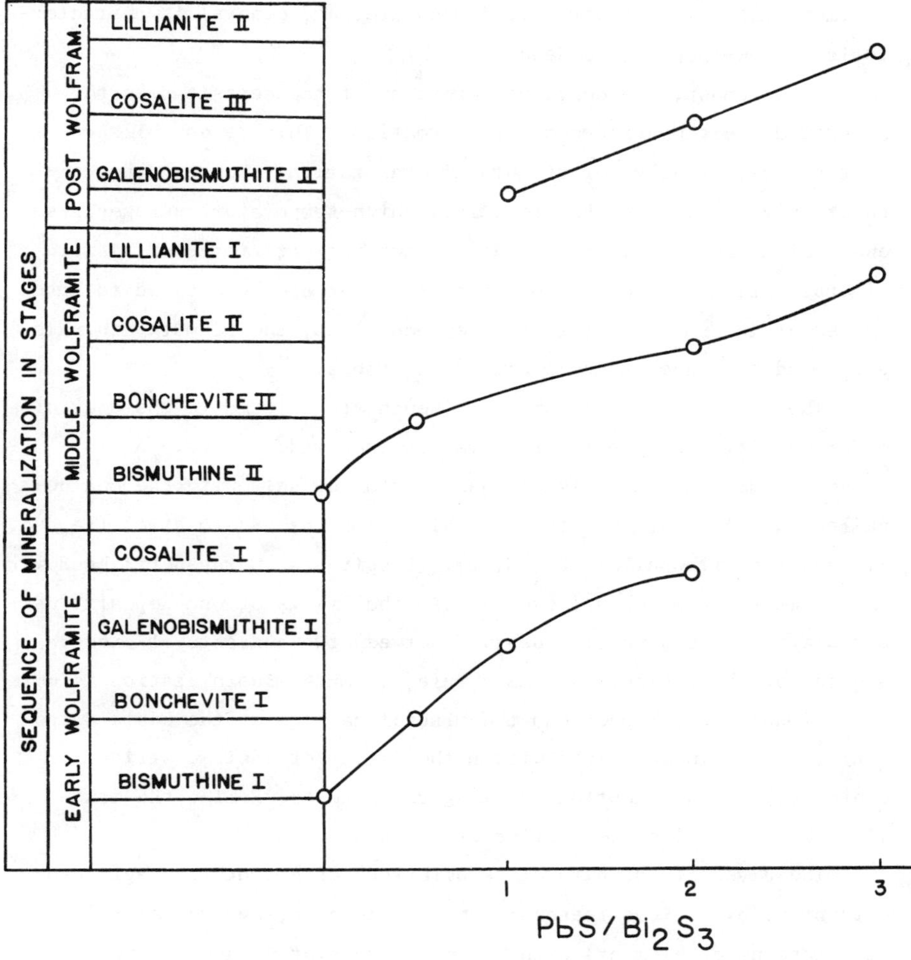

Figure 4. Character of change in composition of minerals of bismuth during process of ore precipitation in the deposit of KARA-OBA (from G. N. Necheljustov).

Cu-Pb-Bi complex sulphides → Cu-Bi complex sulphides → Pb-Cu-Ag-Bi complex sulphides (→ indicates increasing alkalinity and oxidizing potential).

The development of bismuthine, native bismuth, and cosalite is characteristic of skarn and greisen deposits; the development of the complex sulphides of bismuth, lead, copper, and silver (minerals of the rezbanyite and benjaminite type) is characteristic of mid-temperature deposits; and the development of the bismuth-containing antimony minerals (copper ore, jamesonite, and others) is characteristic of low-temperature deposits.

As is known, the depth of formation of the deposit does not determine the temperature of its formation. This is established particularly clearly for bismuth mineralization. Despite the comparatively shallow depths, relatively high-temperature parageneses and a significant role for volatiles are characteristic of bismuth mineralization. Even in those deposits that are considered to have formed at intermediate temperature conditions, the bismuth minerals are found in higher-temperature associations.

The basic concentrations of bismuth are related to the ores of other metals, a complexity that varies over wide limits. As is evident from Figure 5, bismuth associates not only with the chalcophiles (Cu, Ag, Pb, As, etc.) but also with the siderophiles (Fe, Co, Ni) and the lithophiles (Sn, W, U). Despite the paragenetic bond with the elements shown above, even within the limits of one deposit, no correlation dependence is observed between the contents of bismuth and any of these elements. As a rule, bismuth mineralization is more conditionally developed than the mineralizations of the basic components, and it is localized within the limits of their distribution contours. As an exception it is appropriate to examine the monomineralic natural concentration of bismuth.

The diversity of natural associations of bismuth is explained, evidently, by various reasons: on the one hand, by the natural associations of elements with a similar chemistry, and on the other, by a possible generality of derivation in the process of nuclear transmutations. However, these reasons cannot explain the extremely widespread association of bismuth with tungsten and tin. Nevertheless, the fact that natural tin (present in post-magmatic formations connected with the hydrothermal metamorphisis of syenites, granite-porphyries, and alaskites) contains a significant quantity of bismuth, indicates the geochemical bond of tin and bismuth.

It is also imperative to emphasize that although the demonstrated genetic bond of bismuth mineralization with specific magmatic bodies is absent, its association with the acid and moderate-acid intrusive and effusive rocks is clearly observed. This is seen in their spatial and temporal proximity. Indications of the existence of a similar

Group / Period	I	II	III	IV	V	VI	VII	VIII
1								
2						O		
3						S		
4	Cu				As	Se		Fe Co Ni
5	Ag				Nb Sb	Mo Te		Pd
6	Au		Tl	Pb	Ta Bi	W		Pt
7						U		

Figure 5. Elemental association with bismuth in post-magmatic processes (size of letters indicates degree of association).

bond are found in the works of Vernadskii (1916), Fersman (1939), Rankama and Sahama (1950), Goldschmidt (1954), Koptev-Dvornikov et al. (1958), and Ginzburg et al. (1958).

Contemporary presentations on the inter-relationship of magmatism and mineralization give a basis to assume that bismuth, like other heavy metals of which a bond with the granitoides (Sn, W, Pb, etc.) is characteristic, is initially concentrated in the volatile distillates of deep magmatic sources, and then migrates as gaseous-liquid mixtures along magma-ore-evolving fractures--i.e., the source of the bismuth-carrying mixtures is situated in the region of the primary magmatic centers.

Analyzing the geologico-geochemical properties of deposits of high bismuth concentrations in post-magmatic processes (e.g., the bismuth-containing deposits of Central Kazakstan, Central Asia, Eastern Zabaikale, Bolivia, and Spain), it can be seen that bismuth mineralization is related to those areas where a rapid shift of the composition and condition of the ore-forming mixtures occurs. This is indicated by its general geological position (the shallow depth of formation, the comparatively high-temperature, the close bond with magmatism) as well as by the texturo-structural properties of the ores (the broad development of modular and crustified textures, the presence of the collomorphous formations of the ore minerals), and the unusual "tele-accumulated" nature of the ore associations. This explains the spatial limitedness of bismuth mineralization. Despite the presence of the paragenetic bond of bismuth minerals with the minerals of tungsten, tin, molybdenum, copper, arsenic, gold, cobalt, nickel and other elements, even within the limits of one deposit, no correlation dependence is observed between them.

As a rule, the concentration contours of the bismuth ores is narrower than the contours of the ores of other metals associated with it, and the alteration of the degree of concentration is more distinct. Often within the limits of an ore body (at a distance of several meters) the degree of concentration varies by 3--5 orders of magnitude. These properties are characteristic, to a lesser degree, of the deposits where bismuth is present in the form of micro-inclusions within the intrinsic minerals of one or several ore-forming sulphides, or enters them isomorphously. The degree of concentration of bismuth in similar deposits is lower, but in most cases, it is these which present the greatest practical interest. For this reason, a study of the types of location of bismuth in them, and of the regularities of the paragenesis and the distribution of bismuth minerals is of primary interest, since it would permit a resolution of the problem of the incidental derivation of bismuth.

Vernadskii (1915) further emphasized that: "In the question of bismuth extraction the matter concerns not only clarification of the segregations of its ores, their study and properties--but also the extraction of the small quantities of bismuth which accompany the ore of gold, silver, lead, and copper. It concerns a change of the extraction of these metals by which the bismuth accompanying them would not be lost, but, on the contrary, would be obtained in the form of a secondary product."

In this study, the behavior of bismuth in the oxidation zone of ore deposits and in sedimentary rocks has not been examined, since a recently published work by Sakarova and Ratnikova (1965) is directed to this question.

Literature

Ahlfeld, F. Uber die Verbreitung der Verbreitung der Wismuts in der Zinnprovim Boliviens. Zeitschr f. prakt Geol., 1933, H. 9.

Andrianov, K. A., Khaidyk, L., and Khananashvili, L. M. Concerning the facility of the elements to form polymers with the inorganic chains of molecules. Uspekh Khimii, 1965. 12 iss.

Anikeeva, N. F. The role of various associations of chemical elements in the geochemical and metallogenetic specialization of magmas. The Metallo-genetic Specialization of the Magmatic Complexes. Moscow: Nedra, 1964.

Babko, A. K. A Physico-chemical Analysis of the Complex Compounds in Solutions. Kiev: Akademia Nauk, Ukrainian USSR, 1955.

Babko, A. K., and Golub, A. M. A study of the halide complexes of bismuth. A Collection of Articles on General Chemistry, Vol. 1. Akademia Nauk, USSR, 1953.

Bandy, M. C. The Ribaue - Alto Logonha Pegmatites district, Portuguese East Africa. Rock and Minerals, 1951, 26, N 9-10.

Barabanov, V. F. The Mineralogy of the Tungsten Deposits of Eastern Zabaikal., Vol. 1 L.G.U., 1961.

Berg, G. The Geochemistry of the Deposits of Useful Minerals. Moscow-Leningrad: Gos. Nauchn. - Tekhn. gorn - geol - neft., 1933.

BjØrlykke, H. The mineral paragenesis and classification of the granite pegmatites of Iveland Setesdal, Southern Norway. Norsk. geol. tidsskr., 1937, 14, H. 3--4.

Bobrov, V. A. The intrusive complexes of Eastern Mongolia and a comparison of them with the intrusive complexes of Zabaikal. Materials on the Petrology of the Granitoides of Zabaikal. Moscow: Gosgeoltekhizdat, 1962.

Bokii, G. V. An Introduction to Crystallo-chemistry. Moscow: M.G.U., 1954.

Brooks, R. R. Apparent geochemical association of bismuth and thallium. Nature, 1961, 189, N 4768.

Brooks, R. R., and Ahrens, Z. H. Some observations on the distribution of thallium, cadmium and bismuth in silicate rocks and the significance of covalency on degree of association with other elements. Geochim. et Cosmochim. Acta, 1961, 23, N 1--2.

Brown, H. A table of relative abundances of nuclear species. Rev. Mod. Phys., 1949, 21.

Busev, A. I. The Analytic Chemistry of Bismuth. Moscow: Akademia Nauk., USSR, 1953.

Cherbina, V. V. Methods of the clarification of the forms of transfer of the chemical elements in the geochemical processes. Geoximia, 1962, No. 2.

Chuxrov, F. V. The Mineralogy and zonality of Eastern Conrad. Trudy IGEM., Akademia Nauk., USSR., 1960. 50 iss.

Cissarz, A. Quantitativ-spectralanalytische Untersuchung eines Mansfelder Kupfer sehieferprofils. Chem. Erde, 1930, 5.

Clarke, F. W., and Washington, H. S. The composition of the Earth's crust. U.S. Geol. Surv., 1924, N 127.

Dana, D., et al. The System of Mineralogy. Moscow: I.L., 1951-1954.

Dunin-Barkovskaya, E. A. The paragenesis of scheelite and bismuthine in the ores of the arsenic-bismuth deposit in Uzbekistan. Uzbek geol. z, 1966, No. 1.

Dymkov, Yu. M. The Uranium Mineralization of the Rudnij Mountains. Moscow: Gosatomizdat, 1960.

Ehmann, W. D., and Huizenga, I. R. Bismuth, thallium and mercury in stone meteorites by activation analysis. Geochim. et Cosmochim. Acta, 1959, 17, p. 125-135.

Ezhov, A. I. On the geochemical specialization of the granitoides of the Salginsky region (Central Kazakhstan). The Metallogenetic Specialization of the Magmatic Complexes. Moscow: Nedra, 1964.

Fersman, A. E. Geochemistry, Vol. IV., 1939.

Fersman, A. E. The Pegmatites, Vol I: "The Granite Pegmatites." Akademia Nauk, USSR, 1940.

Genkin, A. D., Zhuravleva, N. N., and Smirnova, E. I. Moncheite and Kotulskite--the new minerals, and the composition of maichenerite. Zapiski Vses. min. ob-va., 1963, 92, 1 iss.

Ginzburg, A. I. The types of rare-metal deposits genetically connected with various intrusive complexes. Materials for the conference of All-Union Petrographers. Questions of the Magmatism and Metallogenus of the USSR. Tashkent: Akademia Nauk, Uzbek USSR, 1958.

Godovikov, A. A. On the impurities of silver, bismuth and antimony in galena. Geol. Rudn. Mestorzhd., 1962, No. 2.

Goldschmidt, V. M. The principles of distribution of chemical elements in minerals and rocks. J. Chem. Soc., 1937.

Goldschmidt, V. M. Geochemistry. Oxford, 1954.

Gorokhova, U. N. On the content of rhenium in the molybdenites of the Kadzaranskii cupro-molybdenum deposit. Trudy IMGRE, 1960, 4 iss.

Gorzhevskij, D. I. To the question of several pyritic deposits with the manifestation of nickel and bismuth minerals (Central Asia). Uchenye Zapiski Lvovsk gos. un-ta., seria geol., 1955.

Graner, F., Olin, A., and Sillen, L. G. Studies of the Hydrolysis of Metal Jons. Acta chem. scand., 1956, 10.

Green, G. Geochemical table of the elements of 1953. Bull. Geol. Soc. Am., 1954, 64, 1001-1012.

Grinberg, A. A. An Introduction to the Chemistry of the Complex Compounds. Goskhimizdat, 1951.

Haberlandt, H. The significance of the dispersed elements in geochemical research. The Rare Elements in the Igneous Mountain Rocks and Minerals. Moscow: I.L., 1952.

Hamley, G. E., and Nichol. G. Trace elements in pyrite, pyrrhotite and chalcopyrite of different ores. Econ. Geol., 1961, 56, N 3.

Hintze, C. Handbuch der Mineralogie, Bd. 1. Berlin u. Leipzig, 1933.

Hoehne, K. Chemie Erde. 1934-1935, 9.

Ipatiev, I. S. On the morphological properties of zircon from the granites of the Omchikandinskij massif. Nauchn. soobshch. Yakutskogo fil. Akademia Nauk, USSR, 1960, 4 iss.

Ivanov, V. V. The Mineralogico-geochemical Features and Indium-bearing of the Tin Deposits of Yakutie. Moscow: Nauka, 1964.

Karasik, M. A. On several properties of the Magnitogorskij ore field and the perspectives of its expansion. Trudy Gorno-geol. in-ta Uralskogo fil. Akademia Nauk, 1959, 40 iss.

Kakharov, A. K. Towards the distribution of the small elements in the scarn-magnetite deposits of the south-western part of the Kuraninskii Range. Uzbek geol. z., 1959, No. 1.

Kidd, D. F., and Haycock, M. H. Mineragraphy of the ores of Great Bear Lake. Bull. Geol. Soc. Am., 1935, 46, N 6, p. 879-9609.

Klepper, M. R. The Sangdong tungsten deposit, Southern Korea. Econ. Geol., 1947, 42, N 5.

Koptev-Dvonrikov, V. S., Grigoriev, I. E., et al. The intrusives of small-depth granite formation. The behavior in their rocks of the element-impurities and the criteria of the genetic bonds of ore-formation with them. Materials for the Conference of All-Union Petrographers. Questions of Magmatism and Metallogenus. USSR. Tashkent: Akademia Nauk., Uzbek USSR, 1958.

Koptev-Dvornikov, V. S., Polkvoi, O. S., and Markova, N. G., et al. The paleozoic intrusive complexes of Betpakdal. Trudy IMGRE, Akademia Nauk, USSR, 1966, 44 iss.

Krasnikov, V. I., Kirillov, G. I., and Menshikov, V. S., et al. On the possible reasons for the combined location of the elements in rocks and cres. Vestnik nauchnoi informatsii Zabaikalskogo otd., geogr. ób-va USSR, 1965, No. 4.

Krauskopf, K. The sedimentary deposits of rare metals. Problems of Ore Deposits. Moscow: I.L., 1958.

Krebs, G. The inorganic high polymers. The Inorganic Polymers. Moscow: I.L., 1961.

Lacroix, A. Sur un groupe des niobotantalates cubiguer, radioctifs, des pegmatites du Vakinankarata. Bull. Soc. Branc. Mineral., 1912, 35, N 2.

Leden, I., and Chatt, F. Chem. Soc., 1955.

Leontiev, A. N., and Bojko, T. F. On the greisenised granite domes of the Altai. Trudy IMGRE, 1959, 3 iss.

Levitskii, O. D. The Geology of the Ore Deposits of Zabaikalie. Moscow: Nauka, 1964.

Liaxovich, V. V. The accessory minerals and the rational nomenclature of the granitoides. Sov. geol., 1963, No. 9.

Liaxovich, V. V., and Chervinskaja, A. D. The accessory minerals in the granitoides of Tyrny-Auza and their petrogenetic significance. Trudy IMGRE, 1961, 7 iss.

Liaxovich, V. V., Noneshnikova, V. I., and Chervinskaja, A. D. Some data on the accessory minerals of the granitoides. Trudy IMBRE, 1959, 3 iss.

Liaxovich, V. V., and Noneshnikova, V. I. The accessory minerals of the granite intrusions of Western Tuva and the rocks connected with them. Trudy IMGRE, 1961, 7 iss.

Liaxovich, V. V., and Noneshnikova, V. I. On the influence of the late processes of the content of the accessory minerals in the granitoides. Trudy IMGRE, 1960, 4 iss.

Liaxovich, V. V., Zolotariev, B. P., Podionov, D. A., and Soboliev, S. F. The accessory minerals in the granitoides of Gornii Altai. Trudy IMGRE, 1959, 2 iss.

Magakian, I. G. The Dispersed and Rare-earth Metals. Erevan: Akademia Nauk, Armen. USSR, 1957.

Malakhov, A. A., and Nazirova, R. The endogenic formations of bismuth. The Endogenic Ore-formations of Uzbekistan. Tashkent: Akademia Nauk, Uzbek USSR, 1966.

Maleev, E. F. The inter-relationship of mineralization and vulcanism in Zakarpatie. Sov. Geol., 1963, No. 1.

Malevskii, A. Yu. On the isomorphous entry of thallium into galena. Dokl. Akademii Nauk, USSR, 1955, 169, No. 6.

Mamedov, E. M., and Efendiev, G. X. On the new minerals of the Paragajskii Deposit. Dokl. Akademii Nauk, Azerb. USSR, 1963, No. 10.

Matias, V. V. Uranium-containing microlite from the sodium-lithium pegmatites. The Geology of the Deposits of the Rare Elements. VIMS, 1961, 9 iss.

Meliksetian, B. M. The accessory minerals in the rocks of the Megrinskii pluto. Izv. Akademia Nauk, Armen. USSR, serija geol. geograf. nauk., 1960, 13, No. 2.

Mrnja, F. The polyascendent and monascendent zonality on the Jakhimovskii veins (the Rudnii mountains, Czeckoslovakia). Problems of Post-Magmatic Ore-Formation, Vol. 1. Prague, 1963.

Naboko, S. I. The alteration of rocks in the zones of active vulcanism. Trudy Labor. vulkanol., 1958, 13 iss.

Nakovnik, N. I. The greisens. The Altered Near-Ore Rocks and their Research Significance. Moscow: Gosgeoltekhizdat, 1954.

Naumova, E. N., and Sidorenko, G. A. Cosalite in the ferrous ores of the Chokadam-Bulakski Deposit. Trudy Min. muzeya, USSR, 1965, 16 iss.

Necheljustov, G. N., Popova, N. N., and Mintser, E. F. The distribution of the element-impurities in the process of hypogenic mineral-formation in the lead-zinc and cupramolybdenum deposits of Karamazar. Trudy IMGRE, 1961, 1 iss.

Necheljustov, G. N., Popova, N. N., and Mintser, E. F. On the isomorphism of selenium and tellurium in galena. Geokhimija, 1962, No. 11.

Nemets, D. The rare elements of the scarns of Western Moravia as an indicator of their derivation. Geokhimia, 1964, No. 3.

Nemoto, T., Hayakawa, M., and Takahashi, K. Report on the geological, geophysical, and geochemical studies of showa-shinzan, usn volcano. Geol. Surv. Gapan Rept., 1957, 170.

Nenadkevich, K. A. An outline of the research on the bismuth ores of Zabajkalie. Gos. in-ta narodnogo obrazovania, 1922, kn. 1.

Nesterova, Ju. S. Towards the question of the chemical composition of the galenas. Geokhimia, 1958, No. 7.

Nikolskii, A. P. On the accessory minerals of heavy fractions in the granitoides of the Altai. The Petrology and Mineralogy of Several Ore Areas of the USSR. Moscow: Gosgeoltekhizdat, 1952.

Noddack, J. a. W. Die Hangfigkeit der Chemischen Elemente. Naturwissanschaft, 1930, 18.

Noddack, J. a. W. Svensk kem., Tid., 1934, 46.

Oftedal, I. Untersuchungen uber die Nebenbestandteile von Erzmineralien norwegischer zinkbenderfurender Vorkommen. Skr. Norske. Vid. Akad. Oslo. Mat.-naturv. Kl., 1940, N 8.

Oftedal, I. On the occurrens of tellurium in Norwegian galenas. Norsk: Geol. Tidsskrift, 1959, 39, Ht. 1.

Olin, A. Studies on the hydrolysis of metal gons. Acta chem. scand., 1959, 13.

Ontoev, D. O. The properties of bismuth mineralization in some tungsten deposits of Eastern Zabaikalie. Trudy Min. muzeya, Akademia Nauk, USSR, 1964, 15 iss.

Ontoev, D. O. Nissenbaum, P. N., and Organova, N. I. The nature of the high contents of bismuth and silver in the galenas of the Bukukinskij deposit and some questions on the isomorphism in the system PbS--Ag_2S--Bi_2S_3. Geokhimija, 1960, No. 5.

Padera, K., Bouska, V., and Pelican, G. Rezbanyit aus Dobeing in der Osteslowakei, CSR. Chem. Erde, 1955, 17, N 4.

Pauling, G. A. The Nature of the Chemical Bond. Cornell University Press, 1960, N 4.

Pecherskij, D. I. Some data on the geochemical properties of the Lisiinskij granitoide massif. Materially po geol. i polezn. iskopaemym Severo-Vostoak USSR, 1961, 15 iss.

Penbield, S. G., and Ford, W. E. On stibiotantalite. Amer. J. Sci., 1906, 22, N 127.

Polynkva, T. N., and Poraj-Koshchits, M. A. The stereochemistry of the compounds of trivalent antimony. Zhurnal strukt. Khim., 1966, 7, No. 1.

Popova, N. N., Necheljustov, G. N., and Razina, I. S. The Rare Elements in the Molybdenum-Tungsten deposit of Central Kazakhstan. Moscow: Nauka, 1966.

Preobrazhenskii, I. A. The granites of Man-Khambo. The circumpolar Urals. Trudy IGN, Akademia Nauk, USSR, 1940.

Preuss, E. Beitrage Ger spektralanalytischen Methodik, Bestimmung von Zn, Co, Hg, In, Te, Ge, Sn, Pb, Sb, Bi durch fractionierte Destillation. Zeitsehr. Angew. Mineral., 1941.

Rananaev, N. A., Medvedeva, G. A., et al. A Qualitative, Chemical, Fractional Analysis. Sverdlovsk, 1962.

Rankama, K. The Isotopes in Geology. Moscow: I.L., 1963.

Rankama, K., and Sahama, Th. Geochemistry. Chicago, 1950.

Reed, I. W., Kigoshi, K., and Turkevich, A. A concentration of some heavy elements in meteorites by activation analysis. Geochim. et Cosmochim. Acta, 1960, 20.

Remi, G. A Course in Inorganic Chemistry, Vol. I. Moscow: I.L., 1963.

Sakharova, M. S. On the bismuth sulphosalts of the Ustarasaisky deposit. Trudy Min. Muzeya, Akademia Nauk., USSR, 1955, 7 iss.

Sakharova, M. S., and Ratnikova, G. I. Bismuth. The Metals in the Sedimentary Strata. Moscow: Nauka, 1965.

Schmitz-Du Mont, O. On the high-polymer coordination compounts. The Inorganic Polymers. Moscow: I.L., 1961.

Solovieva, F. I. Natural bismuth in the Pre-Cambrian of Krivoi Rog. iz Akademia Nauk, Ukrain. USSR, 1962, No. 2.

Tarasov, V. V. General features in the structures of the poly-conductors, the high-polymers and glasses. Fiz-Khim., 1958, 32, No. 9.

Terner, F. A comparative characterization of the main ore deposits of Central Bolivia. Problems of the Endogenic Deposits. Moscow: I.L., 1964, 2 iss.

Turovskii, S. D. On the finding of natural tin in Northern Kirgizia. Trudy In. Geol., Akademia Nauk, Kirgiz. USSR, 1956, 8 iss.

Tvalcheridze, T. A. The gold-bismuth-tellurium association of the Dambludskij and Zodskij deposits. Geol. Sbornik, Kavk. In. Min., 1959, No. 1.

Urey, H. C. The abundances of the elements. Phys. Rev., 1952, 88, 248.

Vernadskii, V. I. On the study of the natural productive forces of Russia--A note to the commission for the researching of the natural productive forces of Russia. Bull. Akademia Nauk., 1915.

Vernadskii, V. I. Bismuth in the Earth's core. Bull. Akademia Nauk., 1916, 10, No. 15.

Vernadskii, V. I. On the utilization of the chemical elements in Russia. Notes and Speeches. Petrograd, 1922.

Vinogradov, A. P. The regularities of the distribution of the chemical elements in the Earth's core. Geokhimia, 956, No. 1.

Vinogradov, A. P. The average content of the chemical elements in the main types of igneous mountain rocks of the Earth's core. Geokhimia, 1962, No. 7.

Vlasov, K. A., and Kutukova, E. I. Emerald Mines. Moscow: Akademia Nauk., USSR, 1960.

Vlodavets, V. I. The dispersed elements in vulcanic products. Trudy Labor. Vulkanol., 1958, 13 iss.

Warren, H. V. A gold-bismuth occurrence in British Columbia. Econ. Geol., 1936, 31, N 2.

Xetchikov, L. N. Towards the question of the content of bismuth in galena. Bull. of the Far-East Section, Akademia Nauk., USSR, (Komarov), 1958, 9 iss.

Yurk, Yu. Yu. Bismuth minerals in the Ukraine. Proc. Akademia Nauk of the Ukraninian USSR, 1939, No. 5.

Yanulova, M. K. The Mineralogy of the Scarno-Baryto-Polymetallic Deposit of Karagaila (Central Kazakhstan), Vol. I. "Hypogenic mineralization." Alma-Ata: Akademia Nauk., Kazakh USSR, 1962.

Yanulova, M. K., and Potok, O. I. On the bismuth mineral of the Karagailinskii deposit. Bull. Akademia Nauk., Kazakh USSR, Ser. Geol., 1956, 25 iss.

Zabin, A. G., Mukhitdinov, G. P., and Kazakova, M. E. The Paragenetic associations of the accessory minerals of rare elements in the exocontact phenytised rocks to the intrusion of the myaskites of the Vishnevii Mountains. Trudy IMGRE, 1960, 4 iss.

Zalaskova, N. E., and Kukharchik, M. V. Bismutho-microlite--a new variety of microlite. Trudy IMGRE, 1957, 1 iss.

Zuckert, R. Die Paragenesen von gediegen Silber und Wismut mit den Kobaltnickelkiesen und der Uranpechblende zu St. Joachimstahl in Bohmen. Mitt. Abt. f. Geol., 1926, N 1.

FOOTNOTES

1. According to the new system based on carbon and accepted in 1960 (Spravocnik Khimiz [Handbook of Chemistry] Vol. I, 1962), the atomic weight of bismuth is 208.98.
2. In connection with the postulation of the multiplicity of the bonds, these compounds were taken earlier for monomers.
3. This question, as applied to the magmatic rocks of Southern Fergana, is specially examined in an article by A. S. Velikii (see the present collection).
4. Bismuth has not been discovered in the monzonites and granosyenites of pluton.
5. This question is examined in greater detail in the article by O. E. Jusko-Zaxarova and V. V. Ivanov (see the present collection).
6. An article by G. N. Necheljustov et al. is devoted to the bismuth-bearing of the scarn deposits (see the present collection).
7. An article by G. N. Necheljustov et al. is devoted to a detailed description of the bismuth-bearing of the greisen deposits (see the present collection).
8. The regularities of bismuth's distribution in the cassiterite-sulphide deposits are described in more detail in an article by V. V. Ivanov and O. E. Jusko-Zaxarova (see the present collection).

17

BISMUTH AND ANTIMONY IN GALENA AS INDICATORS OF CERTAIN CONDITIONS OF ORE DEPOSIT FORMATION

A. A. Malakhov

This article was translated expressly for this Benchmark volume by Mary P. Stuart, University of Kansas, Lawrence, from the original Russian article in Geokhimiya **11**: 1055–1068 (1968).

The present study examines the question of the effect of geological and physico-chemical factors on the concentration of antimony and bismuth in galena. For the solution of the problem, 204 analyses of galena were used from 84 deposits of various origin, primarily endogenous (Table 1).

The most important geological factors that may affect the variation of the concentration of antimony and bismuth in galena are the content of these elements in the rocks and the degree of reactivity of the enclosing rock [12]. A moderate amount of antimony and bismuth in the rocks is not very distinctive. The concentrations gradually increase from the ultrabasic rocks (0.1 g/T Sb, 0.007 g/T Bi) to acidic granitoid (0.26 g/T Sb, 0.01 g/T Bi) and slate (2 g/T Sb, 0.01 g/T Bi) [13, 14]. There is no reason to expect a significant effect on the variation of the content of these elements in galena as a function of a moderate concentration in the enclosing rocks. For clarification of the role of the second factor, the deposits were combined into two groups, one located in carbonaceous (carbonate) rock (I), and the other in alumino-silicate and silicate media (II). The corresponding points for the galena were plotted on two orthogonal diagrams with a logarithmic scale (Figures 1 and 2).

Among the physico-chemical factors that directly affect both the variation of the concentration and the form of the entry of antimony and bismuth in galena are: the rate of temperature and pressure decrease [15], the mass of the solutions and of the dissolved substances, the form of the transfer of the ore elements, the alkalinity and

Table 1. The content of antimony and bismuth and the ratio Sb:Bi in galenas

No. in order	No. of sample	Ore Deposit (Occurrence)	Ore Formation	Content, g/T *** Sb	Bi	Sb:Bi
I. Deposits and Ore Manifestations in Carbonate Rock						
a) Scarn-hydrothermal						
1	1340	Koshmansay, Chatkalsky range	scarn-chalcopyrite-galena sphalerite	<10	8,000	0.0006
2	1339	Same	same	<10	4,800	0.001
3	368	Miskan, Chatkalsky range	same	<10	4,800	0.001
4	355	Same	same	<10	4,000	0.001
5	90	Koshmansay, Chatkalsky range	same	<10	4,000	0.001
6	352	Miskan, Chatkalsky range	same	<10	3,900	0.001
7	366	Same	same	<10	3,600	0.001
8	1246	Altyn-Topkan	scarn-sphal	<10	3,300	0.001
9	371	Miskan, Chatkalsky range	scarn-chalco pyrite-galena	<10	3,000	0.002
10	1344	Koshmansay, Chatkalsky range	same	<10	1,500	0.003
11	2643	Orlinaya Gorka (Eagle Hill)	scarn-magnetite	<10	1,500	0.003
12	1355	Koshmansay, Chatkalsky	scarn-chalcopyrite-galena sphalerite	<10	1,200	0.004
13	1968	Karkhona	scarn-galena sphalerite	<10	800	0.006
14	150	Leadville, CO	scarn-chalcopyrite-galena sphalerite	2300	114,700	0.02[3]
15	161	Akchagyl, Kazakhstan	same	400	18,500	0.02[3]
16	369	Miskan, Chat-Kalsky range	same	120	5,600	0.02

Table 1. cont'd

No. in order	No. of sample	Ore Deposit (Occurrence)	Ore Formation	Content, g/T *** Sb	Bi	Sb:Bi
17	217	Dzhurdzhurek, Chatkalsky range	same	125	4,900	0.03
18	1943	Altyn-Topkan, level 1450	scarn-sphalerite-galena	200	5,500	0.03[4]
19	70611	Mishikkol	scarn-sphalerite-galena	<100	2,400	0.03
20	164	Snowy Range, USA	scarn-chalcopyrite-galena-sphalerite	500	12,500	0.04[3]
21	1934	Altyn-Topkan, 1450	scarn-sphalerite-galena	400	8,000	0.05[4]
22	1625	Chetka	same	10	60	0.17[8]
23	168	Akchagyl, Kazaknstan	scarn-chalcopyrite-galena-sphalerite	600	9,500	0.06[3]
24	353	Miskan, Chatkalsky range	same	250	4,900	0.05
25	av. of 3 anal.	Paybulak (ore pipe 4)	scarn-sphalerite-galena	260	2,800	0.09
26	-	Zakhkan	same	550	5,500	0.10
27	296	Paybulak	same	210	1,960	0.11
28	1	Paybulak	same	150	1,250	0.12
29	70579	Mishikkol	same	300	2,300	0.13
30	2093	Altyn-Topkan, 1700 level	same	240	1,740	0.13[4]
31	2969	Altyn-Topkan, main ore-run (ore-bed*)	same	380	2,300	0.16[4]
32	1379	Chetka	same	10	3 30	0.30[8]
33	2967	Altyn-Topkan, main ore-run(ore-bed*)	same	500	1,600	0.31[4]
34	5	Paybulak	same	300	850	0.35
35	av. of 3 anal.	Kurusay 2	same	410	850	0.44
36	-	Trak, Bambl, Norway	scarn-magnetite	500	1,000	0.50[11]
37	1619	Chetka	Scarn-sphalerite-galena	2000	3,800	0.52[8]
38	1288	Altyn-Topkan, 1786 level	same	traces	<10	0.60

Table 1. cont'd

No. in order	No. of sample	Ore Deposit (Occurrence)	Ore Formation	Content, g/T *** Sb	Bi	Sb:Bi
39	379	Paybulak	same	600	940	0.63
40	av. of 4 anal.	Kurusay 1, level 4	same	740	1,080	0.67
41	2008/1	West Dzhangalyk	scarn-magnetite	820	980	0.83
42	-	Korsegard, Oslo	scarn-sphalerite	300	300	1.00[11]
43	-	Paybulak (ore pipe 3)	scarn-sphalerite-galena	240	210	1.10
44	6	same	same	1500	1,300	1.10
45	593	Shevchukovskoe	same	320	260	1.28
46	av. of 4 anal.	Kurusay 1, level 5	same	450	300	1.50
47	av. of 7 anal.	Paybulak (ore pipe 2)	same	920	577	1.60
48	av. of 4 anal.	Kurusay I, level 5	scarn-sphalerite-galena	710	360	1.97
49	1043	Chokadambulak	scarn-magnetite	100	50	2.00
50	av. of 4 anal.	same	same	70	30	2.30
51	av. of 2 anal.	Paybulak (ore 1)	scarn-sphalerite-galena	1300	500	2.60
52	84a	same	same	1100	400	2.70
53	262	Okurdavan	same	1300	460	2.82
54	1915	Chokadambulak	scarn-magnetite	30	10	3.00
55	23	Paybulak	scarn-sphalerite-galena	170	40	4.20
56	10	Gavasay	scarn-magnetite	100	traces*	5.00[6]
57	11	same	same	100	traces*	5.00[6]
58	2577	Chokadambulak	same	50	10	5.00
59	292	Paybulak	scarn-sphalerite-galena	500	96	5.20
60	68	same	same	240	40	6.00
61	1	Altyn-Topkan	same	2500	400	6.20[4]
62	468	Altyn-Topkan, level 1800	same	408	60	6.67[4]

Table 1. cont'd

No. in order	No. of sample	Ore Deposit (Occurrence)	Ore Formation	Content, g/T *** Sb	Bi	Sb:Bi
63	109	Paybulak	same	220	30	7.30
64	3150	Altyn-Topkan	same	750	100	7.50[4]
65	187	Southern Darbaza	same	400	<100	8.00[1]
66	168/299	Paybulak	same	400	100	8.00
67	181/302	same	same	400	50	8.00
68	7	same	same	1000	100	10.00
69	762	Altyn-Topkan, level 1740	same	460	40	12.00[4]
70	5579	Altyn-Topkan, main ore-run (ore-bed)	same	600	40	15.00[4]
71	145	Aktash	same	1100	60	18.00
72	335	Central Kansay	same	1600	80	20.00
73	58	Paybulak	same	230	10	23.00
74	754	Shevchukovskoe	same	120	traces	24.00
75	157/249	Paybulak	same	150	<10	30.00
76	12	Gavasay	scarn-magnetite	600	traces*	30.00[6]
77	446	Altyn-Topkan, main ore-run (ore-bed)	scarn-sphalerite-galena	200	traces	40.00[4]
78	683	Altyn-Topkan, level 1800	same	440	10	44.00[4]
79	749	Tashgeze	same	2200	40	55.00[4]
80	225	Altyn-Topkan, main ore-run (ore-bed)	same	1300	20	75.00[4]
81	3570	Altyn-Topkan, main ore-run (ore-bed)	scarn-sphalerite-galena	800	<10	160.00[4]
82	1259	Central Kansay	same	1000	traces	200.00[8]
83	2935	Altyn-Topkan	same	100	<10	200.00[4]

b) Hydrothermal (metasomatic and vein deposits)

No. in order	No. of sample	Ore Deposit (Occurrence)	Ore Formation	Sb	Bi	Sb:Bi
84	1419	Molodezhnoe, Chatkalsky	quartz (cassiterite)-galena	1000	1,000	1.00
85	1426	same	same	3500	2,100	1.67
86	1424	same	same	3000	1,000	1.00

Table 1. cont'd

No. in order	No. in sample	Ore Deposit (Occurrence)	Ore Formation	Content, g/T *** Sb	Bi	Sb:Bi
87	315	Aygyrbulak	quartz-calcite-sphalerite-galena	600	200	3.00
89	33	Korolevo	calcite-galena	100	10	10.00
90	45	same	same	200	20	10.00
91	319	Aygyrbulak	quartz-calcite-sphalerite-galena	520	40	13.00
92	44	Korolevo	carbonate-galena	300	20	15.00
93	21	Kan, Alaysky range	carbonate-sphalerite-galena	60	traces	20.00[6]
94	1253	Molodezhnoe, Chatkalsky	quartz (cassiterite)-galena	3100	100	31.00
95	318	Aygyrbulak	quartz-calcite-sphalerite-galena	640	20	32.00
96	17	Iokundzh, Vakhshsky range	calcite-sphalerite-galena	100	traces	33.00[6]
97	2391	Uchochak 4	quartz-calcite-sphalerite-galena (with Ars.)	1700	40	42.50
98	1254	Molodezhnoe, Chatkalsky	quartz (cassiterite)-galena	3100	60	51.00
99	-	Karaotek	carbonate-sphalerite-galena	300	<10	60.00[6]
100	-	Eskikan, Fergana	calcite-galena	300	<10	60.00[6]
101	22	Kan, Alaysky range	carbonate-sphalerite-galena	300	traces	100.00[6]
102	3	Kadainskoe, Zabaykale	quartz-carbonate sphalerite-galena	2000	10	200.00[6]

Table 1. cont'd

No. in order	No. of sample	Ore Deposit (Occurrence)	Ore Formation	Content, g/T *** Sb	Bi	Sb:Bi
		c) Telethermal layered deposits of disputed origin				
103	2269	Karkara, Kungey Alatau	lead	50	20	2.50
104	-	Ameliya	upper valley of Mississippi, USA layered deposits in limestone	50	3.5	14.30[10]
105	-	Lova	same	50	<3	16.50[10]
106	-	Blekston A	same	120	3.5	34.28
107	-	Blekston B	same	150	<3	75.00
108	-	Hickory Hill	same	300	<3	150.00
109	-	Tennyson A	same	600	<3	150.00
110	-	Hancock	same	350	<3	175.00
111	-	Tennyson B	same	500	<3	250.00
112	-	Ivey	same	550	<3	275.00[10]
113	3519	Sumsar, Fergana	lead	1600	10	320.00
II.		Hydrothermal Deposits and Ore shows Aluminosilicate and Silicate Rock				
		a) In granitoids				
114	-	Skoldevik, Norway	quartz-molybdenum	100	3,000	0.003[11]
115	2353	Aktyuz, Zailisky range	quartz-galena sphalerite	10	400	0.01
116	2357	same	same	10	240	0.02
117	94/55	Bukukinskoe, Zabaykale	quartz-molybdenum	400	18,500	0.02[7]
118	146/55	same	same	500	12,500	0.04[7]
119	-	Laven, Norway	in pegmatite	50	1,000	0.05[11]
120	177/56	Bukukinskoe, Zabaykale	quartz-molybdenum	600	9,500	0.06[7]
121	2331	Kenkol	barite-fluorite-quartz-sphalerite-galena	10	100	0.15[5]

Table 1. cont'd

No. in order	No. of sample	Ore Deposit (Occurrence)	Ore Formation	Content, g/T *** Sb	Bi	Sb:Bi
122	1269	same	same	10	100	0.10[5]
123	13	Naugarzan	barite-fluorite-quartz-galena	10	100	0.10[6]
124	1985	same	same	10	70	0.14
125	1990/1	same	same	10	60	0.17
126	476a	Kalmakyr	sericite-quartz (molybdenum)-chalcopyrite	180	1,000	0.18
127	37	Sadonskoe, Northern Caucasus	quartz-sphalerite-galena	200	600	0.33[6]
128	2470	Kenkol	barite-fluorite-quartz-galena-sphalerite	112	300	0.37[5]
129	2448	same	same	70	150	0.46[5]
130	1389	same	same	130	260	0.50[5]
131	1308	Kenkol	barite-fluorite-quartz-galena-sphalerite	70	120	0.60[5]
132	19	Granitogorskoe, Kirgizsky range	quartz-chalcopyrite-sphalerite-galena	400	600	0.67[6]
133	36	Sadonskoe, Northern Caucasus	quartz-sphalerite-galena	1500	2,200	0.68[6]
134	72	Boordu, Kirgizsky range	quartz-chalcopyrite-sphalerite-galena	750	800	0.93
135	1235	Kenkol	barite-fluorite-quartz-sphalerite-galena	40	40	1.00[5]
136	1263	same	same	traces	traces	1.00[5]
137	261	Boordu, Kirgizsky range	quartz-chalcopyrite-sphalerite-galena	3300	1,200	2.75
138	13	same	same	30	10	3.00
139	48c	same	same	600	190	3.15
140	18	Granitogorskoe, Kirgizsky range	same	400	100	4.00[6]

Table 1. cont'd

No. in order	No. of sample	Ore Deposit (Occurrence)	Ore Formation	Content, g/T *** Sb	Bi	Sb:Bi
141	1265	Kenkol	barite-fluorite-quartz-sphalerite-galena	48	10	4.80[5]
142	1259	same	same	108	20	5.40[5]
143	2339	same	same	60	10	6.00[5]
144	-	Eykakhalmen, Norway	in pegmatite	100	10	10.00[11]
145	1293	Kenkol	barite-fluorite-quartz-sphalerite-galena	120	traces	24.00[5]
146	42b	Boordu, Kirgizsky range	quartz-chalcopyrite-sphalerite-galena	750	10	75.00
147	1984/1	Naugarzan	barite-quartz-fluorite with (galena)	440	10	88.00
148	3440	Takob, Turkestansky range	same	1100	10	110.00
149	2099	Kanszhol	barite-fluorite-quartz-sphalerite-galena	1900	10	190.00
150	70	Boordu, Kirgizsky range	quartz-chalcopyrite-sphalerite-galena	2250	10	225.00
151	2081	Kandzhol	barite-fluorite-quartz-sphalerite-galena	1200	<10	240.00
		b) In moderate and acidic effusives in subvolcanic formations				
152	1508	Chukurdzhilga	carbonate-quartz-sphalerite-galena	120	7,900	0.02
153	2983	Berezovskoe, Altay	carbonate-quartz-chalcopyrite-sphalerite	traces	100	0.10[2]
154	4656	Belousovskoe, altay	quartz-galena-chalcopyrite-sphalerite	130	1,100	0.11[2]

Table 1. cont'd

No. in order	No. of sample	Ore Deposit (Occurrence)	Ore Formation	Content, g/T *** Sb	Bi	Sb:Bi
155	1245	Ridderovskoe	same	90	670	0.13[2]
156	304	Budo, Japan	calcite-quartz-galena-sphalerite	100	500	0.20[3]
157	2040	Barite Hill (Mt.)	barite-(sulphide)	500	2,300	0.40
158	180	Zolotushinskoe, Altay	quartz-galena-sphalerite-chalcopyrite	200	500	0.40[2]
159	2388	Ridderovskoe, Altay	quartz-galena-chalcopyrite-sphalerite	traces	traces	1.00[2]
160	860	Grekhovskoe, Altay	chalcopyrite-galena-pyrite-sphalerite	200	200	1.00[2]
161	370	Tara, Japan	calcite-quartz-galena-sphalerite	300	200	1.50[3]
162	752, t.2	Kanimansur	barite-quartz-sphalerite-galena	500	30	1.67
163	669	Zambarak	quartz-sphalerite-galena	1200	600	2.00
164	76	Taryekan	same	110	50	2.20
165	1566	Chukurdzhilga	carbonate-quartz-sphalerite-galena	200	90	2.22
166	3628	Berezovskoe, Altay	same	400	100	2.66[2]
167	2972	same	same	400	100	2.66[2]
168	304	Khosokura, Japan	calcite-quartz-galena-sphalerite	900	300	3.00[3]
169	av. of 4 anal.	Taryekan	quartz-sphalerite-galena	320	70	4.57

Table 1. cont'd

No. in order	No. of sample	Ore Deposit (Occurrence)	Ore Formation	Content, g/T *** Sb	Bi	Sb:Bi
170	1520	Chukurdzhilga	carbonate-quartz-sphalerite-galena	620	120	5.16
171	3506	Agalkhar, Pamir	same	50	10	10.00
172	1719	Karatashkotan	same	1900	190	10.00
173	2557K	Taryekan	quartz-sphalerite-galena	530	50	10.60
174	2557	Taryekan	quartz-sphalerite-galena	600	50	12.00
175	65-4b	Kanimansur	barite-quartz-sphalerite-galena	600	40	15.00
176	1717	same	same	90	10	18.00
177	8	Lashkerek	quartz-chalcopyrite-sphalerite-galena	60	traces	20.00[6]
178	1500	Agalkhar, Pamir	quartz-sphalerite-galena	230	10	23.00
179	638	Zambarak	same	1420	60	23.00
180	2028	same	same	810	30	27.00
181	564	same	same	2700	100	27.00
182	2103/1	same		330	10	33.00
183	1516	Chukurdzhilga	carbonate-quartz-sphalerite-galena	200	10	40.00
184	589	Zambarak	quartz-sphalerite-galena	2400	50	48.00
185	1767	same	same	270	5	54.00
186	135	Takyrekan	same	300	5	60.00
187	603	Zambarak	same	300	5	60.00
188	2103/2	Zambarak	same	380	10	76.00
189	3499	Chatyrtash, Pamir	same	400	10	80.00

Table 1. cont'd

No. in order	No. of sample	Ore Deposit (Occurrence)	Ore Formation	Content, g/T *** Sb	Bi	Sb:Bi
190	1522	Chukurdzhilga	carbonate-quartz-sphalerite-galena	400	10	80.00
191	2556	Taryekan	quartz-sphalerite-galena	500	10	100.00
192	2035	Zambarak	same	500	10	100.00
193	2036	same	same	500	10	100.00
194	2006	Eastern Karamazar	same	500	10	100.00
c) In sandstone, shale, metamorphic and crystalline schist						
195	1230	Charchar	barite-quartz galena	10	1,500	0.007
196	1140	Chavata 2, Chatkalsky range	quartz-calcite-sphalerite-galena	20	40	0.24
197	2264	Charchar	barite-quartz galena	20	150	0.13
198	-	Lassedalen, Norway	quartz-carbonate-galenite	300	100	3.00[11]
199	-	Khestekletten, Norway	sphalerite-pyrite	1000	100	10.00[11]
200	-	Gottes Khilfe, Norway	calcite-silver	1000	50	20.00[11]
201	0	Alte Anne, Norway	same	300	10	30.00[11]
202	av. of 4 anal.	Barskaun, Terskey Alatau	sphalerite-pyrite	800	25	32.00
203	-	Kuveemenskoe, Central Chukotka	quartz-chalcopyrite-sphalerite-galena	1500	20	75.00[9]
204	-	Promezhutochnoe, Central Chukotka	same	5000	50	100.0[9]

NOTES TO TABLE 1

*If after the name of the deposit, the names of the country or range are not indicated, the Kuraminsky range is implied.

**The name of an ore formation is determined by the major non-metallic and economically important ore minerals. Non-metallic and ore minerals are listed in order of increasing quantity.

NOTES TO TABLE 1 cont'd

***The present authors: Nos. 8, 11, 13, 38, 41, 54, 103-113, 115, 116, 124, 125, 147, 148, 149, 151, 157, 165, 170, 171, 172, 176, 178, 181, 182, 183, 188, 189, 190, 191, 192, 193, 194, 195, 197, 202. The following analyses were most graciously provided by: E. F. Bagrovaya - 49, 50; R. L. Dunin-Barkovsky - 134, 138, 139, 146, 150; A. Kakhkharovy - 3, 4, 5, 6, 7, 9, 16, 17, 24; S. Ya. Klempert - 25, 27, 28, 34, 39, 43, 44, 47, 51, 52, 55, 59, 60, 63, 67, 68, 73, 75; A. S. Kudryavtsev - 162, 164, 169, 173, 174, 185, 186; M. Mansurovy - 89, 90, 92; M. I. Moiseeva - 35, 40, 46, 48, 97, 137, 163, 179, 181, 184, 187; R. Nazi rova - 1, 2, 10, 12, 84, 85, 86, 94, 98, 196; Z. M. Protodyakova - 45, 53, 71, 72, 74; M. F. Finko - 19, 29.

I am very grateful for the use of these analyses.

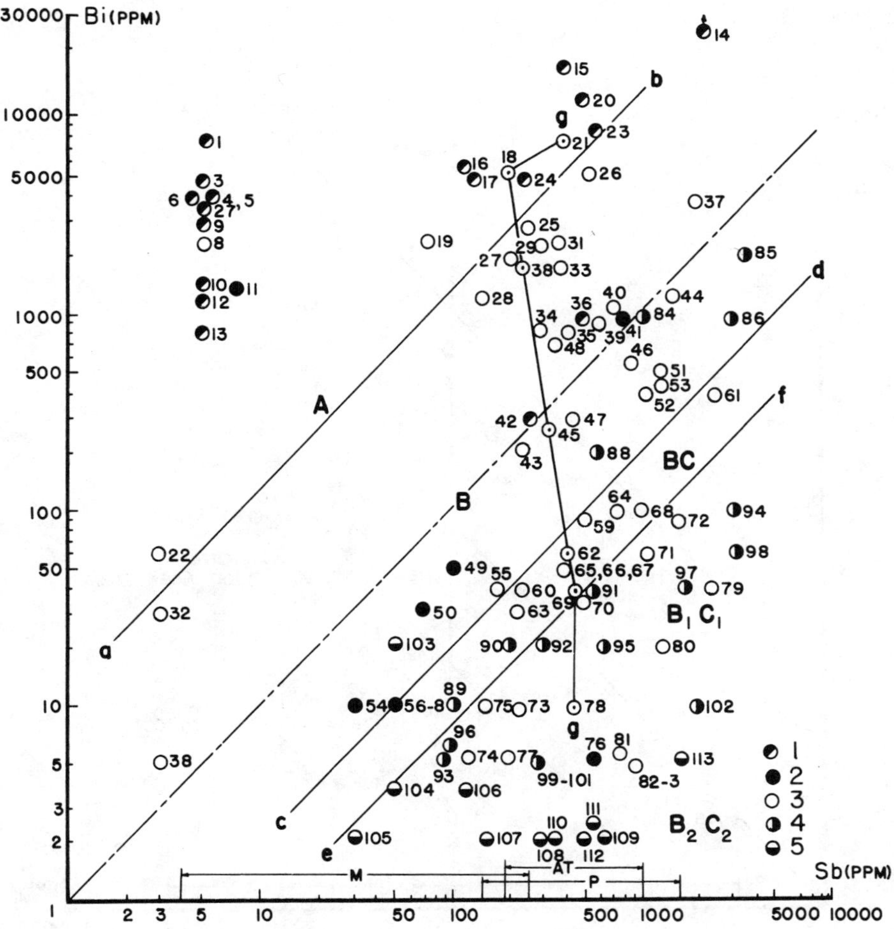

Figure 1. Diagram Sb:Bi, the distribution of the figured points for galena from deposits in carbonate rock

Key: 1-scarn-chalcopyrite-galena-sphalerite; sphalerite; 2-scarn-magnetite; 3-scarn-sphalerite-galena; 4-vein; 5-layered deposits. Fields of galena: A-high temperature; B-moderate temperature; BC-moderate-low temperature; C-low temperature. The lines of the borders between fields: ab (Sb:Bi=0.05), cd (Sb:Bi=5.0), ef (Sb:Bi=13.0). Line gg joins the data points of galena from various horizons of Altyn-Topkan. The range of content of antimony in galenas: AT - Altyn-Topkan; P = Paybulak; M - Miskan.

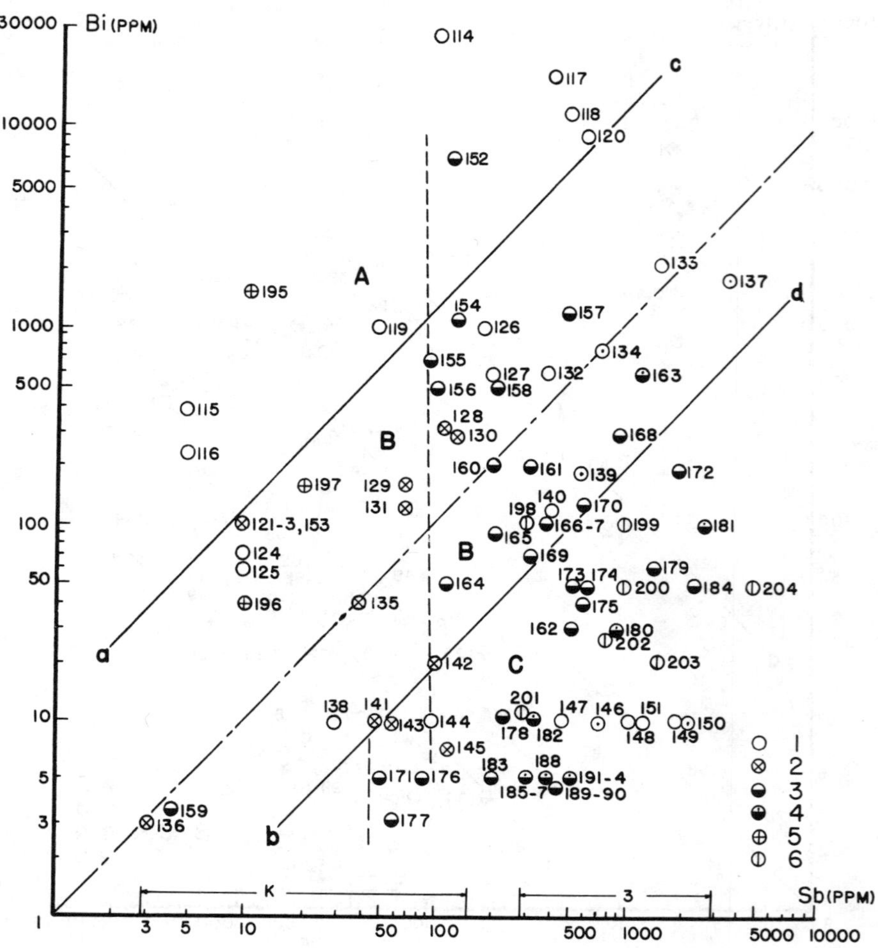

Figure 2. Diagram Sb:Bi, the distribution of the contents for galena from deposits in aluminosilicate rock

Key: 1-beds in granitoids; 2-Kenkol; 3-in effusive and subvolcanic formations; 4-Zambarak; 5-in sandstone; 6-in shale and gneiss. The fields of the galena - A, B, C (as on Figure 1). The lines of the borders between the fields - ab, cd. To the right of the broken line are distributed the data points of galena from deposits in shale, metamorphic schist, in effusive and subvolcanic formations.

acidity of the solutions in the micro-inclusions of the minerals, etc.

For determining the pressures, data about the depths of formation of the deposits were used. In assigning the galena to one or another temperature group, both direct and indirect criteria were used. The direct criteria include data from the thermometric analyses of the galena, its thermoelectric effect, the presence of micro-inclusions of mineral thermometers (argentite, matildite), and the decomposition structure of solid solutions. The indirect criteria were: thermometric analysis of sphalerite, pyrite, quartz, fluorite, barite, and calcite from galena-containing paragenetic mineral associations, the presence in them of mineral thermometers (chalcosite, hessite, calaverite, and others). The indirect criteria were: the known temperature ranges of the formation of the intact deposits, their mineral composition, the peculiarities (features) of the circum-ore changes, and in the deposits of deep-seated origin, the possible geothermal gradient. As a result, the following arbitrary fields were isolated: A-high temperature (200-300°); B - moderate temperature (140-220°); BC, C, C_1 - low temperature (100-160°) and telethermal (60-110°) galena.

During the crystallization of the galena, part of the antimony and the bismuth enters it in the form of isomorphous admixtures (impurities), and part forms syngenetic microinclusions of independent minerals. Micro-inclusions of argentite, chabazite-matildite, galenobismuth, beergerite, bournonite, grayish copper ores, silver thiosalts and other minerals have been identified in the galena of Central Asia [16] and in other regions of the U.S.S.R. [2, 3, 6, 7, 17]. The first three [18] and beergerite [7] form solid solutions with galena and determine the isomorphous entry of bismuth, antimony, silver, and tellurium. The composition of the admixtures of bismuth and antimony was examined by Van Hook, Nesterova [6], I. Kostov, A. Yu. Malevsky, and other workers, from a physico-chemical viewpoint. It is examined in particular detail in the work of Godovikov [3]. These data were considered during the study of the problem described in this article.

GALENA FROM DEPOSITS IN CARBONACEOUS (CARBONATE) ROCK

In the compilation of the diagram Sb:Bi (see Figure 1), 112 chemical analyses from 43 deposits were used. The diagram is divided into four sections, A, B, BC, C.

Section A of high temperature galena occupies the lefthand corner of the diagram Sb:Bi above the line ab. Here are plotted the points of 25 galena samples from deposits of scarn-chalcopyrite-galena sphalerite formations and their deeper level deposits of scarn-sphalerite-galena formations, which formed at hypabyssal depths of the order of 1.7-4.0 km [19-21] and at pressures of 270-1000 kg/cm^2. The characteristic features of these deposits are their cupriferous nature (chalcopyrite), the predominance of sphalerite over galena, the presence in the galena of microinclusions of bismuthic minerals, the high content of bismuth in the galena (almost always higher than 1000, 114,700 g/T), the wide fluctuation of the content of antimony (from traces to 2300 g/T), the predominance of bismuth over silver, and the low value of the ratio of antimony to bismuth (from 0.0006 to 0.05).

Section B of moderate temperature galena is located between lines ab and cd. Thirty-eight points are plotted in this section, representing galena from 16 deposits of primarily scarn-sphalerite-galena, scarn-magnetite, and vein quartz-galena with cassiterite formations. In the ores of these deposits there is very little chalcopyrite, and the Sb:Bi ratio varies from 0.7 to 10. For the galena, significant variations in the content of antimony (traces - 2000 g/T) and bismuth (traces - 5500 g/T) are characteristic; the ratio Sb:Bi is from 0.06 to 5.00.

Section BC of medium temperature galena is bounded by lines cd and ef. There are thirteen points plotted here, selected from five deposits of scarn-sphalerite-galenite, two of scarn-magnetite and vein carbonaceous (carbonate)-galenite formations. The galena contains little bismuth (traces to 400 g/T) and more antimony (100-3000 g/T). Chalcopyrite is present in the ores in very small quantities.

Section C_1-C_2 of low temperature galena is located in the lower righthand corner of the Sb:Bi diagram. Near the abscissa (antimony) are plotted the points (subsection C_2) of galena from stratified telethermal deposits, which formed at shallow depth (0-300 m.) and at pressures of less than 70 kg/cm^2. Above (subsection C_1), galena from vein and metasomatic deposits are represented; these were formed in a range of depths 0.5-1.2 km [20] and pressures 70-150 kg/cm^2. Bismuth in telethermal galena was found in traces (less than 4 g/T), but antimony in quantities from 30 to 1700 g/T. The galena from scarn-hydrothermal, metasomatic and vein deposits contains very little bismuth (5-100 g/T) and antimony (60 to 3100 g/T).

Chemical analysis of small amounts of aqueous extracts of the residues of mineral-forming solutions from the minerals of the lead-zinc mineralization of the carbonaceous (carbonate) rock of Central Asia [23-25] and in the telethermal deposits of the United States and Spain [26] indicates the existence of basic differences.

In the hypabyssal deposits and in the deposits changing from high temperature scarn to hypabyssal, the precipitation of aluminosilicates (660-300°C) occurs from solutions which are supersaturated with chlorides and fluorides, and the crystallization of sulphides from solutions which are highly saturated with carbonic acid and with minor levels of chlorides.

In the medium temperature lead-zinc deposits, formed mainly at near-surface depths and at depths up to hypabyssal, the sulphides (sphalerite, galena) precipitate from solutions in which the anions (HCO_3, Cl) dominate over (SO_4), and the cations (Na, K) over Ca, Mg). The ratio of HCO_3 to Cl varies significantly, and sodium predominates over potassium [27].

The composition of solutions from microinclusions in galena from low temperature stratified deposits is characterized by: the predominance of the Cl anion over HCO_3 where there is a negligible concentration of SO_4, the predominance of the Na cation over the sum (Ca + Mg) a low ratio of potassium to sodium, a high solution concentration during the precipitation of the sulphides, and a sharp reduction of this concentration in the last state of mineralization [26, 27].

The study [24, 25] of hydrothermal solutions from inclusions of minerals of the deposits of the family of scarn-lead-zinc formations in the Kurusaysky ore field (heavily diluted during the process of analysis) indicates the repeated shift of alkalinity and acidity and the oxidizing and reducing properties of solutions even within one stage of mineralization. The highest temperature stage of the precipitation of the sulphides (pyrite, pyrrhotite, sphalerite) occurs primarily in alkaline and reducing media, and the low temperature stage (galena) in an acidic medium.

It is possible to formulate the hypothesis that the variation of the pH of solutions depends mainly on their interaction with the enclosing medium and on temperature. In view of this, in aluminosilicate rock the precipitation of galena occurred in less alkaline conditions than under carbonaceous (carbonate) conditions. In the high and moderate pressure zones, the solutions had an acidic, neutral, and weak-alkaline character during the precipitation of galena. In zones of lower pressures and temperatures, they acquired a more alkaline character, even more alkaline during the formation of telethermal sulphides. Along the same lines, the reduction of the potassium to sodium ratio in the solutions can be observed along with an increase of the ratio of antimony to bismuth in galena and a reduction of the content of bismuth in it. This is a natural (i.e., regular) variation of the composition of the minerals in the galena.

This regular increase with depth of the content of bismuth in galena from the Freiberg veins was noted by Leutwein and Herrmann [28], for the deposit of Altyn-Topkan by Melnichenko, for the Kurusay deposit by Moiseyeva, and for a series of scarn-sphalerite-galena deposits in the Kuraminsky mountains by Malakhov [16]. On the Sb:Bi diagram this is illustrated by the line gg (Figure 1), connecting the plotted points of galena (Numbers 21, 18, 30, 69, 78) from Altyn-Topkan. Located toward the top are the plotted points of galena (rich in bismuth) from the deeper levels, and toward the bottom are the points representing galena from the shallow upper levels and the surface of the deposit.

Many investigators relate the reduction of antimony and the increase of bismuth in galena with an increase in the temperature of crystallization [7, 16, 28, 29].

The principal accumulation of bismuth in galena of the first generation scarn-hydrothermal deposits of the Far East was described by Khetchikov [17]. The same phenomenon was also observed in galena from the scarn deposits of Central Asia. An extremely wide range in the concentration of bismuth (from traces to 114,000 g/T) and a very narrow range for antimony is observed in the galena from the main generation in these deposits. Thus, in the Paybulak deposit (19 analyses) the content of antimony in galena varies from 140 to 1600 g/T. In the Altyn-Topkan deposit (17 analyses) antimony varies from

10 to 250 g/T. The range increases from the higher temperature deposits to the low temperature deposits (see Figure 1).

GALENA FROM DEPOSITS IN ALUMINOSILICATE AND SILICATE ROCKS

Ninety-two points for galena from 41 deposits are plotted in the Sb:Bi diagram (see Figure 2). Three sections, A, B, and C may be singled out.

Section A--high temperature galena: plotted here are data from seven galenas from pegmatite, deposits of quartz-molybdenum, and distinctive quartz-galena-sphalerite formations, which formed at hypabyssal depths and at high pressure (>270 kg/cm^2). In the ores of these deposits there is more sphalerite than galena; in the galena very large concentrations of bismuth (from 9500 to 30,000 g/T) and antimony (from 5 to 500 g/T) occurs; the ratio Sb:Bi does not exceed 0.06. In the galena from sandstone ore deposits there is little antimony (from 5 to 200 g/T), more bismuth (from 100 to 1500 g/T), and the ratio Sb:Bi reaches 0.13.

Within the limits of this section there are two anomalous points which represent moderate temperature galena. The very high content of bismuth (7900 g/T) in the galena from Chukurdzhilga is explained by the fact that there galena mineralization is partially superimposed on the veins of an earlier quartz-chalcopyrite-bismuth formation. The reason for the high content of bismuth in the ore deposits of Charchar has not been explained.

Section B of moderate temperature galena is indicated between the lines ab and cd. In this section are plotted 43 points from 19 vein and metasomatic deposits of quartz-galena-sphalerite-chalcopyrite, quartz-sphalerite-galena, pyrite-polymetallic, carbonate-quartz-galena-sphalerite, barite, barite-sphalerite-galena, and barite-fluorite-quartz-sphalerite-galena formations (table). The content of bismuth and antimony in the galena generally varies within the range 20 - 1000 g/T; galena from higher temperature deposits contains more bismuth than antimony. Galena, which was formed at lower temperatures, contains more antimony than bismuth. This is clearly demonstrated upon comparing galena from the polymetallic deposits of Altay with galena from the lead-zinc veins of Karamazar. The moderate temperature origin of the galena from Karamazar confirms the thermometric data. Thus, for the deposit of Boordu, six thermal-emf readings of galenas give the low values of 68 to 77.1 mv, which is characteristic of moderate temperature galena.

Section C of low-medium and low-temperature galenas is located to the lower right of line cd. In this section are plotted 40 points of galena from 15 deposits of quartz-sphalerite-galena, barite-fluorite-quartz-sphalerite-galena, calcite-quartz-sphalerite-galena, and barite-quartz-sphalerite-galena formations, which were deposited in effusive rock and which were formed under relatively low pressure. The range of depths of the ore deposits of the subvolcanic deposits of the Kuraminsky mountains examined in the present study was determined to be 0.3-1.2 km [21].

The low and medium temperature range of the precipitation of sphalerite and galena of the Kuraminsky Mountain deposits in effusive rock was established from the complex chemical admixtures in the sulphides [16]. For the Karamazarsay basin, however, the range was established on the basis of thermometric analyses of vein minerals [30]. The galena usually contains from 100 to 2700 g/T of antimony and bismuth in the galena is observed in granodiorite and crystalline schists; the ratios are lower in galena from deposits in effusive rock, where a wide variation in the content of antimony and a minor one for bismuth can be observed (Zambarak, 11 analyses, from 270 to 2700 g/T Sb, from 5 to 600 g/T Bi). In moderate temperature galena, negligible variations in the content of antimony and bismuth are observed (Kenkol, 12 analyses, from traces to 120 g/T Sb, from traces to 300 g/T Bi).

The distribution of the plotted points for galena (from deposits in schists in the righthand part of the diagram and from deposits in sandstone in the lefthand part) is attributed to the effect of the enclosing rock. The distribution of the points of galena from subvolcanic deposits in effusive rock in the lefthand part of the diagram can be explained by: the characteristic conditions of their formation at shallow depths, comparatively rapidly changing pressures and temperatures, the considerable mass of the solutions, the penetration to significant depths of vadose waters, the comparatively elevated partial pressure of the oxygen, and the acidic nature of the solutions, etc.

The technique of using ratios [22, 31] (in our case the ratio Sb:Bi, Figure 1, Table 1) indicates the dependence of this ratio on the temperature of crystallization of the galena in natural conditions. Taking into account the nature of paragenetic galena-containing associations of minerals, the composition of the chemical admixtures in galena, the data from thermometric analyses, the depths of the formations, and the fact that the deposits belong to ore formations, it is possible to lay out the boundaries between high and medium temperature galena along the line ab (Sb:Bi = 0.05), between medium and medium-low temperature galena along the line cd (Sb:Bi = 5-6), and between medium-low and low temperature galena along the line ef (Sb:Bi = 13).

Unfortunately, we did not finish the work on the determination of the lattice parameters in galena from deposits which formed at different depths, but it was conducted on sphalerites from 29 of the deposits examined here, mainly from sphalerite-galena paragenetic associations. The chemical composition, the parameters of the crystal lattices, and the specific gravity were determined for the sphalerites. The dependence of the variation of the parameter of the crystal lattice on the quantity of isomorphic admixtures [2] was established. Data for sphalerites formed at various depths are plotted in Figure 3. From their distribution, it is evident that the least dense sphalerites (3.95-4.03) with the largest lattice parameters (a_o = 5.408 - 5.424 A) were formed in a region of hypabyssal and changing depths, and vice versa, the most dense (4.04 - 4.08) and compact (a_o = 5.408 - 5.414 A) sphalerites were formed in surface-layer or near-surface zones.

CONCLUSIONS

An appreciable effect of the enclosing rocks on the variation of the concentration of antimony and bismuth in galena was not observed, with the exception of galena from deposits in sandstone (little antimony) and shale (much antimony).

The arrangement in the righthand part of Figure 2 of the data points for the galena from deposits that are connected with the formation of volcanic-plutonic complexes indicates the specific physicochemical conditions of the formation of these deposits.

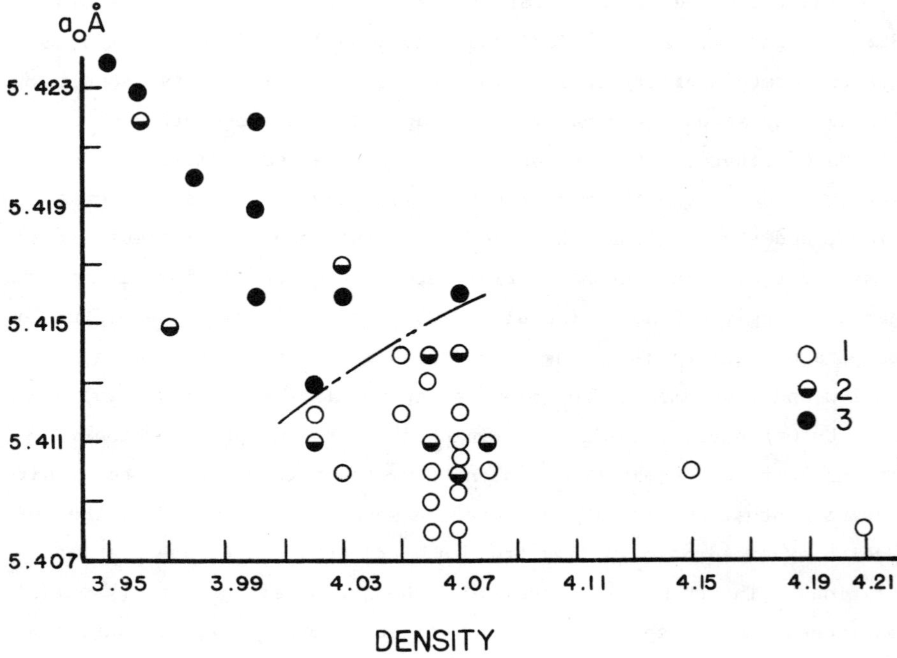

Figure 3. Diagram of the density of sphalerites
Key: Points for sphalerites from deposits which formed in: 1-surface layer near surface depths (Banska Shtyavnitsa, Barite Hill (Mt.*), Gudas, Zambarak, Lashkerek, Takob, Taryekan, Chukurdzhilga); 2-scarn-hydrothermal, surface-layer, transitional, and hypabyssal depths (Altyn-Topkan, Kansay, Mishikkol, Paybulak, Stari Trg, Turangly, Chalata); 3-hydrothermal vein from transitional and hypabyssal depths (Barskaun, Boordu, Kank, Obizarang, Chalkuryukakdzhilga). The broken line - border between the fields of points of sphalerite from surface layer and transitional depths.

The functional relation between the variation of temperature and pressure on the one hand and the variation of the concentration of antimony and bismuth in the galena on the other is clearly reflected in the diagram. In deeper and higher temperature deposits, wide fluctuations of the content of bismuth and antimony are observed, and the reverse relation is observed in low temperature and surface-layer or near-surface deposits.

With depth, the quantity of the admixture of bismuth in the galena is increased.

The rare encounter of micro-inclusions of bismuthic minerals isomorphous with galena indicates the wide miscibility of these compounds under natural conditions.

It is possible to hypothesize that the temperatures of crystallization of ore formation of galena (and other minerals) in various sections of the deposit have different values.

The ratio of antimony to bismuth in the galena indicates the temperatures and pressures which existed at the time of their crystallization. Very small values (Sb:Bi$<$0.06) are characteristic of high temperature galena which was formed at a pressure higher than 250 kg/cm^2, and large values (Sb:Bi$>$6.0-13.0) are typical for low-temperature galena which crystallized out at relatively low (150-70 kg/cm^2) and low ($<$70 kg/cm^2) pressures.

The variation of the crystal lattice parameters and densities of the natural sphalerites and galena which crystallized out at a range of depths 0.3-5.0 km (pressure up to 1250 kg/cm^2) depends only on the entry of isomorphous chemical compounds into their lattice.

BIBLIOGRAPHY

1. Arapov, Yu. A. The mineralogy and geochemistry of the Kansaysky Deposit, Working Papers of the Tadzhiko-Pamirsk Expedition, issue no. 49, 1934.
2. Veyts, V. I., Pokrovskaya, I. V. and Bologov, G. P. The Mineralogy of the Polymetallic Deposits of Altay Ore, Alma-Alta: The Academy of Sciences of the Kazakhstan Soviet Socialist Republic, vol. 1, 1957.
3. Godovikov, A. A. The Minerals of the Bismuth-Galenite Series, Novosibirsk: Science, 1965.
4. Yenikeyev, M. R. The galena from Southwest Karamazar, The Notes of the Uzbekistan Division of the All-Union Mineralogy Association, issue no. 13, Tashkent: Academy of Sciences of the Uzbekistan Soviet Socialist Republic, 1961.

5. Moiseyeva, M. I. The mineralogical and geochemical features of the quartz-fluoride-sulphide deposit of Kenkol, Questions of Mineralogy and Geochemistry, Tashkent: Science, 1964.

6. Nesterova, Ju. S. On the Question of the Chemical Composition of Galena, Geochemistry, no. 7, 1958.

7. Ontoev, D. O., Nissenbaum, P. N., and Organova, N. I. The nature of large amounts (contents) of bismuth and silver in the galena of the Bukukinsky Deposit and certain questions about isomorphism in the system $PbS-Ag_2S-Bi_2S_3$, Geochemistry, no. 5, 1960.

8. Popov, V. S. On the geochemistry of the ores of Central Kansay in Karamazar, Scientific Notes, S.A.I.G.I.M.Sa. issue no. 3, Tashkent, 1960.

9. Sidorov, A. A. The gold-silver mineralization of Central Chukotka, Moscow: Science, 1966.

10. Marshall, R. R., and Joensuu, O. I. Econ. Geol. 56, no. 4, 1961.

11. Oftedal Ivar. Untersuchungen uber die Nebenbestandteile von Ermineralien norwegischer zinkblandefuhrender Vorkommen. Norsk. Videnskap. Akad. Skrifter. Mat. Naturwiss. Kl., no. 8.

12. Shcherbina, V. V. On the question of the zonality of ore deposits, Working Papers of the Conference on the Problem of Postmagmatic Ore Formation, vol. I. Prague, 1963.

13. Onisi, H., Sandell, E. B. Observations on the geochemistry of antimony (stibium), The Geochemistry of Rare Elements, Moscow: Foreign Literature, 1969.

14. Vinogradov, A. P. Moderate amounts of chemical elements in the major types of igneous rock in the earth's crust, Geochemistry, no. 7, 1962.

15. Shcherbina, V. V. On the effect of pressure on isomorphic substitution (replacement), Problems of the Crystallochemistry of Minerals and Endogenous Mineral Formation, Leningrad: Science, 1967.

16. Malakhov, A. A. Minor elements in the sulphides of magnetite, polymetallic, and fluoride deposits, The Essential Features of Magmatism and the Metallogenesis of the Chatkalo-Kuraminsky Mountains. Tashkent: The Academy of Sciences of

the Uzbekistan Soviet Socialist Republic, 1958.

17. Khetchikov, L. K. On the question of the content of bismuth in galena, Report of the Far-Eastern Branch of the Academy of Sciences of the U.S.S.R., issue no. 9, 1958.

18. Ramdohr, P. Die Erzmineralien und Yhre Verwachsungen. Berlin: Akademie Verlag, 1960.

19. Kushnarev, I. P. The depth of formation of the endogenous deposits of the Kuraminsky structural-facies zone and the role of the erosional section in their distribution, The Geology of Ore Deposits, no. 6, 1961.

20. Bragin, I. K., Goldshmidt, E. Kh., Kalabina, M. G., and Malakhov, A. A., The depths of formation and the vertical amplitude of the mineralization of the Chatkalo-Kuraminsky mountains, Report of the Academy of Sciences of the U.S.S.R., vol. 169, no. 6, 1966.

21. Bragin, I. K., Kalabina, M. G., Goldshmidt, E. Kh., and Malakhov, A. A. On the depths of the formation of certain endogenous ore formations of the Uzbekistan, The Endogenous Ore Formations of Uzbekistan, vol. 2. Tashkent: Fan, 1968.

22. Malakhov, A. A. On the typology of chemical complexes of elements in minerals according to the example of sphalerite, Geochemistry, no. 5, 1966.

23. Sazonov, V. D. On the question of the determination of the temperatures of sedimentation (deposition) of zinc blende according to the content of iron, Working Papers of the Geological Institute, vol. 4, Academy of Sciences of the Takzhik, Soviet Socialist Republic, 1961.

24. Sazonov, V. D., Geochemical and physico-chemical characterizations of the processes of hypogenetic mineralization of the Kurusaysky ore field, Working Papers of the Geological Institute, issue no. 6, Academy of Sciences of the Takzhik, Soviet Socialist Republic, 1962.

25. Sazonov, V. D. Certain physico-chemical properties of hypogenetic mineral-forming solutions according to the example of the Kurusaysky ore field. Prague: Conference on the Problem of Postmagmatic Ore Formation, vol. I, 1963.

26. Hall, W., and Friedman, I. Composition of fluid inclusions cave in rock fluorite district, Illinois and Upper Mississippi District, Econ. Geol. 58, no. 6, 19-3.

27. Khodakovsky, I. L. The characterization of hydro-thermal solutions based on the data of the study of gaseous-liquid inclusions in minerals, Mineralogical Thermometry and Barometry, Moscow: Science, 1965.

28. Leutwein, F., and Herrmann, A. G. Kristallochemische und Geochemische Untersuchungen uber Vorkommen und Verteiligung des Wismuts im Gleiglanz der Kiesigblendigen Formation des Freiberger Gangreviers, Geologie, no. 8, 1954.

29. Tischendorf, G. Paragenetische und tektonische Untersuchungen auf Cangen des fluobarytischen Bleiformation Freiberts, Freiberger Forschungsh, vol. 18, 1955.

30. Bragin, I. K. The temperature conditions of the formation of the vein minerals of the Zambarak-Taryekansky ore field, The Uzbekistan Geol., no. 6, 1964.

31. Malakhov, A. A. On the application of the ratio method to the elements-admixtures of multi-component minerals in metallogenetic and mineral-geochemical research, The Notes of the Uzbekistan Division of the All-Union Mineralogical Society, no. 14. Tashkent: The Academy of Sciences of the Uzbekistan Soviet Socialist Republic, 1962.

32. Kliyentova, G. P., and Malakhov, A. A. The variation of the crystal lattice parameter of sphalerite as a function of the content of iron, cadmium and manganese, The Radiological Mineralogy of Raw Materials, 5, Moscow: "Nedra" ("The Depths"), 1966.

18

THE MINERALOGY OF BISMUTH IN THE ZONE OF OXIDATION

F. V. Chukhrov, V. M. Senderova, and L. P. Ermilova

This article was translated expressly for this Benchmark volume by Kathleen Ottinger, University of Kansas, Lawrence, from the original Russian article in Kora Vyvetrivaniia **3**:5–25 (1960).

INTRODUCTION

The hypergenic minerals of bismuth are among the least studied minerals of the oxidation zone. The diagnostic complexity of these minerals is explained by:

1. The great similarity of their external appearance when their Bi content is quite different.

2. The existence of different mineral types giving the same or similar qualitative reactions (the carbonates--bismutite, beyerite; the oxides--bismite, sillenite).

3. The primary latent crystal structure.

4. The practical impossibility of employing optical methods for simple diagnostic determinations.

5. The frequent presence of fine, often outwardly homogeneous, mixtures. Additionally, the chemical aspects of the hypergenic minerals of bismuth have not been adequately studied. The number of reliable analyses of these minerals is very small. This paper presents the results of a study of the pseudomorphs of the hypergenic minerals of bismuth as they relate to the primary minerals.

ANALYTICAL TECHNIQUES

To obtain the desired information, each sample was investigated using chemical, X-ray, thermal, and optical methods. The methods of chemical analysis were modified somewhat in the process of studying the minerals and the mixtures described below. Essentially the process is as follows:

The mineral is decomposed in nitric acid, the solution is evaporated to dryness, and the dry residue is treated with nitric acid. Bismuth, lead, copper, iron, aluminum, calcium, magnesium, and the basic form of molybdenum are found in the diluted nitric acid solution. Tungsten and some molybdenum occur in the sediment. The nitric acid solution is then saturated with H_2S, and the resulting sulphides of bismuth, lead, copper, and molybdenum are separated by filtration from iron, aluminum, calcium, and magnesium. They are then treated with sulphurous ammonia. Molybdenum is transferred to the solution, with bismuth, lead, and copper remaining in the sediment. The residue that contains bismuth, lead, and copper is then dissolved in nitric acid. The nitric acid solution of the analytical sub-group of bismuth, lead, and copper is divided in the following way: bismuth and lead are separated from copper by ammonium hydrate and carbonic ammonium in the presence of a coagulator (alumo-ammoniac alum) leaving the copper in the solution. The sediment containing bismuth and lead is dissolved in nitric acid, and from the cooled nitric acid solution, bismuth is deparated from the lead by cupferron. Tungsten and molybdenum are determined from another aliquot: the mineral is decomposed in nitric and hydrochloric acids with subsequent evaporation until dry, then the nitric acid treatment is repeated; tungsten and some molybdenum are measured in the sediment; molybdenum is determined in solution by a volumetric method. Chlorine is determined by coagulation with soda in a nickel crucible. Water determination is carried out in Penfield tubes.

BISMOCLITE

Bismoclite (BiOCl) was first described by Mauntain (1935). The first samples of bismoclite came from Steinkopf in Namaqualand, South Africa where it was discovered in pegmatites in the form of columnar or fibrous aggregates. According to X-ray data, the bismoclite from Namaqualand is identical with BiOCl samples produced artifically (i. e. both have tetragonal symmetry). Bismoclite was also reported by Schaller (1941) in samples from Goldfield, Nevada (U.S.A.). Bismoclite deposits were also reported from the U.S.S.R.

Bismoclite has been found by the authors in the oxidation zone of the Karaoba deposit (Betpak-Dala) in quartz veins containing tungstenite, bismuthine, cosalite, natural bismuth and other minerals. The quartz veins occur primarily in granites and partly in quartz porphyries.

Bismoclite is found in a mass of veined quartz, mainly in the form of pseudomorphs after the primary bismuth minerals of bismuthine and cosalite; it is rarely seen, but if present, occurs as pellicles and fine crusts on quartz. In many samples bismoclite replaces bismuthine and cosalite, the replacement beginning from the periphery of the material or along the fractures in the samples.

Anglesite is comparatively common with the bismoclite pseudomorphs of cosalite. Bismutite $Bi_2(CO_3)O_2$, which is found with the bismoclite, is quantitatively very subordinate to it. The size of the largest occurrences of bismoclite reaches dimensions of 5 x 3 x 2 cm. The mineral is present in dense sub-crystallic aggregates of a light to dark gray and yellow-green color. Under the microscope, bismoclite is isotropic or weakly anisotropic and has a muddy-green color.

Analyses of bismoclite from the Karaoba deposit were first carried out by V. M. Senderova; data from these analyses are given in Table 1. Data for the bismoclite analyses from Namaqualand and Nevada samples are presented for comparison.

Analyzed samples of bismoclite exhibit fine admixtures of clay material, iron oxides and anglesite; the general content of which averages about 3% in gray bismoclite, about 6% in yellow-green bismoclite, and about 8.5% in light gray bismoclite. In the yellow-green bismoclite an admixture of koechlinite (Bi_2MoO_6) is possible.

Allowing for the mechanical admixtures, the chemical data correspond well to the formula of bismoclite--BiOCl. The somewhat smaller content of chlorine in yellow-green bismoclite can be explained by partial replacement of the chlorine by the hydroxide group. At the present time such a replacement is known as daubreeite BiO(Cl,OH), in which the chlorine content is 7.5%. The existence of a compound with the structure BiO(Cl,OH) was proven experimentally by Bannister and Hey (1935). Spectral analyses of

Table 1. The chemical structure of bismoclite

Components	Gray Bismoclite		Yellow-green Bismoclite	
	weight %	molecular quanity	weight %	molecular quantity
Bi_2O_3....	84.94	0.182	84.12	0.181
Sb_2O_3....	0.89	0.003	0.00	-
As_2O_5....	0.00	-	0.00	-
CuO......	0.84	0.011	1.37	0.017
PbO......	1.10	0.005	0.38	0.002
CaO......	0.00	-	0.00	-
MgO......	0.00	-	0.00	-
Al_2O_3....	0.14	0.001	0.78	0.007
Fe_2O_3....	0.04	0.0003	0.07	0.0004
WO_3......	0.00	-	0.00	-
MoO_3.....	0.00	-	3.50	0.024
Cl.......	12.84	0.362	10.78	0.304
F........	-	-	-	-
CO_2......	0.00	-	0.19	0.004
SO_3......	0.15	0.002	0.09	0.001
H_2O^+.....	0.43	0.024	0.66	0.037
H_2O^-.....	0.20	0.011	0.10	0.006
	0.50	-	0.37	-
Sum......	102.07	-	102.41	-
$-O=Cl_2$...	2.90	-	2.43	-
$-O=F_2$....	-	-	-	-
Sum......	99.17	-	99.98	-
Specific weight	7.45	-	7.26	-

Light gray Bismoclite		Namakv Valende	Nevada
weight %	molecular quantity	weight %	weight %
79.47	0.171	88.49	88.53
-	-	-	-
-	-	-	-
-	-	-	-
5.30	0.024	traces	-
traces	-	-	-
0.00	-	-	-
0.27	0.003	-	-
0.06	0.0004	0.12	-
0.30	0.001	-	-
0.00	-	-	-
12.00	0.338	13.00	12.51
0.08	0.004	-	-
-	-	-	-
2.16	0.027	-	-
2.48	0.138	0.45	1.58
0.25	0.014	0.42	0.20
0.45	-	0.77	-
102.82	-	103.25	102.82
2.71	-	2.93	2.82
0.03	-	-	-
100.08	-	100.32	100.00
7.24	-	7.36	-

the bismoclite samples indicate the presence of silver, tellurium, and strontium.

The heating curve of bismoclite indicates endothermic peaks with upper temperature limits of 120, 600-640, and 835-860°C. A typical heating curve for bismoclite is shown in Figure 1.

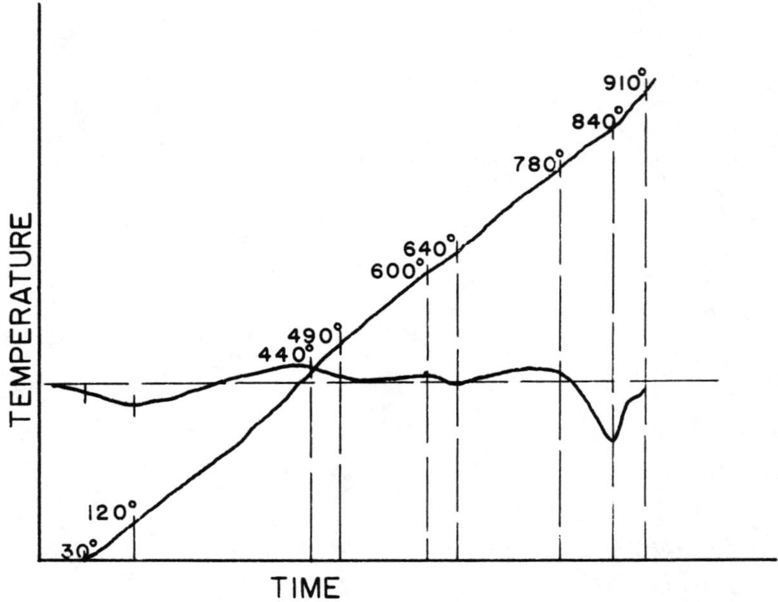

Figure 1. The heating curve of bismoclite.

X-ray analysis of gray, yellow-green, and light gray samples of bismoclite shows no substantial difference between them. Interplanar distances, determined by N. N. Sludska, are given in Table 2.

X-ray study of a series of other pseudomorphs of the hypergenic minerals of bismuth by bismuthine or cosalite from the Karaoba deposit shows that these pseudomorphs are often formed by mixtures of bismoclite and bismutite. One such pseudomorph of a light yellow coloration was subjected to chemical analysis (Table 3).

According to the analytical data, the sample studies gave values of BiOCl 0.169, $Bi_2(CO_3)O_2$ 0.076, $(Mo,W)O_3$ 0.28, and Bi_2O_3 0.021%. The mineral studies may be considered as a mixture of bismoclite with a subordinate amount of bismutite and a mineral of the koechlinite type.

Table 2. The interplanar distances of bismoclite

Gray Bismoclite		Yellow-Green Bismoclite		Light Gray Bismoclite	
I	d	I	d	I	d
4	7.36	3	7.26	-	-
-	-	-	-	-	3.72
3	3.67	4	3.65	3	3.64
9	3.42	8	3.40	7	3.38
-	-	1	3.13	-	-
1	2.95	1	2.92	3	2.92
9	2.75	9	2.75	8	2.73
10	2.66	10	2.65	10	2.65
4	2.46	5	2.44	6	2.44
7	2.21	7	2.19	7	2.19
1	2.15	1	2.14	1	2.14
1	2.07	-	-	2	2.05
1	2.03	1	2.00	2	2.01
9	1.949	9	1.940	8	1.931
3	1.886	3	1.872	2	1.867
9	1.832	9	1.830	9	1.820
1	1.725	1	1.714	2	1.720
9	1.696	9	1.688	8	1.682
9	1.669	9	1.663	9	1.657
10	1.574	10	1.572	10	1.565
5	1.528	8	1.523	7	1.519
1	1.483	1	1.480	2	1.469
8	1.378	8	1.372	7	1.374
1	1.359	1	1.352	2	1.351
1	1.347	1	1.339	2	1.337
1	1.303	1	1.298	1	1.297
1	1.290	1	1.287	-	-
1	1.277	1	1.277	-	-
9	1.268	10	1.265	8	1.260
-	-	-	-	6	1.231
9	1.227	9	1.224	6	1.220
-	-	1	1.214	-	-
5	1.201	5	1.199	4	1.197
2	1.178	4	1.173	2	1.171
6	1.170	6	1.165	6	1.164
1	1.145	1	1.145	1	1.138
9	1.126	10	1.123	8	1.120
10	1.101	10	1.098	8	1.097
9	0.070	9	1.068	7	1.066
7	1.062	7	1.060	4	1.059
10	1.034	10	1.035	8	1.038
				8	1.033

Table 3. The chemical structure of the bismoclite and bismutite pseudomorphoses by bismuthine.

Components	Weight %	Molecular Quantities
Bi_2O_3...........	82.37	0.177
Sb_2O_3...........	0.00	-
As_2O_5...........	0.00	-
CuO.............	0.84	0.011
PbO.............	0.58	0.003
MgO.............	0.00	-
CaO.............	0.00	-
Al_2O_3...........	1.02	0.010
Fe_2O_3...........	0.15	0.001
WO_3.............	1.10	0.005
MoO_3............	3.36	0.023
Cl..............	5.96	0.169
CO_2.............	3.35	0.076
SO_3.............	0.00	-
H_2O+............	1.24	0.069
H_2O-............	0.00	-
Undissolved residue	1.07	-
Sum.............	99.70	
Sp. wt..........	7.04	
Components in molecular percentages		
BiOCl...........	57.48	
Bi_2CO_5..........	25.85	
Bi_2O_3...........	7.14	
WO_3+MoO_3........	9.52	

Finding of bismoclite in the Karaoba deposit is one proof of the occurrence in this area of an ancient oxidation zone formed under the conditions of a hot dry climate.

RUSSELLITE

Hey and Bannister (1938) first described the mineral russellite (Bi_2WO_6) that was examined as an isomorphous mixture of bismuth and tungsten oxides. In russellite samples from Cornwell, Bi_2O_3 varies from 62.3--68.3% and WO_3 from 25.5--32.1%. The molecular ratio of Bi_2O_3 : WO_3 in russellite ranges from 1.23 : 1.00 to 0.90 : 1.00.

In Central Kazakhstan russellite is found in the Karaoba, Western Karaoba, and Baynazar deposits in quartz veins deposited in the granites and rocks of roof rocks (effusive and sedimentary).

The occurrence of russellite is as dense latent crystallic aggregates of a greenish-yellow and yellow color. Russellite is usually observed in pseudomorphs after the primary minerals of bismuth, mainly bismuthine.

Russellite analyses carried out by Senderova are shown in Table 4. Table 5 shows the interplanar distances determined by Sludska for the samples analyzed. For the sake of comparison, the same table shows the standard data for russellite and koechlinite.

The results of studying analyzed samples of russellite by X-ray, thermal, and optical methods are given here. A heating curve of samples from the Baynazar deposit of the Shetsky District was not obtained due to the lack of homogeneous samples. Judging from the powder pattern present for sample 1, russellite is the basic crystal phase present; bismutite is also indicated to be present; on the powder patterns of sample 2, lines of russellite and bismutite are present. Under the microscope, the substance analyzed presents an aggregate of high birefringence yellowish grains with transitions to an isotropic substance; in the secondary material a weakly anisotropic or isotropic densely colored (greenish-gray) substance (bismoclite) is observed. On the periphery of sample 2, fluorite is observed. The presence of the latter probably explains the decreased specific weight of this sample. A heating curve of samples from the Western Karaoba deposit of the Betpak-Dala district was not obtained

Table 4. The chemical composition of russellite

Components	Baynazar		W. Karaoba	Karaoba			Cornwell	
	Greenish-Yellow Sample 1	Bright Yellow Sample 2	Yellow	Bright Green Sample 1	Greenish-Yellow Sample 2	Yellowish-Green Sample 3	1	2
Bi_2O_3.......	73.63	64.00	65.90	77.68	63.13	75.66	68.26	62.3
Sb_2O_3.......	0.00	-	-	0.04	-	0.00	-	-
CuO.........	0.00	0.25	-	0.75	1.11	1.88	-	-
CaO.........	0.00	6.41	2.02	4.47	4.12	3.68		-
MgO.........	0.00	0.29	-	0.00	0.00	0.00	-	-
PbO.........	traces	0.00	0.44	traces	3.62	1.17	-	-
Al_2O_3.......	0.00	2.50	0.65	0.64	0.47	0.67	-	-
Fe_2O_3.......	0.65	0.86	0.40	0.14	0.10	0.12	traces	-
WO_3.........	18.68	14.20	18.24	6.58	12.47	7.72	25.50	32.1
MoO_3........	4.73	0.28	5.90	3.15	10.56	3.70	-	-
As_2O_5.......	-	-	-	-	0.00	0.00	0.26	0.29
Cl..........	0.59	0.25	0.45	0.18	2.00	0.79	-	-
F...........	0.00	3.30	0.00	2.35	0.00	-	-	-
CO_2.........	0.00	3.79	2.63	2.99	0.44	3.18	-	-
SO_3.........	0.00	-	-	0.00	0.08	0.21	-	-
H_2O +.......	1.39	2.31	1.77	1.49	1.36	0.56	4.86	-
H_2O -.......	0.00	3.38	0.71	0.00	0.62	0.63	-	-
undissolved residue...	0.59	0.20	0.85	0.59	0.43	0.37	1.60	1.6
Sum.........	100.26	102.02	99.96	101.05	100.51	100.34	100.48	96.29
$-O=Cl_2$......	0.13	0.05	0.10	0.05	0.45	0.18	-	-
$-O=F_2$.......	-	1.39	-	0.99	-	-	-	-
Sum.........	100.13	100.58	99.86	100.01	100.06	100.16	100.48	96.29
Specific weight....	6.50-6.72	5.43	7.54	6.81	6.92	7.04	7.35	-

Table 4. - cont'd

Components	Baynazar		W. Karaoba	Karaoba			Cornwell	
	Greenish-Yellow Sample 1	Bright Yellow Sample 2	Yellow	Bright Green Sample 1	Greenish-Yellow Sample 2	Yellowish-Green Sample 3	1	2
Molecular Ratios								
MoO_3WO_3.....	0.40:1.00	0.03:1.00	0.52:1.00	1.03:1.00	-	0.79:1.00		
Bi_2O_3 (WO_3+MoO_3)	1.41:1.00	2.18:1.00	1.18:1.00	2.75:1.00	2.50:1.00	2.75:1.00		
H_2O+:(Bi_2O_3 $+WO_3+MoO_3$)	0.29:1.00	0.64:1.00	0.38:1.00	0.37:1.00	0.40:1.00	0.14:1.00		
Components in Molecular Percentages								
Bi_2O_3.......	52.03	24.77	37.26	35.99	26.96	45.43		
(WO_3+MoO_3)..	41.69	19.26	38.88	18.15	18.43	17.99		
BiOCl.......	6.27	2.14	4.25	1.59	19.11	6.71		
Bi_2CO_5......	-	15.90	7.19	16.24	-	0.61		
$CaCO_3$.......	-	8.26	11.76	5.41	-	20.43		
$MgCO_3$.......	-	2.14	-	-	-	-		
CaF_2........	-	26.60	-	19.75	-	-		
$PbCO_3$.......	-	-	-	-	5.46	0.91		
$PbSO_4$.......	-	-	-	-	0.34	0.91		
$CaMoO_4$......	-	-	-	-	24.92	-		
$CuSiO_3$......	-	0.92	0.65	2.87	4.78	7.01		

Table 5. The interplanar distances of russellite

						Russelli	
From Cornwell (acc. to V.I. Mikleev)		From Baynazar				From W. Kar	
		Greenish-yellow Sample 1		Bright yellow Sample 2		Yellow	
I	d	I	d	I	d	I	
–	–	–	4.08	–	–	–	
–	–	–	–	–	–	–	
–	–	2	3.41	–	–	5	3.
–	–	–	–		3.26	–	
10	3.08	9	3.11	10	{	10	3. {2.
			2.92		2.96	3	2.
7	2.68	7	2.70	10	2.70	9	2.
–	–	–	–	1	2.53	–	
–	–	1	2.49	–	–	2	2.
–	–	–	–	–	–	–	
–	–	–	–	–	–	1	2.
2	2.11	3	2.11	4	2.13	6	2.
–	–	–	–	–	–	9	1.
9	1.91	10	1.922	9	1.919	–	
–	–	–	–	–	–	–	
–	–	2	1.805	–	–	6	1.
–	–	–	–	–	–	–	
–	–	1	1.738	–	–	3	1.
–	–	–	–	–	–	–	
–	–	–	–	–	–	–	
10	1.64	10	1.639	–	–	10	1.
–	–	–	–	9	1.622	–	
–	–	4	1.574	–	–	7	1.
5	1.564	–	–	–	–	–	
–	–	–	–	–	–	–	
–	–	–	–	5	1.504	–	
–	–	–	–	–	–	–	

Russellite						Koechlinite	
		From Karaoba				From Schneesburg	
ight green Sample 1		Greenish-yellow Sample 2		Yellowish-green Sample 3		(acc. to Frondel, 1943-6)	
I	d	I	d	I	d	I	d
1	4.08	–	–	–	–	–	–
6	3.72	–	–	2	3.64	–	–
1	3.41	–	–	–	–	–	–
3	3.23	–	–	–	–	–	–
3	3.15	7	3.12	6	3.09	10	3.131
0	2.95	–	–	6	2.92	6	2.733
0	2.73	7	2.71	7	2.71	5	2.683
1	2.54	–	–	–	–	–	–
–	–	–	–	–	–	3	2.473
1	2.36	–	–	–	–	–	–
2	2.29	–	–	–	–	–	–
9	2.13	1	2.10	5	2.12	–	–
0	1.934	–	–	–	–	6	1.936
–	–	8	1.917	7	1.919	8	1.918
1	1.859	–	–	–	–	1	1.879
1	1.810	1	1.817	–	–	–	–
1	1.783	–	–	–	–	–	–
8	1.747	–	–	2	1.738	–	–
4	1.716	–	–	1	1.714	–	–
4	1.686	–	–	–	–	–	–
4	1.645	10	1.635	6	1.651	9	1.647
0	1.616	–	–	6	1.614	7	1.628
–	–	–	–	–	–	6	1.570
–	–	4	1.567	–	–	–	–
–	–	–	–	–	–	–	–
1	1.512	–	–	–	–	–	–
1	1.480	–	–	1	1.480	–	–
3	1.413	–	–	–	–	–	–
7	1.371	2	1.367	3	1.370	3	1.369
2	1.347	–	–	–	–	3	1.346

Table 5. - cont'd

From Cornwell (acc. to V.I Mikleev)		From Baynazar				Russellit	
		Greenish-yellow Sample 1		Bright yellow Sample 2		From W. Kara	
						Yellow	
I	d	I	d	I	d	I	d
–	–	–	–	–	–	–	–
–	–	3	1.362	–	–	4	{1.3 1.3
–	–	–	–	–	–	–	–
–	–	–	–	–	–	–	–
7	1.250	8	1.250	–	–	8	1.2
–	–	–	–	–	–	–	–
–	–	–	–	–	–	–	–
7	2.14	8	1.219	–	–	8	1.2
–	–	–	–	–	–	–	–
–	–	–	–	–	–	4	1.1
–	–	–	–	–	–	–	–
–	–	–	–	–	–	8	1.1
7	1.109	8	1.113	–	–	–	–
–	–	–	–	–	–	–	–
7	1.044	–	–	–	–	8	1.[illegible]

Russellite						
ght green Sample 1	From Karaoba Greenish-yellow Sample 2		Yellowish-green Sample 3		Koechlinite From Schneesburg (acc. to Frondel, 1943-6)	
d	I	d	I	d	I	d
1.297	-	-	1	1.294	-	-
1.275	-	-	-	-	-	-
-	9	1.250	3	1.256	4	1.257
1.243	-	-	-	-	6	1.246
1.224	-	-	4	1.220	5	1.223
-	8	1.217	-	-	4	1.209
1.175	-	-	-	-	-	-
1.154	-	-	-	-	-	-
1.143	-	-	-	-	-	-
1.177	-	-	-	-	5	1.116
-	7	1.110	4	1.110	4	1.107
1.079	-	-	3	1.075	-	-
1.046	7	1.050	3	1.048	4	1.053
					3	1.039
					3	0.971
					5	0.025
					5	0.917
					5	0.908
					4	0.867
					4	0.855
					5	0.834

due to insufficient material. An X-ray powder pattern indicates material similar to that of the sample type from the Baynazar deposit. This suggests that the basic crystal phase present in this sample is russellite. On heating curves of samples from the Karaoba deposit there are deflections with upper limits of: (1) 500°; (2) 660-710°; (3) 720-730°; and (4) 735-760°C; from these the first, second and fourth deflections to some degree may be caused by a bismutite mixture. The type of heating curve obtained is shown in Figure 2. The

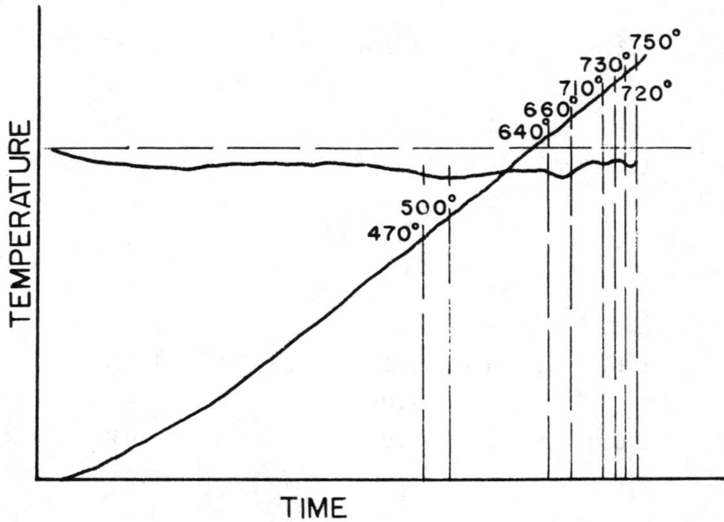

Figure 2. Heating curve for russellite from Karaoba.

lines on the X-ray powder pattern (Table 5) of samples 1 and 3 belong to russellite and bismutite; on that of sample 2 there are only lines of russellite. Under the microscope, the basic mass of the samples analyzed presents a yellowish high birefringence grainy substance, partially mixed with a densely colored greenish or yellow-greenish, almost isotropic substance. In sample 2 a small quantity of chrysocolla is indicated.

In agreement with data from the chemical analyses, results obtained by X-ray and optical methods indicate the existence of fine mixtures of russellite with bismutite, bismoclite, calcite, fluorite, chrysocolla, and other minerals.

In a few pseudomorphs after bismuthine, molybdenum oxide predominates over tungsten oxide. This is the case, for instance, in samples from the Batystau and Dzhaur deposits of the Shetsky District. Chemical and X-ray data for these samples are given in Tables 6 and 7, respectively.

It is possible that the samples described represent a mixture of russellite and koechlinite with a predominance of the latter. Due to the great similarity of the X-ray powder patterns of these minerals, no simple solution to this question exists.

Chemically, the closest analog of russellite is koechlinite from Schneeburg in Saxony. The analytical data for this mineral yields a molecular ratio for Bi_2O_3 : MoO_3 of 1.06 : 1.00, which reduces to a formula of Bi_2MoO_6.

According to Zemann (1956), koechlinite is rhombic (pseudotetragonal) and has a schistose lattice with a_o = 5.50, b_o = 16.24, and c_o = 5.49. Russellite, according to Hey and Bannister, has a tetragonal lattice with a_o = 5.42 and b_o = 11.3. A comparison of the interplanar distances indicates that nearly all the lines of russellite coincide with the more intensive lines of koechlinite, but on the X-ray powder patterns of the latter a series of other lines of weak and average intensity occur. The finding is in agreement with the fact that koechlinite is of lower symmetry than russellite. All these data speak against the existence of isomorphour mixtures of russellite and koechlinite, although they do not exclude any isomorphism between molybdenum and tungsten.

Sillen and Lundborg (1944), having made a study of the system Bi_2O_3---WO_3, concluded that these oxides form several phases of definite composition with the general formula $Bi_{1-x}W_xO_{1.5(1+x)}$. Apart from the pure oxides Bi_2O_3 and WO_3, phases of the following composition are obtained:

(1) $Bi_{0.95}W_{0.05}O_{1.58} = 0.475Bi_2O_3 \cdot 0.050WO_3 = 9.5Bi_2O_3 \cdot WO_3$;

(2) $Bi_{0.855}W_{0.145}O_{1.718} = 0.42Bi_2O_3 \cdot 0.145WO_3 = 2.95Bi_2O_3 \cdot WO_3$;

(3) $Bi_{0.833}W_{0.167}O_{1.751} = 0.417Bi_2O_3 \cdot 1.167WO_3 = 2.50Bi_2O_3 \cdot WO_3$;

(4) $Bi_{0.60}W_{0.40}O_{2.1} = 0.30Bi_2O_2 \cdot 0.40WO_3 = 0.75Bi_2O_3 \cdot WO_3$;

(5) $Bi_{0.4}W_{0.6}O_{2.4} = 0.20Bi_2O_3 \cdot 0.60WO_3 = 0.33Bi_2O_3 \cdot WO_3 = Bi_2(WO_4)_3$

Table 6. The chemical composition of the pseudomorphoses of russellite and koechlinite by bismutine (weight %)

Components	Batystav	Dzhaur
Bi_2O_3	74.75	74.97
CuO	0.60	-
CaO	0.18	1.34
MgO	0.00	0.00
PbO	0.12	0.50
Al_2O_3	0.49	0.18
Fe_2O_3	0.09	00.05
MoO_3	10.33	14.88
WO_3	9.00	5.18
Cl	0.48	-
CO_2	1.50	2.00
H_2O+	0.68	0.70
H_2O-	0.86	0.70
Sum	99.08	100.50
$-O=Cl_2$	1.11	
Sum	98.97	
Sp. Wt.	7.88	
Molecular Ratios		
$MoO_3:WO_3$	1.84:1.00	3.81:1.00
$Bi_2O_3:(WO_3+MoO_3)$	1.44:1.00	0.24:1.00
$H_2O+:(Bi_2O_3+WO_3+MoO_3)$	0.14:1.00	0.11:1.00
Components in Molecular Percentages		
Bi_2O_3	44.41	44.79
(MoO_3+WO_3)	38.81	41.01
BiOCl	4.89	-
Bi_2CO_5	9.09	5.99
$CaCO_3$	1.05	7.57
$PbCO_3$	0.35	0.63
$CuCO_3 \cdot Cu(OH)_2$	1.39	-

Table 7. The interplanar distances of the mixtures of russellite and koechlinite forming pseudomorphoses by bismutine

Batystav		Dzhaur		Batystav		Dzhaur	
I	d	I	d	I	d	I	d
-	-	2	3.67	9	1.641	10	1.645
5	3.43	3	3.42	7	1.618	3	1.614
9	3.14	10	3.13	6	1.576	7	1.574
-	-	2	3.02	1	1.478	1	1.478
7	2.95	6	2.92	-	-	1	1.411
9	2.73	9	2.72	5	1.374	5	1.371
-	-	2	2.47	5	1.297	2	1.292
-	-	1	2.26	8	1.257	10	1.252
8	2.13	7	2.13	8	1.225	10	1.223
10	1.934	10	1.931	5	1.151	2	1.155
4	1.812	4	1.812	8	1.115	10	1.114
6	1.752	-	-	7	1.079	2	1.075
-	-	5	1.741			8	1.050
-	-	2	1.714				

The experiments of Sillen and Lundborg showed that there is no isomorphism between Bi_2O_3 and WO_3 over wide limits. These investigators were not able to obtain a phase corresponding in composition to russellite. This provides the basis for supposing that russellite may represent a mixture of the various phases obtained by Sillen and Lundborg, or a mixture of one of these phases with some other substance giving fairly weak or wide lines on the X-ray powder pattern. In contrast to the artificial compounds, they will contain water, which enters into the crystallic lattice. Phases 1-4 obtained in the experiments of Sillen and Lundborg, yield substances in which the atoms of the metals form a granocentric cubic or tetragonal lattice; phase 5--$Bi_2(WO_4)_3$--has a monoclinic lattice. The X-ray powder patterns of phases 2, 3, and 4 are practically identical. It is typical that in the X-ray film of some of the preparations of phases 2 and 4, weak lines of another phase are present.

In the study of the system Bi_2O_3---MoO_3 Sillen and Lundborg identified in it two phases that are evidently isomorphous with phases 1 and 2 of the system Bi_2O_3---WO_3. These are:

(1) $Bi_{0.95}Mo_{0.05}O_{0.58} = 0.475Bi_2O_3 \cdot 0.05MoO_3 = 9.5Bi_2O_3 \cdot MoO_3$;

(2) $Bi_{0.85}Mo_{0.05}O_{1.73} = 0.42Bi_2O_3 \cdot 0.150MoO_3 = 2.83Bi_2O_3 \cdot MoO_3$.

Two additional phases yielding complex X-ray powder diffractograms were present as well as the monoclinic form $Bi_2(MoO_4)_3$. The appropriate data are given in Tables 4-6.

It should be noted that compounds of the composition $Bi_2(WO_4)_3$ and $Bi_2(MoO_4)_3$ (tetragonal symmetry) described earlier by Zambonini, Sillen, and Lundborg, were obtained. This omission throws doubt on their existence.

From the experiments of Sillen and Lundborg, it follows that if we consider the russellites as analogs of the compounds Bi_2O_3---WO_3 obtained under artificial conditions, then a significant isomorphous replacement of tungsten by molybdenum must take place with a molecular ratio Bi_2O_3 : (WO_3 + MoO_3) of not less than 2.83 : 1.00. However, inasmuch as a substance of the composition of russellite was not obtained in these experiments, it is impossible to identify the artificially obtained phases and russellite.

An X-ray study of the russellites from Kazakhstan indicates that a similar situation exists as regards the basic lines on the powder pattern. This is typical of those samples which are distinguished by their molybdenum content (see Table 7).

A reason for this may be the similar ionic radii of tungsten and molybdenum and the preservation without essential changes of the crystal lattice parameters with the isomorphous replacement of one of these metals by another. Neither is the possibility excluded that the russellites enriched by molybdenum may contain an admixture of koechlinite, the more intense lines of which practically coincide with those of russellite.

The possibility of the existence of several phases with various ratios of WO_3 and Bi_2O_3 in natural russellites is not excluded, but owing to the small differentiation in their lattice parameters, it is practically impossible to verify this. Similarly, it is not possible to verify in natural material, as with the Bi_2O_3---WO_3 compounds, the presence of Bi_2O_3---MoO_3 compounds with a similar or closely related structure. This is true because although the reflective capacities of tungsten and molybdenum atoms differ, the corresponding lines on the powder patterns must occupy essentially the same position.

As indicated in our experiments, bismuth-tungsten and bismuth-molybdenum sediments may be obtained by precipitation from water solutions. If at room temperature, we add to the basic sulphate solution some bismuth from the nitric acid solution of molybdate or some wolframate from the ammonium solution, the resulting sediment contains Bi_2O_3 and WO_3 or Bi_2O_3 and MoO_3. The quantity of WO_3 and MoO_3 in the sediments obtained from the solutions (with a pH of 1-3) varies from 40-60% (corresponding to 60-40% Bi_2O_3). From those solutions with a pH of 7--8, from 20- to 40% WO_3 or MoO_3 is precipitated (corresponding to 80-60% Bi_2O_3). From the reactions of these solutions at room temperature, we obtain sediments (samples) that are subjected to X-ray study after air drying and also after heating for 100 hours at temperatures of both 60-65° and 100°C. All the air dried sediments appeared to be X-ray amorphous. On the powder patterns of the sediments obtained with a pH of 1--3 and subjected to heating up to both 60-65° and 110°C, very weak lines of russellite and bismite appear. The data computed from the powder patterns are given in Table 8.

The molecular ratio of Bi_2O_3 : WO_3 in the sediments that were examined by X-ray varies from 0.75 : 1.00 to 0.33 : 1.00. These data indicate the existence in the sediment, along with russellite and bismite, of a free amorphous oxide of tungsten.

With the interaction of the basic sulphate of bismuth with the molybdate of ammonium in the water solutions (pH's of 1--3, 4--5, and 7--8), amorphous sediments (precipitates) were obtained on boiling. These were later subjected to heating for 100-300 hours at temperatures of 60 and 75-85°C. X-ray study of the heated sediments showed that those which were obtained at pH's equal to 1--3 and 4--5 contain a crystal phase analogous, on the basis of X-ray powder pattern lines, to one of the phases of the system Bi_2O_3---$WO_3(MoO_3)$ reported by Sillen and Lundborg. This phase is of the composition $Bi_{0.95-0.83}W_{0.05-.17}O_{1.58-1.76}$ or of sillenite or a mixture of these substances (Table 9 gives approximate data).

Table 8. The interplanar distances of bismuth-tungsten sediments obtained at pH = 1-3

The sediment Bi_2O_3 and WO_3 dried out at 60 - 65°C		The sediment Bi_2O_3 and WO_3 dried out at 110°		Russellite (by V. I. Mikheev)		Bismite (by Frondel)	
I	d	I	d	I	d	I	d
10	4.23	7	4.22	-	-	-	-
1	3.29	4	3.31	-	-	10	3.232
1	3.06	5	3.10	10	3.08	-	-
-	-	1	2.79	-	-	1	2.746
10	2.70	10	2.70	7	2.68	9	2.676
1	2.58	2	2.57	-	-	2	2.537
-	-	-	-	-	-	2	2.423
-	-	2	2.33	-	-	-	-
-	-	2	2.24	-	-	1	2.247
3	2.16	7	2.15	-	-	1	2.166
-	-	-	-	2	2.11	1	2.121
-	-	-	-	-	-	1	2.043
1	1.982	3	1.982	-	-	-	-
-	-	-	-	-	-	7.5	1.951
1	1.908	2	1.908	9	1.91	-	-
-	-	-	-	-	-	3	1.873
1	1.861	6	1.859	-	-	-	-
1	1.820	-	-	-	-	-	-
-	-	-	-	-	-	1	1.760
-	-	-	-	-	-	7.5	1.740
-	-	-	-	-	-	5	1.722
-	-	4	1.663	-	-	8	1.670
-	-	-	-	10	1.64	8	1.640
-	-	3	1.586	-	-	1	1.572
-	-	4	1.557	5	1.564	5	1.557
-	-	2	1.493	-	-	4	1.499
-	-	-	-	-	-	3	1.482
-	-	-	-	-	-	3	1.457
-	-	6	1.437	-	-	1	1.433
-	-	-	-	-	-	3	1.406
-	-	-	-	-	-	1	1.390
-	-	-	-	-	-	1	1.377
-	-	-	-	-	-	2	1.361
-	-	-	-	2	1.354	-	-
-	-	-	-	-	-	6	1.342
-	-	-	-	-	-	6	1.325
-	-	4	1.317	-	-	6	1.315
-	-	-	-	7	1.250	-	-
-	-	8	1.227	-	-	-	-
-	-	-	-	7	1.214	-	-
-	-	8	1.199	-	-	-	-
-	-	3	1.166	-	-	-	-
-	-	7	1.152	-	-	-	-
-	-	-	-	7	1.109	-	-
-	-	-	-	7	1.044	-	-

Table 9. The interplanar distances of the bismuth-molybdenum sediment obtained at pH = 1-3

Sediment Bi_2O_3 and MoO_3 dried at 60°C		Sediment Bi_2O_3 and MoO_3 dried at 75-90°		The phase Bi_2O_3-$WO_3(MoO_3)$ by Sillen & Lundborg	Sillenite (by Frondel)	
I	d	I	d	d	I	d
-	-	-	-	-	2	3.60
2	3.53	8	3.54	-	-	-
10	3.20	10	3.20	3.22	10	3.21
-	-	-	-	-	7	2.93
4	2.81	-	-	2.88	-	-
-	-	5	2.77	2.74	8	2.73
1	2.19	-	-	-	3	2.17
9	1.970	7	1.979	1.987	3	1.997
-	-	-	-	1.939	-	-
1	1.872	-	-	-	-	-
-	-	-	-	-	9	1.743
-	-	-	-	1.721	-	-
8	1.703	6	1.696	1.660	4	1.695
-	-	-	-	-	6	1.651
7	1.620	5	1.612	1.609	4	1.618
					6	1.499
					5	1.216
					5	1.198
					5	1.182
					3	1.072
					5	1.028

BISMUTITE

As shown by Frondel (1943), the natural carbonates of bismuth, formerly referred to under the names of hydrobismutite and basobismutite, are identical with bismutite. This is a compound with formula $Bi_2(CO_3)O_2$, often with a small quantity of water that forms a solid solution in the mineral ($Bi_2CO_5 \cdot nH_2O$). This mineral represents a widespread product of the hypergenic change of the primary minerals of bismuth.

Further research by Fischer (1955) established the analog of walterite, considered until recently, as a carbonate of bismuth occurring with walpurgite, an arsenate of bismuth and uranium.

Bismutite in fine mixtures with other minerals was reported by the authors in a series of wolframite-quartz deposits in Central Kazakhstan (Uzunbulak, Akchatau, Baynazar, Northern Conrad, Scorpion, Karaoba, Baykatyn, etc.). The quartz veins are deposited in granites or in their exocontact zones. Bismutite is usually observed pseudomorphic after the primary minerals of bismuth, predominantly bismuthine, more rarely cosalite and native bismuth. The composition of the pseudomorphs is dense and latent-crystallic. Their coloring is yellow and yellow-green and is often uneven.

Table 10 shows the data for chemical analyses of a series of pseudomorphs. The results of the samples analyzed by other methods are given below. For comparison, the standard heating curves given in the work of Beck (1950) are utilized.

SHETSKY DISTRICT, the Uzunbulak, Akchatau, and Baynazar deposits. On the heating curves of the samples from Uzunbulak and Akchatau (Figures 3 and 4), the typically bismutite endothermal effects with upper limits of 490-510 and 750-800°C are given. The lines on the powder patterns of the samples from Uzunbulak and Akchatau belong to bismutite. This accounts for its predominance in the composition of the analyzed material. The powder pattern of the sample from Baynazar may be seen as that of a mixture of bismutite and russellite.

Under the microscope, the samples analyzed show aggregates of high birefracting grains of a yellowish coloring with a small quantity of an almost isotropic, more densely colored greenish-gray substance.

BETPAK-DALA, the Karaoba deposit. On the heating curves (Figure 5) there are endothermal deflections with upper limits of 580-690°C that may be ascribed to bismutite; the deflection with the upper limit of 770°C may be caused by bismutite as well as by an oxide of bismuth. This type of heating curve belongs to bismutite. It is possible that some of the lines result from an admixture of fluorite, sillenite, and quartz. Under the microscope, the basic mass presents an aggregate of high birefracting grains with transitions into an almost isotropic substance. Among minor material, an isotropic, almost anisotropic,

Table 10. The chemical composition of the pseudomorphoses by bismuthine (weight %)

Components	Uzunbulak	Akchatau	Baynazar	Karaoba	N. Conrad Deposit			Scorpion
	Grayish-white	Yellowish-white	Greenish-yellow	Lemon yellow	Yellow	Greenish-gray	Pale yellow	Grayish white
Bi_2O_3........	86.55	86.50	83.50	67.56	77.48	80.34	73.87	80.63
CuO..........	-	-	0.00	0.00	-	-	-	-
CaO..........	1.30	0.70	1.22	10.28	7.26	7.84	10.29	4.73
MgO..........	0.35	0.20	0.15	0.00	0.14	0.14	0.18	0.16
PbO..........	0.00	1.52	0.00	6.84	1.00	0.00	0.90	1.25
Al_2O_3........	0.27	0.40	0.67	0.32	0.28	0.67	0.27	0.23
Fe_2O_3........	0.20	0.12	0.25	0.18	0.20	0.12	0.06	0.12
WO_3.........	0.00	0.00	2.48	0.54	0.38	0.42	0.60	0.00
MoO_3.........	-	-	1.80	0.08	0.24	0.69	1.38	-
Cl...........	0.38	0.53	0.78	0.75	0.35	0.35	0.89	3.88
F............	-	-	0.17	2.38	-	-	5.70	-
CO_2..........	9.20	6.81	5.73	7.84	9.76	7.67	5.22	7.68
SO_3..........	0.58	0.45	-	0.42	0.28	0.37	0.34	0.25
H_2O+.........	1.30	1.33	2.52	2.07	1.05	0.90	2.44	1.50
H_2O-.........	0.15	0.33	0.68	0.00	0.37	0.43	0.76	0.25
Undissolved residue......	0.36	0.64	0.40	0.93	0.54	0.33	0.20	0.36
Sum..........	100.64	99.53	100.35	100.19	99.33	100.27	103.10	101.04

Table 10. - cont'd

Components	Uzunbulak	Akchatau	Baynazar	Karaoba	N. Conrad Deposit			Scorpion
	Grayish-white	Yellowish-white	Greenish-yellow	Lemon yellow	Yellow	Greenish-gray	Pale yellow	Grayish white
$-O=Cl_2$.........	0.08	0.12	0.17	0.17	0.12	0.07	0.20	0.88
$-O=F_2$..........	-	-	0.07					
Sum............	100.56	99.41	100.11	99.02	99.21	100.20	100.50	100.16
Specific wt....	-	-	7.36	5.87	-	-	5.94	-
Components in Molecular Percentages								
Bi_2CO_5.........	81.16	63.59	48.29	9.48	28.94	11.01	21.92	26.40
BiOCl	4.48	6.91	9.40	5.69	3.21	3.06	6.68	33.85
Bi_2O_3..........	-	18.43	23.50	26.82	22.83	40.06	17.11	10.87
(WO_3+MoO_3).....	-	-	9.83	0.81	1.29	2.14	3.21	-
$CaCO_3$..........	7.17	5.53	5.55	31.97	41.37	41.28	8.56	25.77
$MgCO_3$..........	4.04	2.30	1.71	-	0.97	0.91	1.34	1.24
CaF_2...........	-	-	1.71	17.08	-	-	40.11	-
$CaSO_4$..........	3.14	-	-	-	-	1.53	-	0.93
$PbSO_4$..........	-	3.23	-	1.36	1.29	-	1.07	-
$PbCO_3$..........	-	-	-	6.78	-	-	-	0.93

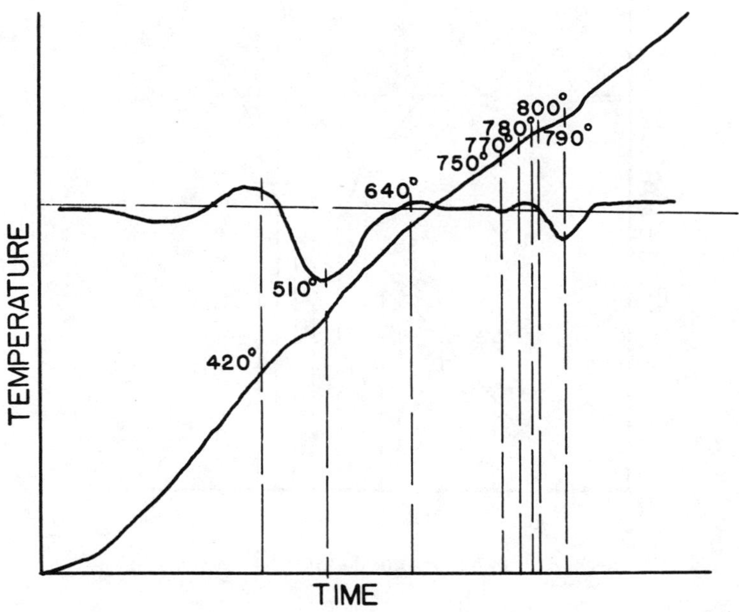

Figure 3. Heating curve of bismutite from Uzunbulak

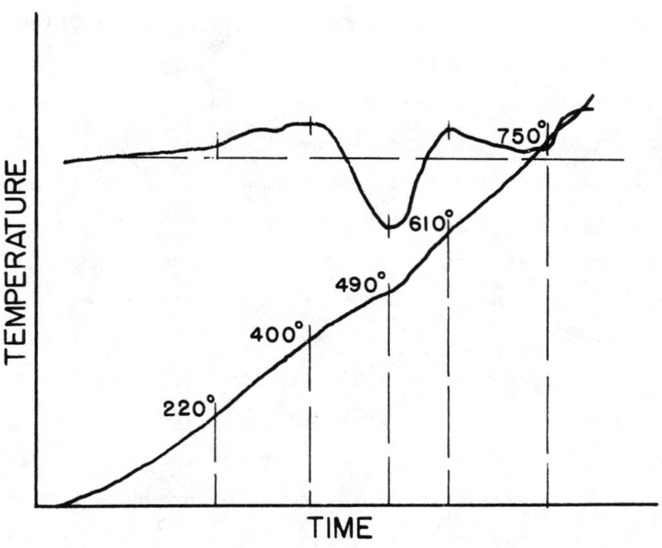

Figure 4. Heating curve of bismutite from Akchatau

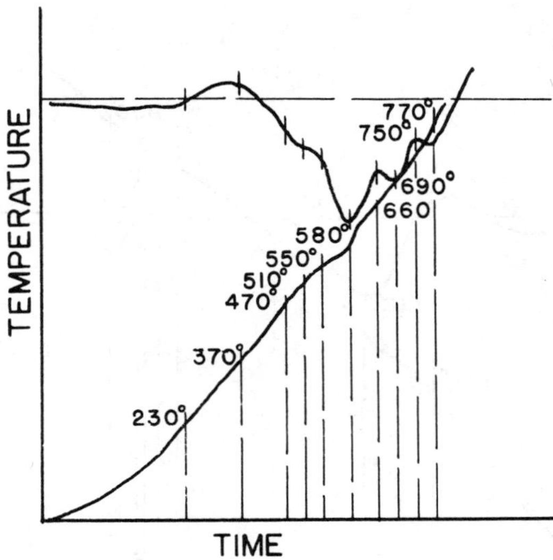

Figure 5. Heating curve of the mixture containing bismutite from Karaoba

substance of a dense greenish coloring is observed. Some edges reveal a gray isotropic substance. In addition, fluorite and calcium carbonate are present.

CONRAD DISTRICT, N. Conrad and Scorpion deposits. On the heating curves (Figure 6, N. Conrad deposit; Figure 7, Scorpion), the deflections with upper limits of 500, 650-690 and 750-790°C, which may be caused by bismutite, are quite pronounced; the last stop is also typical of bismuth oxide. There are also deflections with upper limits of 320°C (oxide of bismuth) and 850-870°C (bismoclite). The deflection for calcium carbonate is absent on the curve since heating ceased at a lower temperature than its dissociation.

On the powder patterns lines of bismutite and bismoclite occur. Some of the lines may be ascribed to bismite, sillenite, fluorite, and calcium carbonate.

Under the microscope, the following are distinguished in the samples analyzed: a grainy, high birefracting substance of a yellowish or brownish coloring, with transitions into an almost isotropic substance; a densely colored isotropic or brownish colored substance with transitions into an almost isotropic substance; a densely colored isotropic or weakly birefracting substance of a greenish-gray or gray coloring; calcium carbonate fluorite.

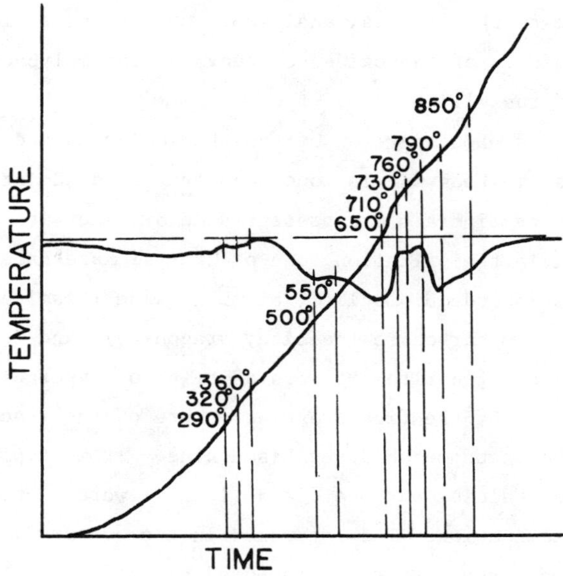

Figure 6. Heating curve of the mixture containing bismutite from the N. Conrad deposit

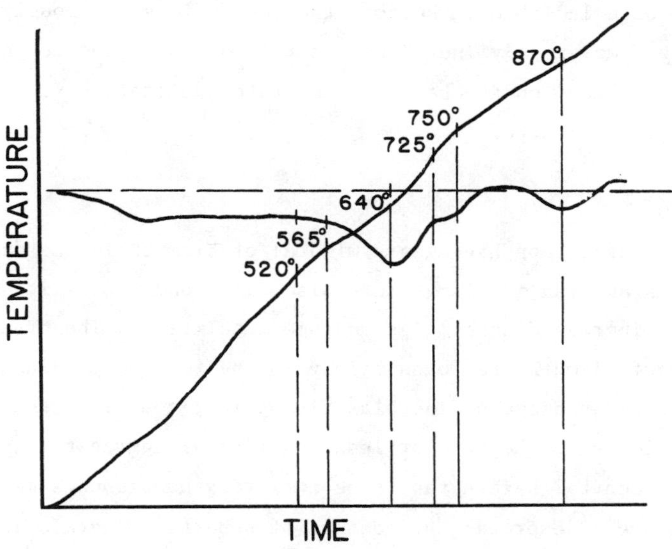

Figure 7. Heating curve of the mixture containing bismutite from the Scorpion deposit

According to the chemical analyses, some of the analyzed samples showed the presence of the oxides of tungsten and molybdenum, probably in the form of russellite.

On the basis of comparison of the chemical data and results obtained by other methods, it is concluded that fine admixtures of bismutite with other minerals are present side by side with relatively pure bismutite in the oxidation zone of Central Kazakhstan; these other minerals include bismoclite, bismuth oxide (bismite) perhaps sillenite, calcium carbonate (calcite, aragonite), and fluorite. Calcium carbonate and fluorite are seen as being of hypergenic formation; this is indicated by the fine granular nature of their occurrence in the mass of the pseudomorph after bismuthine. It is typical that isolated grains of calcium carbonate and fluorite were identified under microscopic examination. They occurred in the outwardly homogeneous pseudomorphs when the latter are composed basically, or to a significant degree, of russellite. In the pseudomorphs studied, the presence of bayerite was not established by the methods applied.

Apart from the samples described above, three samples from the Baynazar deposit (Shetsky district) and a sample from the Baykhatyn (Betpak-Dala) were also subjected to chemical study. The data obtained from these are shown in Table 11.

The data in Table 11 support the conclusions previously cited about the complex polymineral composition of the pseudomorphs after bismuthine which, on the basis of separate qualitative reactions, may be taken as bismutite.

CONCLUSION

The most common hypergenic mineral of bismuth in the deposits of Central Kazakhstan is bismutite. Bismoclite and russellite are observed in increased quantities in some localities. The hypergenic minerals of bismuth are found mainly in the form of pseudomorphs after the primary bismuth minerals--bismuthine, cosalite, and others. Side by side with the more or less monomineral aggregates in the deposits of Central Kazakhstan, fine outwardly homogeneous mechanical mixtures are widespread; the content of separate minerals in such mixtures varies. Besides the bismuth minerals, fluorite, calcium

Table 11. The chemical composition of the pseudomorphoses by bismuthine of a complex composition (weight %)

Components	Baynazar			Baykhatyn
	Sample 1	Sample 2	Sample 3	
Bi_2O_3	67.02	78.59	75.70	77.57
CuO	0.00	0.59	0.50	-
PbO	1.06	0.23	traces	traces
CaO	10.83	6.33	8.50	4.72
MgO	0.00	0.00	0.00	-
Al_2O_3	0.83	0.42	0.50	-
Fe_2O_3	0.33	0.19	0.10	0.12
WO_3	3.15	0.80	1.57	5.06
MoO_3	5.73	0.90	1.08	2.47
Cl	0.48	0.77	0.99	-
F	5.90	3.38	4.58	-
CO_2	4.57	6.57	6.25	7.15
SO_3	0.00	0.20	0.30	-
H_2O+	2.16	1.50	1.26	2.24
H_2O-	1.07	0.79	0.82	0.53
Undissolved residue	-	-	-	0.35
Sum	103.13	101.26	102.15	100.21
$-O=Cl_2$	0.11	0.17	0.22	-
$-O=F_2$	2.48	0.42	1.93	-
Sum	100.54	99.67	100.00	100.21
Specific Wt.	-	-	-	7.12
Components in Molecular Percentages				
Bi_2CO_5	15.14	40.00	32.17	27.00
BiOCl	3.47	6.88	8.11	-
Bi_2O_3	18.85	11.80	10.72	30.45
$(WO_3=MoO_3)$	13.40	2.95	4.07	29.06
$CaCO_3$	9.43	7.87	8.11	29.06
CaF_2	38.46	28.85	34.78	-
$CaSO_4$	-	0.33	-	-
$PbSO_4$	-	0.33	-	-
$PbCO_3$	1.24	0.98	0.87	-
$CuCO_3 \cdot CU(OH)_2$	-	0.98	0.87	-

carbonate (calcite, aragonite) and other minerals, may go into the composition of the mixtures in varying quantities. Such a polymineral composition of pseudomorphs after bismuthine is explained most likely by the complex composition of solutions in the zone of oxidation, or by its change in the course of the lengthy period of the zone's existence.

The polymineral nature of the pseudomorphs of hypergenic bismuth minerals by sulphide minerals is typical also of other districts, for instance Bolivia. According to the data of Ahlfeld and Reyes (1938), in the Bolivian deposits, the products of bismuthine's change, collected under the name of bismuth ochre, often contain (side by side with bismuth oxide) bismutite, and the arsenates and chlorides of bismuth. According to Heinrich (1947), mixtures of bismutite with bayerite are widespread.

The diagnostic results of the minerals, which formed the pseudomorphs after the primary minerals of bismuth, are obtained by a combination of various research methods; X-ray analysis shows the impossibility of utilizing color for diagnosing the hypergenic bismuth minerals. The diagnostics, based on qualitative reactions, were applied with most success to the purest isolates of a simple composition (e.g., the oxides of bismuth, bismutite, bismoclite), but they do not give simple results for mixtures of a complex composition. Diagnoses by microscopic examination of the hypergenic minerals of bismuth in mixtures presents a greater complexity due to their fine aggregate nature and the difficulty of defining the optical constants precisely.

The sharp decrease in the significance of the specific weight (below 6.00) is usually the result of the presence in the substance of increased quantities of fluorite or calcium carbonate (calcite, aragonite) and also the presence of both together. Study of the fine aggregate nature of the pseudomorphs after the primary bismuth minerals does not permit differentiation of their specific and volume weight.

Study of the water from the hypergenic products from the weathering of the primary minerals of bismuth, isolated at a temperature greater than 110°C shows that it evidently enters partly into the composition of bismoclite (the substitution of Cl^{1-} for OH^{1-}).

Part, like the water at a temperature lower than 110°C, is retained by the forces of absorption.

BIBLIOGRAPHY

Ahlfeld, F., and Reyes, M. 1938. Mineralogie von Bolivien. Berlin.

Bannister, F. A., and Hey, M. H. 1935. The crystal structure of the bismuth oxyhalides. Min. Mag., 24, No. 149, 49-58.

Beck, C. W. 1950. Differential thermal analysis curves of carbonate minerals. Amer. Min., 35, No. 11-12, 985-1013.

Fischer, E. 1955. Identität von Waltherit und Walpurgin. Chemie der Erde, 17, H. 4, 341-345.

Frondel, C. 1943a. Mineralogy of the oxides and carbonates of bismuth. Amer. Min., 28, No. 9-10, 521-534.

Frondel, C. 1943b. New data on agricolite, bismoclite, koechlinite, and the bismuth arsenates. Amer. Min., 28, No. 9-10, 536-540.

Heinrich, E. W. 1947. Beyerite from Colorado. Amer. Min., 32, No. 11-12, 660-669.

Hey, M. H., and Bannister, F. A. 1938. Russelite, a new British mineral. Min. Mag., 25, No. 161, 41-55.

Mountain, E. D. 1935. Two new bismuth minerals from South Africa. Min. Mag., 24, No. 149, 59-64.

Schaller, W. T. 1941. Bismoclite from Goldfield, Nevada. Amer. Min., 26, No. 11, 651-654.

Sillen, L. G., and Lundborg, K. 1944. X-ray studies on the systems Bi_2O_3 - WO_3, Bi_2O_3 - MoO_3, PbO - WO_3, PbO - MoO_3. Arkiv F. Kemi, Min., Geol., 17, H. 5, 1-11.

Zemann, J. 1956. Die Kristallstruktur von Koechlinit, Bi_2MoO_6. Heidelberg, Beitr: Min. u. Pettr., 5, H. 2, 139-145.

19

THE BEHAVIOR OF BISMUTH, SILVER, LEAD, AND ZINC IN A PROCESS OF SECONDARY SULFIDE ENRICHMENT

V. N. Naumev, D. N. Pachadzhanov, and T. I. Burichenke

This article was translated expressly for this Benchmark volume by Leonard J. Stanten, University of Kansas, Lawrence, from the original Russian article in Geokhimiya 5:588–591 (1971).

The zone of secondary sulfide enrichment of deposits is interesting to many investigators because it is a development area for hypergenic chemical processes that lead to the formation of sulfide minerals under surface conditions. This type of deposit contains a series of industrially important elements in concentrations significantly greater than that of primary sulfide ores.

The purpose of this investigation is to study the behavior of bismuth, silver, lead, and zinc during the secondary enrichment process. We will use as our example the copper-bismuth deposits of the Eastern Karamazar and the development of conditions effecting the concentrations of these elements in secondarily enriched sulfide minerals. There is significant practical interest in knowing the degree of concentration of elements in the process of secondary sulfide enrichment. From this it is possible to make a judgment about the character of deep-lying primary ores.

The deposits studied are located near the village of Adrasman in the Tadzhik region of the USSR. Situated along both sides of the Karamazar, the deposits occupy an area of approximately 20km.[2] Copper-bismuth deposits are associated with fractures and are grouped in three ore areas: Adrasman (Adrasman and Svintsevy lead fractures), Askazan (turquoise fracture), and Tary-Ekan (Tary-Ekan fracture).

The enclosing (or "country") rocks of the deposit are acid effusives.[1] In a structural sense the lead-bismuth deposits are of the vein type formed as a result of the precipitation of ore minerals from hydrothermal solutions in cavities of fracture origin. Previous investigators have distinguished five formation stages for the sulfide deposits of Eastern Karamazar (Koroleva, Shikhin, Koteneva). Lead-bismuth deposits were formed during a third lead-bismuth stage of mineralization. The main ore minerals are chalcopyrite and a group of natural bismuth minerals which are represented by bismutine, wittichenite, aikinite, emplecitite, cosalite, matildite, and others. Chalcopyrite is widely distributed in the deposits under study. Natural bismuthic minerals are rarely encountered, but sometimes form large pockets. Gangue minerals are represented by quartz and more rarely by ankerite, fluorite, and barite.

Smirnov has described the basic processes leading to the formation of zones of secondary sulfide enrichment of copper deposits. The action of metal rich waters with sulfates of copper and other metals dissolved in them on chalcopyrite causes a replacement of chalcopyrite by chalcocite, covellite, and sulfides of metals with a strong affinity to sulphur. Under these conditions iron rich chalcopyrite is converted to the sulfate form and is carried from the area of reaction by metalliferous waters.

In carrying out this research, the following methods of sample selection and preparation were adopted. In the ore zones of those deposits in outcrops and in underground pit mining, typical samples were selected from primary and secondary ores. Each average sample consisted of a large number of spot ore samples taken over a distance of several tens of meters along the fracture zone. Monomineralic samples of chalcopyrites, chalcosites, and covellites were subjected to quantitative spectral analysis, the results of which are presented in Table 1. For each ore zone, concentration coefficients giving the relationship of the content of an element in chalcocite or in covellite to that of the same element in chalcopyrite were calculated. These coefficients indicate the tendency of the elements to be concentrated in products of secondary sulfide enrichment.

Table 1. Results of analysis of chalcopyrites, chalcosites, and covellites

Deposit	Ore Zone	*	Content, % Bi	Ag	Pb	Zn
		Chalcopyrite				
Tary-Ekan	First Ore	19	0.20	0.099	0.10	0.19
Dzhuzum	Ore Fracture	9	0.22	0.11	0.066	0.039
Almadon and Kaptar-Khona	Almadon Fracture	6	0.36	0.094	0.090	0.034
	Mean		0.28	0.104	0.076	0.037
		Covellites				
Almaly	Central Ore	7	0.38	0.12	0.28	0.19
Dzhuzum	Ore Fracture	9	0.22	0.11	0.066	0.039
Almadon and Kaptar-Khona	Almadon Fracture	6	0.36	0.094	0.090	0.034
	Mean		0.31	0.109	0.141	0.086

Chalcosites

Deposit	Ore Zone	*	Content, % Bi	Ag	Pb	Zn	Concentration Coefficient Bi	Ag	Pb	Zn
Tary-Ekan	First Ore	5	0.84	~1	0.16	≥1	4.2	~10	1.60	≥5
Dzhuzum	Ore Fracture	3	0.80	~1	0.056	0.011	3.6	~10	0.85	0.28
Almadon and Kaptar-Khona	Almadon Fracture	2	1.20	~1	0.052	0.011	3.3	~10	0.58	0.32
	Mean		0.96	~1	0.054	0.011	3.4	~10	0.71	0.30
	Covellites									
Almaly	Central Ore	3	0.48	0.1-0.5	0.15	0.018	1.26	1-5	0.54	0.095
Dzhuzum	Ore Fracture	3	0.41	0.1-0.5	0.037	0.0060	1.86	1-5	0.56	0.15
Almadon and Kaptar-Khona	Almadon Fracture	1	0.55	0.1-0.5	0.068	0.0043	1.53	1-5	0.76	0.13
	Mean		0.46	0.1-0.5	0.090	0.011	1.48	1-5	0.64	0.13

*Number of average samples

Bismuth. As seen in the table, bismuth accumulates in products of secondary sulfide enrichment; especially in chalcosites, it reaches 1.2% and in covellites 0.55%. The concentration coefficient of bismuth is always greater than or equal to one, for covellites 1.48 and for chalcosites 3.4; for chalcosites of the Tary-Ekan deposits, it reaches a maximum value of 4.2

Silver. Semiquantitative determinations of silver in chalcosites and covellites and quantitative determinations in chalcopyrites are given in the table. In the process of secondary sulfide enrichment, a noticeable accumulation of silver was noted. Its content in chalcosites exceeded 1% and in covellites 0.1-0.5%.

Concentration coefficients of silver reach 10 for chalcosites and range from 1-5 for covellites.

Lead and Zinc. In contrast to bismuth and silver, lead and zinc are removed from chalcopyrites in the process of secondary sulfide enrichment. Zinc is removed more intensively than lead. Mean concentration coefficients for lead and zinc, respectively for chalcosites in the Dzhuzum, Almaden, Kaptar-Khona, and Almaly deposits are 0.71 and 0.30; for covellites they are 0.64 and 0.13.

Chalcosites from Tary-Ekan are in exception to this rule. For them, the concentration coefficients of lead and zinc are 1.6 and ≥ 5, respectively - i.e., in the zone of secondary enrichment, there is an intensive concentration of lead and especially zinc. Such a peculiarity was brought about, as will be shown later, by the secondary enrichment processes operating at the given deposit.

The behavior of elements during the secondary sulfide enrichment process depends on a series of factors, of which the main role is played by the migration mobility of the element during the oxidation of primary minerals and the elemental affinity to sulphur. If the elements under study are arranged in decreasing order, according to their concentration coefficients, then for chalcosites and covellites of the Dzhuzum, Almaden, Kaptar-Khona, and Almaly deposits we get the following sequence: Ag-Bi-Pb-Zn. This conforms fully for those elements to the order of decreasing affinity to sulphur.[2] From this fact, it follows that the behavior of elements in the secondary sulfide enrichment process at these deposits is basically determined by exchange reactions between sulfates in ground waters and sulfides of ore bodies.

In the Tary-Ekan deposits the formation of the secondary sulfide enrichment zone took place in a different way. At the time of sample collection (1966), the level of the pit had deepened by 90m. Correspondingly, in this short time interval, the level of metalliferous waters draining into the pit dropped sharply. Chalcopyrites from the primary ores, which were earlier located in the area of stagnant ground waters, entered into the area of intensive oxidation, as a result of which, apart from the usual products (sulfates of copper, iron, and other metals), sulphuric acid was formed in the zone of exidation.[2] The pH of the ground water is lowered. The more the pH is lowered, the more rapidly does the ground water level drop. Consequently the more intensive are the oxidation processes.

Further, on account of bacterial processes that develop in the region of stagnant ground waters, the possibility of free hydrogen sulfide formation is present. If precipitation of heavy metals from metalliferous waters is caused by the hydrogen sulfide of the stagnant zone, it is also necessary that there be a rapid lowering of the ground water level. We did not study the process of free hydrogen sulfide formation. The importance of the formation of hydrogen sulfide in the zone of oxidation of copper-bismuth deposits is indicated by the fact that in the Aksay deposit, a hydrogen sulfide spring flowed from the bore-hole drilled under the Ore Zone fracture. The Ore Zone fracture is a continuation of the ore zone of the Tary-Ekan deposit.

The presence of free hydrogen sulfides leads to the almost complete precipitation of heavy metals in the form of sulfides. In such a case, the degree of element accumulation in the products of secondary enrichment will usually depend on the element's migration capacity in the oxidation process and on the composition of it in ground waters.

In the Tary-Ekan deposits, elemental behavior is determined not by exchange reactions but by precipitation of metals by hydrogen sulfide. If the elements under study are listed in decreasing order according to their concentration coefficients, we get the following series for the chalcosites of the Tary-Ekan deposit: Ag, Ag-Bi-Pb; this corresponds to the decrease of migration capacity of the elements in the oxidation process. Zinc because of the high solubility of its sulfate (544g/l at 20°C) has the highest migration capacity in the oxidation process.[3] Silver and bismuth in an acid medium are also easily removed from oxidizing ore bodies.[4] Lead has the lowest migration capacity, and it forms a poorly soluable sulfate (41mg/l at 20°C).[4]

It should be noted that the precipitation of zinc in the form of a sulfide takes place in a narrow interval of geochemical conditions - i.e., only in an acid medium at a pH of 1.5 to 6.8.[5] Furthermore, the precipitation of zinc by hydrogen sulfide is carried out most fully if copper serves as the carrier.[6] Thus, in the secondary sulfide enrichment zone of the Tary-Ekan deposit, there occur conditions conducive to the precipitation by hydrogen sulfide of zinc and other metals.

The accumulation of zinc in minerals of the secondary sulfide enrichment zone is an exceptionally rare occurrence. Although similar occurrences are described in the literature,[2] in the majority of cases zinc is dispersed in the process of secondary sulfide enrichment, which is characteristic for four of the five copper-bismuth deposits studied.

CONCLUSIONS

1. Zinc, silver, bismuth, and lead enrichment are part of the secondary sulfide enrichment process of the deposits of Eastern Karamazar, but they possess a varying capacity to accumulate in products of secondary sulfide enrichments.

2. The basic control for these elements in the secondary sulfide enrichment process in the Dzhuzum, Almadon, Kaptar-Khona, and Almaly deposits are the exchange reactions between the sulfates of the natural waters and the sulfides of the ore bodies. We observe an accumulation of elements in the mineral zones of the secondary sulfide enrichment areas proportional to the affinity of the elements with sulphur.

3. The experiments have shown that in chalcosites from the Tary-Ekan deposits, all the elements studied are enriched, and the concentration coefficients increase proportional to the movement of the elements in the oxidation process of primary ores.

LITERATURE

1. Naumev, V. N., Pachadzhanov, D. N., Burichenko, T. I., and Kopp, A. N. The behavior of copper, bismuth, silver, lead, and zinc in the oxidation zone of the copper-bismuth deposits of Eastern Karamazar, Geokhimiya, No. 11, 1969.
2. Smirnov, S. S. Oxidation Zone of Sulfide Deposits. Moscow: USSR Academy of Sciences, 1955.
3. Lurle, Yu. Yu. Handbook for Analytic Chemistry. Moscow: Khimiya, 1967.
4. Anonymous. Chemistry Handbook, Vol. 3. Moscow: Khimiya, 1965.
5. Blok, N. N. Qualitative Chemical Analysis. Moscow: Goskhimizdat, 1952.
6. Sandell, E. B. Colorimetric Methods of Determining Traces of Metals. Moscow: Mir (World), 1964.

20

THE PROBLEM OF THE CORRELATION OF SELENIUM AND TELLURIUM WITH BISMUTH AND THALLIUM USING GEORGIA AS AN EXAMPLE

M. E. Dzhandzhgava

This article was translated expressly for this Benchmark volume by Leonard J. Stanten, University of Kansas, Lawrence, from the original Russian article in Acad. Sci. Georgian SSR, Bull. **51**:357–360 (1968).

In the course of a study of the content and distribution of selenium and tellurium in various types of sulfide deposits in Georgia (USSR), we turned our attention to the correlation of these elements with thallium and bismuth.

Literature exists[1] explaining the dependence between the content of selenium and tellurium in galena on one hand and of bismuth, silver, and thallium on the other hand in the high-temperature, polymettalic ore deposits of Karamazar. It is clear that a direct correlation is observed between these elements. It should be pointed out that the authors were not successful in establishing the inclusion of tellurides and bismuthic minerals in experimental galenas. The absence of the elements Te, Ag, and Bi is explained by the fact that isomorphic substitution in galena in relation to admixtures of anions (Se and Te) depends on the presence or absence of admixtures of cations (Bi, Ag, Tl).

A direct correlation between the content of bismuth, silver, and tellurium was noted by R. N. Zarian[2] (1962) in galena from the Armenian deposits. The author explained this by the presence of microscopic intergrowths of tellurides of silver and bismuth and their isomorphic substitution into the structure of galena according to the scheme $2Pb^{2+} \rightarrow Bi^{3+} \rightarrow Ag^{+}$.

We will begin an examination of our data by examining the problem of the correlation of selenium and tellurium with bismuth in galenas and chalcopyrites. The selection of galena and sphalerite was conditioned by the fact that mineralogically and chemically they were specifically studied for their content of bismuth minerals and bismuth.[3] In these sulfides the presence of naturally occurring bismuthic minerals was established (but the possibility of the entrance of bismuth in the form of isomorphic substitution was not excluded); therefore, in our examination of the correlation we will not discuss the problem of the form or the location of bismuth in every instance. The correlation is obviously controlled by the isomorphic capacity of galena for these elements. We will only examine the dependency data between these elements in various types of mineralization.

Selenium and tellurium in the oxide form were determined by a colorimetric method. They were subsequently reduced to elemental selenium and tellurium by dichloride tin and hydrochlorich hydrazine. In the presence of gelatin, elemental selenium and tellurium remain in colloidal solution. Selenium tints the gel a red-orange color; tellurium, a dark brown. To determine the concentration of selenium and tellurium (separately), calibration curves were drawn.

Analyses were carried out by two different methods: (1) the method worked out by the Central Laboratory of the "Uraltsvetmetrazvedka" (Ural Nonferrous Metal Prospecting Organization) trust,[4] and (2) a method of catalytic reduction of selenious acids.[5] The bismuth and thallium content of the research samples were determined by spectral analysis carried out in the geochemical branch of the Geological Institute of the Georgian Academy of Sciences.

It follows from the tables that Bi is usually present in those deposits that contain selenium and tellurium.

Baryta-polymetallic deposits with an extremely low selenium and tellurium content are characterized by a complete absence of bismuth. The absence of these elements in low-temperature lead-zinc deposits with broad development of collomorphic mineral aggregates is noteworthy. It is therefore possible to state that high concentrations of selenium, tellurium, and bismuth are characteristic for deposits of this type.[7]

It should be noted that in shallow, low-temperature, lead-zinc, thallium-bearing deposits, bismuth is not present. The correlation of selenium and tellurium with thallium was studied in marcasites, pyrites, sphalerites, and galenas. On the basis of our samples, an inverse correlation was observed between thallium on the one hand and tellurium and selenium on the other. All of the sulfides that we examined and that contained thallium in significant amounts (0.00n-0.0n, less frequently 0.n) did not for practical purposes contain selenium or tellurium. Furthermore, we analyzed six marcasite deposits from the northern Caucasus in which the thallium was high and in which selenium and tellurium were absent.

When discussing that data showing a negative correlation between selenium and tellurium with thallium in the Georgian deposits, I. M. Yudin's research findings about the Metalliferous Altai should be introduced.[8] He discovered a distinct dependency of the concentrations of selenium, tellurium and thallium with the degree of crystallization of those minerals that were deposited essentially simultaneously. Thus, granular, crystalline pyrites are 3-4 times richer in selenium and tellurium than colloidal pyrite, but which are only one-sixteenth as concentrated in thallium content. I. M. Yudin explained these relationships.[8] Colloidal sulfide particles, which have a negative charge, easily absorb cations of thallium. At the same time selenium and tellurium, which have a negative charge, are not absorbed by the gels of these sulfides, all of which agrees with our research. Thus, high thallium-bearing, low-temperature, lead-zinc deposits with a large development of colloidal mineral aggregates are characterized by a complete absence of selenium and tellurium.

We conclude that on the basis of a study of the Georgian deposits, a direct correlation of the concentration of bismuth, selenium, and tellurium is observed, and an inverse correlation in the concentration of thallium, selenium, and tellurium.

Table 1. Selenium, tellurium, and bismuth content in galenas and chalcopyrites from various types of Georgian mineralization

No.	Samples	Type of mineralization	Content, %		
			Se	Te	Bi
		CHALCOPYRITE			
1	14[B]	Copper-pyrrhotine and	0.004	0.0029	0.005
2	105[a]	iron pyrites (often	0.011	none	traces
3	15	lead or zinc) type of	0.016	0.002	0.02
4	73	mineralization in Lias	0.009	0.0025	0.005
5	75	slate	0.006	0.0032	none
6	91[a]		0.014	0.002	0.05
7	98		0.064	none	traces
8	3		0.0053	none	none
9	4		0.008	0.0025	0.8
10	1		0.0066	traces	none
11	24		0.0017	none	none
12	53		0.0035	traces	traces
13	68	Copper-sulfide minerali-	0.018	0.005	none
14	281	zation of pyrite type in	0.005	0.015	0.017
15	595	a vulcanogenic upper Cre-	0.0015	0.0045	0.061
16	17	taceous stratum	0.0052	0.0112	0.024
17	136		0.004	0.005	none
18	607		0.0058	0.012	0.047
19	148		0.004	0.0025	none
20	619		0.024	0.01	none
21	607[1]		0.0045	0.012	0.047
22	15		0.0032	0.0017	0.003
23	283		0.006	0.0035	none
24	2	Polymetallic, mid-lower	0.005	0.002	none
25-Piece	4	temperature type of	0.0076	none	none
26-of ore	6	mineralization with	0.033	traces	none
27	46	high copper content,	0.0034	traces	none
28	84	and sometimes gold	0.0013	none	none
29-Piece	1	content[1]	0.005	0.02	0.6
30-of ore	4		0.0027	0.0075	none

Table 1. - cont'd

No.	Samples	Type of mineralization	Content, %		
			Se	Te	Bi
		GALENITE			
31	23	Polymetallic, mid-lower	0.0095	0.02	1.05
32	51	temperature mineraliza-	0.007	0.012	none
33	66	tion with high copper,	0.0032	0.0134	none
34	814	and sometimes gold con-	0.0032	0.015	0.04
35-Others	2	tent	none	0.001	none
36-Piece	26		0.002	0.002	none
37-of ore	13		traces	none	none
38	26		0.004	0.0026	none
39	75		traces	none	0.02
40	35		0.002	0.0012	0.25
41	4		none	0.016	none
42	102		0.0024	0.047	none
43	61	Copper pyrrhotine and	0.01	traces	0.03
		iron pyrite (often with	none	none	traces
		lead or zinc) type of	none	none	traces
		mineralization in Lias	0.007	traces	none
		slate			

[1] V. R. Nadiradze notes that bismuth in certain copper-polymetallic ore manifestations is sporadically met in contents - thousandths of parts per hundred, insignificant traces, and weak lines.[6]

REFERENCES

1. Necheljustov, G. N., et al. Geokhimiia, No. 11, 1962.
2. Zarian, R. N., Geokhimiia, No. 3, 1962.
3. Ivanitskii, T. V., and Gvaramadze, N. E., Geokhimiia, No. 2, 1960.
4. Blium, I. A., et al. Bull., No. 8 (194) of the All Union Scientific Research Institute of Metrology and Standardization, 1958.
5. Taraian, V. M., and Avakian, G. G. Zavodskaia Laboratoriia (Factory Laboratory), No. 8, 1961.

6. Nadiradze, V. R. Reports of the Georgian Academy of Sciences, Vol. XXII, No. 1, 1959.

7. Dzhanzhgava, M. I. Geokhimiia, No. 5, 1966.

8. Yudin, I. M. Proceedings of the USSR Academy of Sciences, Vol. 155, No. 2, 1964.

21

NATIVE BISMUTH AS A GEOLOGICAL THERMOMETER IV. THE EXISTENCE OF A STABILITY FIELD OF METALLIC BISMUTH AT TEMPERATURES ABOVE ITS BOILING POINT

G. R. Kolonin

This article was translated expressly for this Benchmark volume by Leonard J. Stanten, University of Kansas, Lawrence, from the original Russian article in Akad. Nauk SSSR Sibirsk. Otdeleniye Inst. Geologii i Geofiziki Trudy **5**: 5–9 (1967).

On the basis of a study of a great variety of genetic types of natural bismuth deposits and also of the morphological peculiarities of a number of samples, it has been shown that this element can be separated from solution in a liquid drop form when the temperature exceeds its melting point (Dymkov 1963; Godovikov and Kolonin 1964a, 1964b, 1965; Kolonin 1966). This implies that under certain conditions at temperatures above 271° (the melting point of Bi), there occur times at which the melt solutions cause natural bismuth to be more stable in comparison with the state in which it was introduced (e.g., BiO, $BiCl_4$, BiS, etc.).

Therefore, the possibility of the separation of natural bismuth from solutions at temperatures above 271° and of the Eh and pH conditions under which such a phenomenon could occur can be found on the basis of precise calculations of the high-temperature conditions of the type reaction $Bi + H_2O \rightleftarrows BiO^+ + 2H^+ + 3\bar{e}$ (Kolonin 1965a).

A similar calculation is presently impossible for the ions BiO^+, $BiCl_4^-$, etc. due to a lack of thermodynamic data. Such data would make it possible to compute the isobaric potential of these ions at high temperatures. However, with the existence of data on the isobaric potential of natural Bi at temperatures above its melting point, it might be possible to evaluate the immediate influence of the melting of bismuth on the equilibrium constants of those reactions that lead to its formation as well as the thermodynamic probability of their occurrence.

A series of calculations of the isobaric potential of bismuth immediately before and immediately after melting is presented below. Also, ΔZ of liquid bismuth (here and subsequently, the generally accepted designations of thermodynamic functions are used) at T = 600°K (327°C) is given.

Accordingly, for standard conditions and for any high temperature, it is possible to write (Karapet'ian 1953):

$$\Delta Z_{298} = \Delta H_{298} - 298\ \Delta S_{298};\ \Delta Z_T = \Delta H_T - T\ \Delta S_T,$$

then $$\Delta Z_T - \Delta Z_{298} = \Delta H_T - \Delta H_{298} - (T\ \Delta S_T - 298\ \Delta S_{298}).$$

Consequently,

$$\Delta Z_T = \Delta Z_{298} + \Delta H_T - \Delta H_{298} - (T\ \Delta S_T - 298\ \Delta S_{298}).$$

For metallic[4] Bi $\Delta Z_{298} = 0$ and (5) $\Delta H_{298} = 0$, $\Delta S_{298} = S_{298} =$ 13.6 cal/degr. mol. In this manner

$$(6)\quad \Delta Z_T = \Delta H_T - (TS_T - 4053);$$

i.e., for a calculation of the ΔZ of bismuth at any temperature, it is necessary to know the quantities of ΔH_T and S_T of bismuth at that temperature. The latter can be found using the well-known formulas:

$$\Delta H_T = \Delta H_{298} + \int_{298}^{T} \Delta Cp\,dT;$$

$$(7)\quad S_T = S_{298} + \int_{298}^{T} \Delta Cp\,\frac{dT}{T}$$

In our calculations we used the following data from Kubaschewski and Evans (1956) for bismuth: ΔH melting = 2600 cal/mol, ΔS melting = 4.83 cal/deg. mol.,

$$(8)\quad Cp = 5.38 + 2.60 \cdot 10^{-3}T \text{ and}$$

$$Cp \text{ liquid} = 7.60 \text{ cal/deg. mol.}$$

For crystals of metallic Bi at the melting point (544°K), we get:

$$(9)\quad \Delta H_{544} = \int_{298}^{544} (5.38 + 2.60 \times 10^{-3}T)dT = 1594 \text{ cal/mol.};$$

$$\Delta S_{544} = \int_{298}^{544} (5.38 + 2.60 \times 10^{-3}T)\frac{dT}{T} = 17.5 \text{ cal/deg. mol.};$$

$$\Delta Z_{544} = 1594 - (17.5 \times 544 - 13.6 \times 298) = -3873 \text{ cal/mol.}$$

Examining the melting of bismuth as a balanced process, a condition at which $\Delta H = T\Delta S$ and $\Delta Z = 0$, we have no right to expect any perceptible change of ΔZ during the course of melting.

In fact:

$$(10)\quad \Delta Z \text{ melting} = 2600 - 544 \times 4.83 = 0$$

Subsequently, assuming that the specific heat of melted Bi, as a first approach, remains equal to 7.60 cal/deg. mol., it is possible to find the value of ΔH_{600} and S_{600}, and from them calculate

(11) $\Delta Z_{600} = 4620 - (23.1 \times 600 - 13.6 \times 298) = 5287$ cal/mol.

The data produced are given in Figure 1 and Table 1.

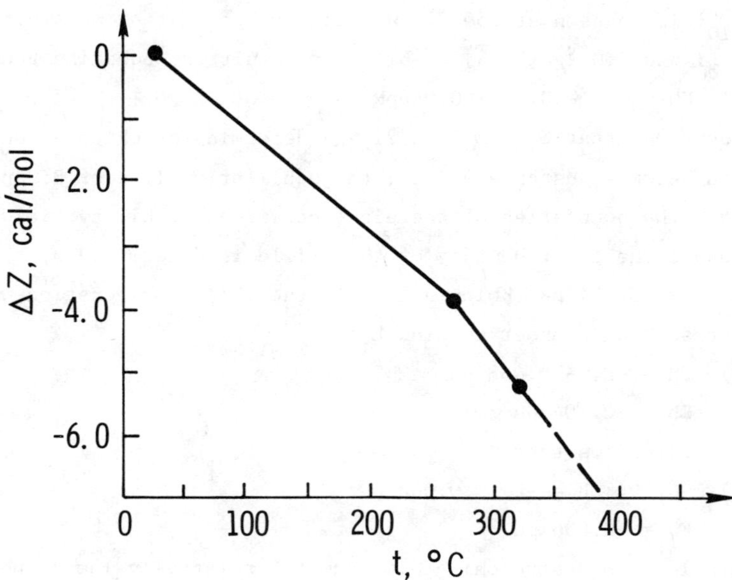

Figure 1. The change of the isobaric potential of metallic Bi with temperature

Table 1. Thermodynamic characteristics of bismuth at high temperatures

Temperature	ΔH, cal/mol	S, cal/mol.deg	ΔZ, cal/mol
544	1594	17.5	-3873
600	4620	23.1	-5287

Using this data, it is possible to evaluate the influence of the change of ΔZ of elemental bismuth on the equilibrium constants for the reactions which lead to its formation.

As an example, we will examine the reaction introduced above (Figure 2). Using Nernst's formula for 250°C, it is possible to formulate the following equation of the dependency of Eh on pH:

(12) $Eh = Eo + \frac{0.104}{3} \ell g\ [BiO^+] - 0.068\ pH$

where E° is the normal electrode potential of the reaction at the selected temperature. It is impossible to calculate E^o due to the fact that ΔZ_{BiO^+} is unknown at 250°C. We will accept that at an arbitrary boundary Bi and $BiO^+ \ell g\ [BiO^+] = -6$. As a result the equation becomes:

(13) $Eh = -E. - 0.21 - 0.068pH$

We will consider that $E_1 = E + 0.21$ and determine at which value of E^1_o bismuth becomes unstable - i.e., the equilibrium line of Bi and BiO^+ leaves the boundaries of the electrochemical stability field of H_2O. Because the lower boundry of this field is determined by the equation Eh = -0.104 pH (Kolonin 1956b), the following equation will have to be solved in order to find E^o_1:

(14) $Eh = -E_1 - 0.068\ pH$

$Eh = -0.104\ pH$

From this it follows that

(15) $-0.104\ pH = -E_1 - 0.068\ pH;$

$E_1 = +0.036\ pH.$

There is such a stretching out from the relatively low slope of the equilibrium lines of Bi and BiO^+ that with an increase of E_1 the field of Bi remains longest in a strongly alkaline region. If we adopt pH = 12, then $E_1 = 0.43$ and $E_o = 0.22$. Knowing E^o from the formula

(16) $\Delta Z = nfE,$

where n is the number of electron participants in the reaction and f is Faraday's constant, equal to 23.06 kcal/v.g.ekv, it is simple to calculate the isobaric potential of the reaction:

(17) $\Delta Z = -3 \times 23.06 \times 0.22 = -14.1$ kcal.

With a change in temperature from 250°C to 271°C, this quantity does not change substantially and can quite possibly serve as a characteristic reaction with the participation of liquid metallic bismuth near its melting point. In this manner, a little above 271°C, a condition for the disappearance of the field of metallic Bi will be

(18) $\Delta Z_p = \Delta Z_{BiO^+} - \Delta Z_{H_2O} - \Delta Z_{Bi} < -14.1$ kcal.

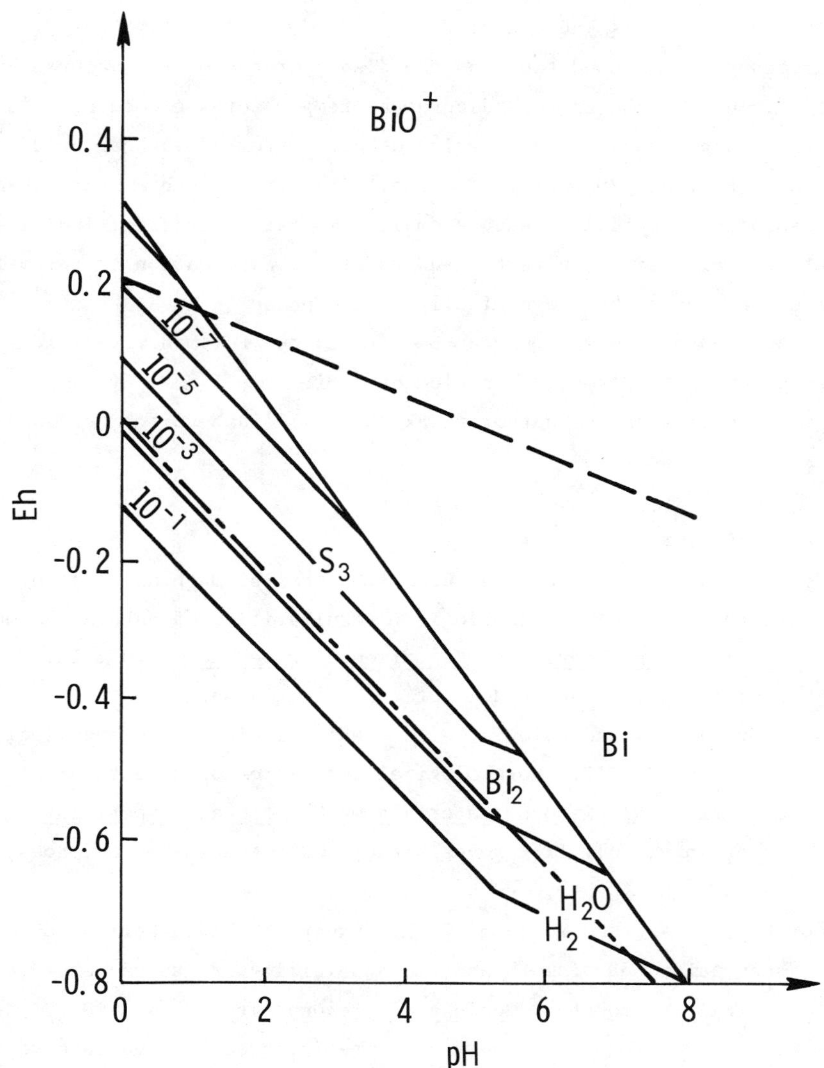

Figure 2. Stability field of Bi and Bi_2S_3 at 250° and of various activity values of a sulfasalt (Kolonin 1965a)

The dotted line shows the boundary of the fields of Bi and of the ion BiO^+ at 250°C.

Putting into this relationship the quantity produced above ΔZ_{Bi} = 3.9 kcal/MOLE and the value calculated by us according to the formula of Nikolaev and Dolivo-Dobrovol'ski (1961) for 250°C ΔZ_{H_2O} = -48.8 kcal/MOLE, we find

(19) $\Delta Z_{BiO^+} < -66.8$ kcal.

This means that a condition for the disappearance of the natural bismuth field from the Eh - pH diagram at temperatures exceeding Bi's melting point should be a lowering of ΔZ_{BiO} from -34.5 kcal at 25°C to -66.8 kcal at 270-280°C. The impossibility of such a sharp change of isobaric potential in such a small temperature interval indicates that natural bismuth, notwithstanding its transformation to the melted state, remains a thermodynamically stable phase.

Analogous calculations can be carried out for other reactions that produce the formation of elemental Bi.

In conclusion the author thanks N. A. Il'iasheva for her valuable suggestions.

LITERATURE CITED

Godovikov, A. A., and Kolonin, G. R. Natural bismuth as a geological thermometer. I: Morphological peculiarities of natural bismuth, Materials on Genetic and Experimental Mineralogy, Vol. II. Novosibirsk: USSR Academy of Sciences, 1964a.

Godovikov, A. A., and Kolonin, G. R. Natural bismuth as a geological thermometer. II: Morphological and microscopic peculiarities of artificial bismuth, Materials on Genetic and Experimental Mineralogy, Vol. II. Novosibirsk: USSR Academy of Sciences, 1964b.

Godovikov, A. A., and Kolonin, G. R. Experimental investigation of separations of bismuth and the possibility of its use as a geological thermometer, Geology of Ore Deposits, No. 3, 1965.

Dymkov, Yu. M. Natural bismuth over-precipitated from various veins under the influence of tertiary basalts, Annals of the All-Union Mineral Society, Vol. 92, No. 3, 1963.

Karapetiants, M. Kh. Chemical Thermodynamics, Goskhimizdat, 1953.

Kolonin, G. R. Several physico-chemical conditions of the formation of natural bismuth and bismutine (calculated data), Reports of the USSR Academy of Sciences, Vol. 163, No. 1, 1965a.

Kolonin, G. R. Conditions of the formation of natural bismuth and several problems of the physical chemistry of hydrothermal solutions, Dissertation, Novosibirsk, 1965a.

Kolonin, G. R. Natural bismuth as a geological thermometer. III: The problem of the duplication of bismuth, Materials on Genetic and Experimental Mineralogy, Vol. IV. Novosibirsk: USSR Academy of Sciences, 1966.

Kubaschewski, O., and Evans, El. Metallurgical Thermochemistry, London, 1956.

Nikolaev, V. A., and Dolive-Dobrovolsky, V. V. Bases of the theory of the processes of magmatism and metamorphism. Gosgeoltekhizdat, 1961.

Ramdor, G. Ore minerals and their intergrowth. Foreign Literature Press, 1962.

22

NATIVE BISMUTH AS A GEOLOGICAL THERMOMETER IN THE ORE OF THE SMIRNOVSK DEPOSITS (SOUTHERN MARITIME TERRITORY)

G. P. Vasilenko, M. I. Efimova, and V. N. Kolesnikov

This article was translated expressly for this Benchmark volume by Leonard J. Stanten, University of Kansas, Lawrence, from the original Russian article in Voprosy Geologii, Geokhimii i Metallogenii Severo-zapadnogo Sektora Tikhookeanskogo Poiasa, *I. N. Govrov, ed., Vladivostok, 1970, pages 311–313.*

In the present-day literature there exists two points of view on the formation temperature of natural bismuth. Ahlfeld, describing the tin-tungsten deposits of Tasna (Bolivia), discussed the possibility of drops of natural bismuth separating from solution at a temperature of about 400°C--i.e., higher than its melting point (271°C). Similar examples are given in the work of Chukhrova, Levitsky, Dolomanova, Dymkova, Ermilova, Ontoeva, and others. Fersman showed by phase diagrams that the separation of natural bismuth in pegmatites can occur at temperatures of 600-400°C (geophases D-E-F-G) from magmatic, pegmatitic and pneuma to hydrothermal solutions.

Turnor, studying the tin ore deposits at Llalgua and Potosi (Bolivia) came to the conclusion that the formation temperature of natural bismuth was lower than 269°C. Shtemprock also rejected the possibility of the separation of drops of natural bismuth in solution in quartz-cassiterite veins of the Ore Mountains. He stated that in this case there should be gravitational concentration of bismuth in the lower parts of the vein. Barsanov and Pogonia proposed that the position of natural bismuth on the phase diagrams should be moved to the low temperature region. It is worth noting that in this respect, Maukher relates the formation interval for natural bismuth in the Ore Mountains to the H-1 phases (400-300°C and 300-200°C). The separation of sulfosalts of bismuth and colloidal solutions occurs primarily in the lower boundary region.

Godovikov and Kolonin's experimental data on the synthesis of bismuth from melts and also by a hydrothermal route indicates the possibility of the separation of natural bismuth in solution over a broad temperature interval above as well as below its melting point. Here the occurrence of a rounded, drop-shaped or irregular shaped drops is evidence of their separation from solution at temperatures of 400-500°C. With a drop in the temperature of the surrounding medium, the liquid inclusions of bismuth crystallized with the formation of characteristic parquet-type duplicates. Inclusions of a dendritic or spear-shaped form, (i.e., with more or less distinct crystallographic characteristics) point to a bismuth formation temperature and drops of metallic bismuth on the stopper of the autoclave and in the lower parts of the titanium insert indicate the absence of a gravitational concentration.

In the ore of the Smirnov tin-sulfide deposit, it is possible to distinguish three morphological groups of natural bismuth inclusions. To the first group are assigned the drop-shaped, oval, and xenomorphic inclusions of natural bismuth in arsenopyrite and stannin. This mineral association is especially characteristic for greisen formations. It is interesting that in pyrite, widely distributed in greisens, inclusions of natural bismuth were not discovered, although spectral analysis of the pyrite by Osipova and others showed a high bismuth content (0.287%). We probably have here an intricate mechanical admixture of natural bismuth in the pyrite because Bi-F isomorphism is improbable in view of the great difference in their ionic radii.

The size of the inclusions of bismuth in arsenopyrites and stannin does not exceed 0.0/mm--rarely 0.00/mm. Therefore we did not examine the cleavability described by Godovikov and Kolonin.

At high magnification we noticed numerous branch-like intrusions of natural bismuth from the inclusions into the enclosing mineral. This phenomenon can be explained as follows. Bismuth, on hardening, increases its volume by approximately 3%, whereby its density decreases--i.e. it becomes more plastic. A squeezing out (expansion) of the bismuth occurs along microfractures that occur in the enclosing mineral (usually in stannite) as a result of crystallization pressure.

The second morphological group consists of joint fillings

(separations) of natural bismuth and bismutine in stanninite pyrrholite, sphalerite, and chalcopyrite. Cubanite is not infrequently present in this association. The form of the bismuth inclusions is acutely angled, tabular, rudely laminated. Their dimensions are significantly greater than the inclusions of the first morphological group, ranging from 0.1-0.5 mm long and reaching 1.5 cm in isolated cases. The structure of such fillings (separations) is large grained, but often among the individual large grains, a great quantity of smaller individual grains are seen filling gaps between the large ones. This gives the aggregate a distinctive block-mosaic structure. For large grains laminated duplicates of the first order are characteristic. The form of replicates is quite varied; most often they are lancet-shaped, spindle-shaped belts, the ends of which are slightly bent. Along the periphery of large individuals, it is sometimes possible to observe a system of evenly parallel belts with distinct tips. The presence of replicates, in Kolonin's opinion, is a criterion of the separation of natural bismuth, initially in a liquid drop-like form. These same drops occur as a result of crystallization pressures.

The structural relationships of bismuth and bismutine with the sulfides enclosing them are quite complex. For example, bismutine is isolated among these minerals in the shape of cross-cutting veins and lenses and is apparently a later mineral in the paragenesis given. The separation of bismuth from solutions, judging by the morphology of its inclusions, takes a long time. On the one hand, it is possible to observe in stannite and chalcopyrite separate occurrences of natural bismuth. From these forms it is possible to make a judgement about its initial separation in the form of drops of solution with subsequent crystallization; on the other hand, natural bismuth and bismutine form individual and cross-cutting veins in stannite, pyrrhotine, and chalcopyrite. Apparently, the separation of bismuth from solution continued even after the precipitation of the basic mass of sulfides at a temperature close to 271°C.

Finally, we assign to the third morphological group colloidal separations of natural bismuth precipitating together with sulfobismutites of copper and lead. As a rule, inclusions of natural bismuth occur along the periphery of anhedral separations of bismuthic sulfosalts among a mass of stannite, chalcopyrite, and ankerite. Replicating and other structural peculiarities whose size did not exceed 0.1 mm were not observed in inclusions of natural bismuth.

Summarizing the above facts with Kolonin's experimental and numerical data as well as with the results of the authors' thermobarometric investigations, it is possible to make the following deductions:

1. Morphological and structural peculiarities of inclusions of natural bismuth and data about homogenization of gas-liquid inclusions of the remaining solutions in quartz are evidence that the separation of bismuth in the Smirnov ore deposit occurs over a wide temperature range (410-190°C), lying below as well as above the melting point of bismuth. Pressure in the fracture system was lowered from 185-190°C to 72 atm.

2. No less than three generations of natural bismuth separations occur. The first two generations, associated with quartz-arsenopyrite and sulfidic mineral parageneses, separate at temperatures above 271°C. The formation temperature of colloidal inclusions of natural bismuth lies below 271°C. These colloidal formations are not a product of epigenetic reduction of the complex sulfosalts of bismuth because a chemical reaction between them at the concluding stage of mineral formation is hardly possible.

3. The simultaneous separation of natural bismuth and bismutine points to the complex character of the oxidation-reduction conditions occurring in zones of enriched H_2S solutions. The boundary of the stability field of Bi and Bi_2S_3 on an Eh-pH diagram lies above the electrochemical stability field for H_2O at average values of pH. Kolonin's data have shown that under these conditions, natural bismuth in the liquid state remains a thermodynamically stable phase.

AUTHOR CITATION INDEX

SUBJECT INDEX

About the Editors

ERNEST E. ANGINO is professor of geology and chairman of the Department of Geology at the University of Kansas. He received the B.S. in mining engineering from Lehigh University in 1954, the M.S. in geology in 1958 and the Ph.D. in geology in 1961, both from the University of Kansas. He was on the staff of Texas A&M University and the Kansas Geological Survey for several years before joining the faculty at Kansas in 1972. Professor Angino has broad interests in sedimentary and aqueous geochemistry. His more detailed studies have focused on trace element and brine geochemistry.

DAVID T. LONG is an assistant professor of geology at Michigan State University. He received the B.A. in geology from Monmouth College, Illinois, in 1969, the M.S. in geology (hydrogeochemistry) from the University of Illinois at Chicago Circle in 1973, and the Ph.D. in geology (geochemistry) from the University of Kansas in 1977. Dr. Long has an appointment with the Marine Science Section of the United States Geological Survey. His research interest is the migration of chemical speciation of the elements as particularly related to low-temperature ore deposits and health and disease.